18. COLLOQUIUM DER
GESELLSCHAFT FÜR PHYSIOLOGISCHE CHEMIE
VOM 5. BIS 8. APRIL 1967 IN MOSBACH/BADEN

WIRKUNGSMECHANISMEN DER HORMONE

BEARBEITET VON P. KARLSON

MIT 84 ABBILDUNGEN

SPRINGER-VERLAG
BERLIN · HEIDELBERG · NEW YORK
1967

ISBN-13: 978-3-540-03750-7   e-ISBN-13: 978-3-642-88715-4

DOI: 10.1007/978-3-642-88715-4

© by Springer-Verlag Berlin • Heidelberg 1967

Library of Congress Catalog Card Number 67-26114

Titel Nr. 4346

Vorwort

Das 18. Mosbacher Colloquium, über welches der vorliegende Band berichtet, war den Wirkungsmechanismen der Hormone gewidmet. Es war die Idee der Veranstalter, in diesem Rahmen möglichst viele Aspekte des Problems zu zeigen; eine enzyklopädische Vollständigkeit hinsichtlich der zahlreichen Hormone wurde nicht angestrebt. Das wäre bei der beschränkten Zeit auch unmöglich gewesen. Einen kurzen Überblick über den behandelten Themenkreis und den wissenschaftlichen Ertrag des Colloquiums habe ich im Schlußwort gegeben.

Sinn eines Colloquiums ist das Gespräch, die Diskussion; der Vortrag soll nur die Basis dazu liefern. In diesem Sinne haben wir auch im vorliegenden Band der Diskussion einen breiten Raum eingeräumt, und ich glaube, daß manche dieser Diskussionsbemerkungen die Vorträge erheblich ergänzen und vertiefen.

Die Organisation der Veranstaltung lag weitgehend in den Händen von Herrn Professor Dr. AUHAGEN. Die Tonbandaufnahmen der Diskussion, die für die schriftliche Fixierung der Diskussion von besonderer Bedeutung waren, wurden in technisch hervorragender Weise von Herrn Dr. WINTER, Stuttgart, besorgt; die hohe Qualität dieser Aufzeichnungen hat meine Arbeit als Herausgeber sehr erleichtert, und ich möchte Herrn Dr. WINTER deshalb auch an dieser Stelle danken. — Weiterhin gilt unser aller Dank den Vortragenden sowie den Diskussionsrednern, die zum Gelingen des Colloquiums beigetragen haben, insbesondere auch unseren Gästen aus dem europäischen Ausland und den USA.

Marburg a. d. Lahn, im Juni 1967 PETER KARLSON

Inhalt

The Role of Adenosine 3′,5′-monophosphate in Hormone Action *

By E. W. Sutherland, R. W. Butcher**,
G. A. Robison and J. G. Hardman

*Departments of Physiology and Pharmacology
Vanderbilt University School of Medicine
Nashville, Tennessee, USA*

With 9 Figures

Experiments carried out in recent years have shown that many hormones act by means of an intracellular second messenger (Fig. 1).

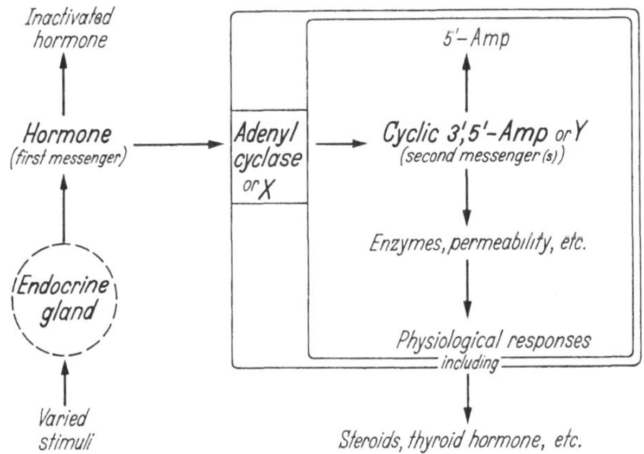

Fig. 1. Schematic representation of the second messenger concept (from Sutherland and Robison, 1966)

Hormones, the first messengers, are released by a variety of stimuli. This area will not be discussed in detail except to point out

* The research described in this communication was supported in part by grants HE 08332, AM 07642 and MH-11468 from the National Institutes of Health, United States Public Health Service.
** Investigator, Howard Hughes Medical Institute.

that the release of certain hormones such as insulin and possibly the anterior pituitary hormones may be mediated by second messenger systems. Evidence has been presented (from the work of others) that insulin release from the pancreatic β cell may be mediated by cyclic adenylic acid (Sussman and Vaughan, 1967; Turtle et al., 1967), although data to date do not rule out the possible participation of other second messengers (e.g., cyclic guanylic acid).

Hormone action is modulated by various factors as the first messenger travels to its target organ, or after it reaches the organ. One obvious factor is the rate of inactivation of the hormone. When the first messenger reaches the target cell it reacts with a specific portion of the cell, often the adenyl cyclase system, which when carefully studied has been found in the plasma membrane (Davoren and Sutherland, 1963b). The interaction of the hormone and adenyl cyclase (or X) may lead to an increased or a decreased production of a second messenger, for example cyclic AMP which mediates the action of the hormone. This second messenger does the real work of the hormone by changing enzyme activities, permeability, or other processes (Sutherland et al., 1965).

This report is divided into three sections:

1. A review of some events which led to the discovery of cyclic AMP in biological systems; 2. a section on the biological role of cyclic AMP; and 3. a brief section on some recent work with cyclic guanylic acid.

1. Early events. Some very simple questions were asked in early stages. For example, is glucose pumped out of the liver cell when epinephrine (= adrenaline) is added or does it accumulate and overflow because there is more glucose inside the liver cell? Using liver slices as a test system, it was found that the actions of the hormones were at least in part due to increased formation of glucose from glycogen rather than to an increased rate of release, and that the glycogenolysis was a result of phosphorolysis rather than hydrolysis (Sutherland, 1956).

The next step was to look at the enzyme systems catalyzing the conversion of glycogen to glucose. Three enzyme systems are on this pathway: phosphorylase, phosphoglucomutase and glucose-6-phosphatase. It was determined that phosphorylase was the rate limiting step of the three by several lines of evidence. For example, the addition of glucose-1-P or glucose-6-P to homogenates or slices led to increased glucose output. Also, the addition of crystalline phosphorylase to liver homogenates stimulated the rate of glucose production, but crystalline phosphoglucomutase did not (SUTHER-LAND, 1951).

Although phosphorylase had been identified as the rate limiting step, this did not exclude the possibility that one of the two other enzymes was stimulated and that the phosphorylase equilibrium was shifted by product removal.

To settle this the concentrations of glucose-1-P and glucose-6-P in slices were measured in the absence or presence of epinephrine or glucagon. If either the mutase or phosphatase were stimulated, the concentrations of the hexose phosphates would decrease, while an increase in phosphorylase activity would cause an increase in glucose-1-P and perhaps glucose-6-P as well.

In these experiments it was found that the hexose phosphates were increased in liver slices when epinephrine or glucagon was added. These slices were incubated in a high phosphate medium, and glycogen reacted with phosphate to form glucose-1-P and then the mutase enzyme converted this to glucose-6-P, which then was converted to glucose by the phosphatase present in liver (SUTHER-LAND, 1956).

These experiments are reviewed for three reasons. First, it shows how a series of reactions can at times be analyzed in intact cells. Secondly, these and other experiments led to the conclusion that the phosphorylase system in liver was stimulated by epinephrine and glucagon. This was a slow way of solving this problem but we were handicapped by knowing too much about phosphorylase. Phosphorylase catalyzes a reversible reaction which favours the synthesis of glycogen when studied in a test tube under certain conditions, so it was thought that stimulation of this enzyme might favour the synthesis of glycogen rather than the breakdown which is seen after epinephrine and glucagon. At the time it was not clearly recognized that this reaction is almost irreversible because of the high levels of inorganic phosphate in cells (LARNER et al., 1959).

Thirdly, we could have been misled in our conclusions because we didn't know about the UDPG transglucosylase which LELOIR later discovered (LELOIR and CARDINI, 1957). An inhibition of this enzyme could have caused similar results, and indeed cyclic AMP

1*

4 E. W. SUTHERLAND et al.:

is capable of stimulating the inactivation of this enzyme in some tissues, and it seems not unlikely that this may occur in liver. In any event, it was concluded from these and other experiments that phosphorylase activation was occurring.

The next step was to study the levels of phosphorylase activity in liver slices (Fig. 2). When liver slices were homogenized at zero time and phosphorylase measured, the activity of the enzyme was found to be very high. However, after incubation of the slices for

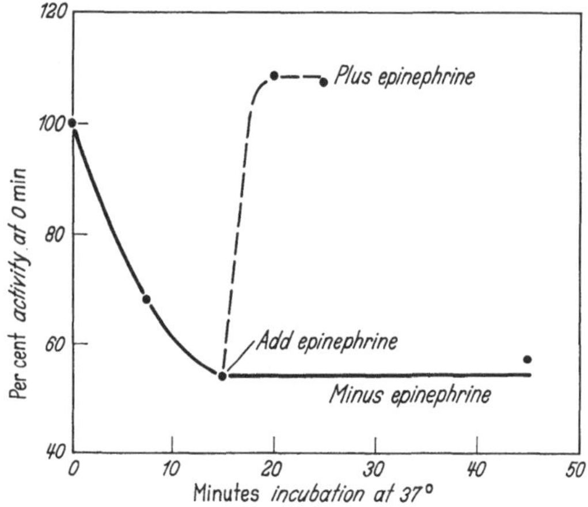

Fig. 2. Effect of epinephrine on phosphorylase concentration in dog liver slices. Seven 15 gm portions of slices were incubated in 30 ml portions of medium containing 0.15 M glycylglycine + 0.001 M phosphate buffer (pH 7.4). The slices were homogenized at the indicated times and phosphorylase activity was determined (from RALL et al., 1956)

15 min, the phosphorylase activity was decreased. The addition of epinephrine (or glucagon) to the slices led to a very rapid reactivation of the enzyme (RALL et al., 1956).

These experiments indicated that the level of phosphorylase activity in liver slices represented a hormonally controlled balance between inactivation and reactivation of the enzyme. Having localized an action of the hormones to phosphorylase, the mechanism of the interconversion was studied. Although at this time the reactivation process could not be elicited in cell free systems, the inactivation reaction did occur in homogenates and purified fractions. Therefore, in order to study the changes in the phosphorylase molecule, liver phosphorylase and the inactivating enzyme were purified (WOSILAIT and SUTHERLAND, 1956).

Experiments with the purified enzymes demonstrated that the enzymatic inactivation of phosphorylase did not involve a large change in molecular weight, nor was the product of the reaction, inactive phosphorylase, fully activated by 5′-AMP as was muscle phosphorylase *b*. A search for small fragments released during the inactivation reaction was made and inorganic phosphate was found to be a product (Fig. 3).

Liver phosphorylase was incubated with or without the inactivating enzyme for the times shown, and the phosphorylase activity

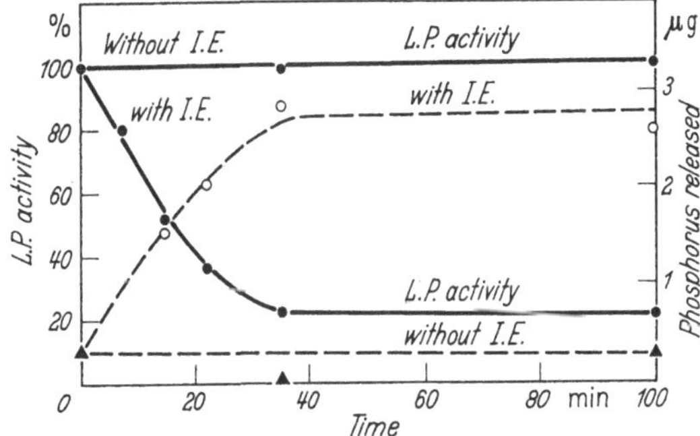

Fig. 3. Appearance of inorganic phosphate in the supernatant fluid of liver phosphorylase precipitated with trichloroacetic acid coincident with enzymatic inactivation (from SUTHERLAND, 1956)

(solid lines) and TCA soluble inorganic phosphate (broken lines) measured. The two parameters were exactly complementary — as phosphorylase activity decreased, phosphate release increased. In this experiment 1 mole of inorganic phosphate was released per 120,000 grams of protein — that is, about 2 moles of Pi per mole of enzyme (6 moles still remain). Since inactivation involved phosphate release, it seemed reasonable for activation to involve the addition of phosphate to the phosphorylase molecule.

Experiments with homogenates fortified with ATP and Mg^{++} suggested that a kinase was involved in the conversion of inactive phosphorylase, now called dephosphophosphorylase, to phosphorylase. Such an enzyme was located in liver homogenates and was purified. It required magnesium ions and ATP for the activation of the substrate enzyme (RALL et al., 1956).

Thus, phosphorylase in the liver had been shown to be a site of action for the glycogenolytic hormones. The enzyme had also been shown to exist in both active and inactive forms of the enzyme, the balance of which was regulated by a phosphatase and a kinase.

The next problem was how the hormones affected the phosphorylase balance (Fig. 4). Homogenates of liver fortified with $MgSO_4$, and ATP were incubated with or without epinephrine and

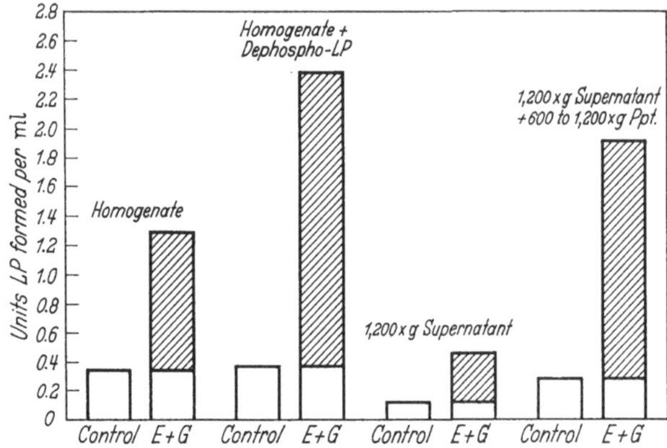

Fig. 4. The effect of epinephrine and glucagon on active phosphorylase (LP) formation in a whole and fractionated cat liver homogenate. 0.14 ml of homogenate or homogenate fraction was incubated with 2×10^{-2} M. Tris buffer (pH 7.4), 2.5×10^{-3} M $MgSO_4$, and 1.7×10^{-3} M ATP in the presence and absence of 0.4 of 1-epinephrine plus 1.0 of glucagon. The final volume was 0.20 ml. Dephospho-LP (4.2 units per ml) was added where indicated. The supernatant fraction used in this experiment was the 1200 x g supernatant fraction. LP activity was assayed before and after 10 min incubation at 30°. The bars represent the amount of LP formed during the incubation period; the cross-hatched portions of the bars represent the increased LP formation above that of the control (from RALL et al., 1957)

glucagon for 10 min and then the active phosphorylase measured (RALL et al., 1957). The addition of the hormones resulted in a 4 to 7 fold stimulation of phosphorylase activation. The addition of dephosphophosphorylase served a double purpose, for in addition to enhancing the response of the homogenate system, it also suggested that the effect was not occurring in intact cells. However, a surprising thing occurred when the homogenate was centrifuged. Although the kinase, phosphatase, and phosphorylase were all soluble enzymes, low speed supernatant fractions were essentially unresponsive to the hormones. However, if the low speed precipitate was added to the supernatant fraction a large hormone effect occurred.

Results obtained in investigations into the participation of the particulate fraction in the hormone response are shown in Fig. 5. Particulate fractions were incubated for five minutes with Mg^{++},

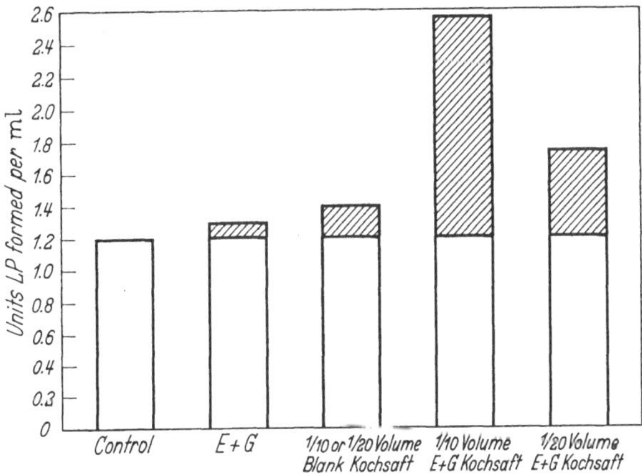

Fig. 5. The effect of preparations of the boiled extract on the formation of LP. Two 25 ml portions of a suspension of washed particles in 0.25 M sucrose, derived from about 25 gm of cat liver slices, were incubated with 2×10^{-2} M Tris buffer (pH 7.4), 2.5×10^{-3} M $MgSO_4$, and 1.7×10^{-3} M ATP in a final volume of 30 ml. One vessel contained 150 of both 1-epinephrine and glucagon (E + G), while the other contained no added hormones (blank). After shaking for 5 min at 30°, the flasks were heated in boiling water for 3 min and then chilled to 0°. The flask contents were centrifuged at 15,000 x g for 15 min in the cold. 0.02 ml and 0.01 ml aliquots of the supernatant fluids (boiled extracts) were incubated with 0.13 ml of an 11,000 x g supernatant fraction of a dog liver homogenate in 4×10^{-2} M Tris buffer (pH 7.4), 2.5×10^{-3} M $MgSO_4$, and 1.7×10^{-3} M ATP. The final volume was 0.20 ml. Control experimental vessels contained either water or 1.2 of both 1-epinephrine and glucagon in place of the boiled extracts. LP activity was assayed before and after 5 min incubation at 30°, and the bars represent the amount of LP formed during this incubation period (from RALL et al., 1957)

ATP, and with or without epinephrine and glucagon, and then boiled. The boiled extracts were then added to a supernatant fraction of a liver homogenate (containing the kinase and phosphatase) which had been fortified with purified inactive phosphorylase, and the conversion to active phosphorylase was measured.

The control level of phosphorylase was 1.2 units/ml, and the addition of epinephrine and glucagon to the soluble system had only a small effect. Likewise, the control heated supernatant, i.e., the particles incubated in the presence of ATP and Mg but without the hormones, had only a small effect. However, the heated supernatant from the particles incubated with hormones strikingly stimulated phosphorylase activation in a concentration-dependent fashion. Thus, particulate fractions of liver homogenates incubated with Mg and ATP produced a heat-stable factor which then stimulated phosphorylase activation in supernatant fractions of cell homogenates, and the production of this heat-stable factor was stimulated by low concentrations of glycogenolytic hormones (Rall et al., 1957; Rall and Sutherland, 1958).

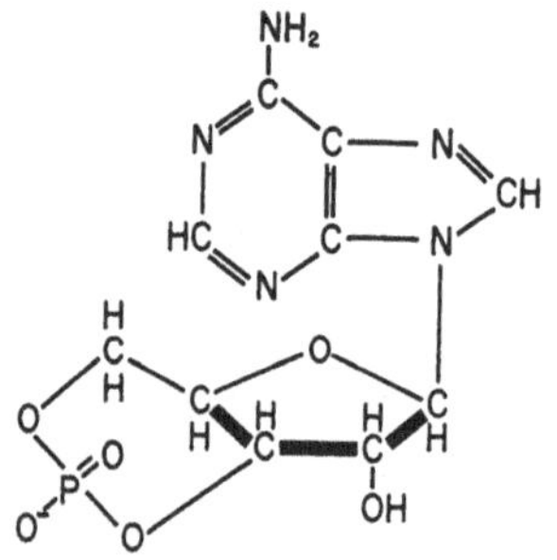

Fig. 6. Cyclic AMP

The next step was to identify the heat stable factor. The compound was extensively purified using ion exchange procedures and crystallized, and was found to contain adenine, ribose, and phosphate in the ratio 1:1:1 (Sutherland and Rall, 1958). Incubation of the purified compound with a variety of phosphatase preparations was without effect on either phosphate release or activity. Fortunately, at about the same time a group of organic chemists at Washington University in St. Louis led by David Lipkin were working on the structure of a compound isolated from a $Ba(OH)_2$ digest of ATP. Both groups wrote to Dr. Leon Heppel asking for a purified enzyme preparation to help in identifying the compound, and he recognized that the tentative structures were the same. He suggested that the compounds might be identical, and samples were exchanged. They were indeed identical, and the identification was greatly accelerated by the efforts of Dr. Lipkin and his co-workers (Lipkin et al., 1959).

The compound was shown to be adenosine 3′, 5′-monophosphate, or as it is familiarly known, cyclic AMP (Fig. 6).

The compound has an interesting structure — the cyclization of the phosphate between 3 and 5 greatly enhances the stability of the glycosidic bond, something which was most unexpected. The phosphate ring is in the main resistant to enzymatic attack except by a specific phosphodiesterase (SUTHERLAND and RALL, 1958). The ribose moiety is not flat as chemists once thought, and this caused some delay to Dr. LIPKIN and our group since some molecular models indicated that this compound was impossible.

The ability of cyclic AMP to stimulate phosphorylase activation was used as the basis of a very sensitive assay for the compound which has been used throughout these studies (RALL and SUTHERLAND, 1958; BUTCHER et al., 1965).

The cellular components known to be involved in the control of cyclic AMP levels in tissues are illustrated in Fig. 7. The particulate

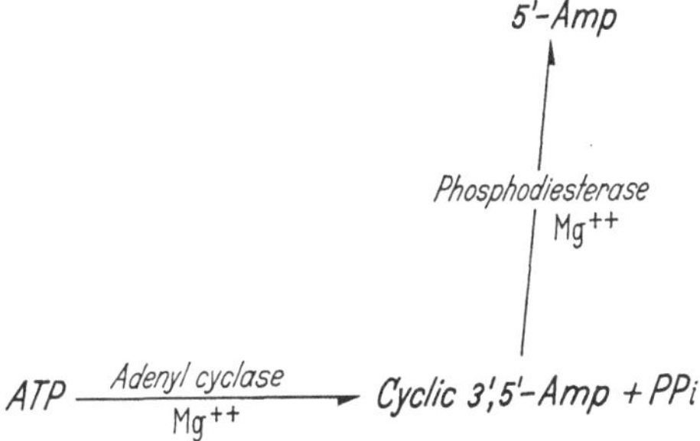

Fig. 7. The relationship of adenyl cyclase and the cyclic 3′, 5′-nucleotide phosphodiesterase to cyclic AMP levels

enzyme system adenyl cyclase, which is hormonally regulated, catalyzes the cyclization of ATP in forming cyclic AMP. Isotopic studies showed that cyclic AMP was derived from 8-^{14}C-ATP without loss of specific activity (RALL and SUTHERLAND, 1958). It was later found that it is the phosphate of ATP that is incorporated into cyclic AMP, and that the byproduct of the cyclization is pyrophosphate (RALL and SUTHERLAND, 1962). Where carefully studied, the adenyl cyclase has been found to be associated with membranes. The intracellular distribution of adenyl cyclase was initially puzzling because the activity was associated with the low speed, or nuclear fraction (RALL et al., 1957; RALL and SUTHERLAND, 1958). However, it had been shown that under certain conditions of homogenization and centrifugation, membranes were also found in the low speed fractions. This problem was studied in the avian erythrocyte and using special homogenization coupled with density centrifugation it was possible to separate adenyl cyclase activity from nuclei. It was found that adenyl cyclase was concentrated at a density between 1.18 and 1.21, where membranes are found in gradient density centrifugation (Davoren and Sutherland, 1963b). More recently, cell membranes free of nuclei possessing hormone sensitive adenyl cyclase activity have been prepared from turkey erythrocytes (Oye and Sutherland, 1966).

Adenyl cyclase has not been amenable to purification. It is very labile, and the particulate nature of the enzyme system precludes the use of most techniques used in enzyme fractionation. Highly washed, hormone-sensitive preparations have been obtained from several tissues, but as yet highly purified samples are not available (Sutherland et al., 1962).

Adenyl cyclase is widely distributed, e.g., it has been found in all the tissues of the dog which have been investigated with the exception of red blood cells. The CNS has the highest level of activity, followed by skeletal muscle, heart and spleen. Other tissues, like the adrenal cortex and corpus luteum, are also rich in the enzyme. Adenyl cyclase has also been detected in a variety of other species and phyla. Beef, sheep, and pig brain (as well as a variety of human tissues) have been found to be active, and in addition, the flat worms, segmented worms, bird tissues, fly larvae, minnows, and bacteria are also equipped with adenyl cyclase (Sutherland et al., 1962).

The other enzyme system is a phosphodiesterase which inactivates cyclic AMP by opening the ring at the 3′ position with the production of 5′-AMP. The phosphodiesterase is Mg^{++} dependent and is strongly inhibited by methyl xanthines which may explain a portion of the pharmacological action of these agents (Sutherland and Rall, 1958; Butcher and Sutherland, 1962). It is possible that certain hormones (e.g. insulin, which lowers intracellular cyclic AMP in certain tissues) (Butcher et al., 1966), may affect the phosphodiesterase, as reported recently by Dr. Senfts' group from Germany (Schultz et al., 1966). The catecholamines, however, exert their activity in broken cell preparations by stimulating adenyl cyclase and did not affect the phosphodiesterase where studied (Sutherland and Rall, 1960).

Like the adenyl cyclase, the cyclic nucleotide phosphodiesterase is widely distributed, the highest activities being in the central nervous system, followed by the adrenal and kidney. In most mammalian tissues the intracellular distribution of the enzyme is mixed, being partly soluble and partly (around 50%) particulate. The reason for the mixed distribution is unclear. However, to date the two activities have been catalytically indistinguishable (BUTCHER and SUTHERLAND, 1962).

2. **The biological role of cyclic AMP.** The wide distribution of the adenyl cyclase system and the phosphodiesterase suggested that cyclic AMP might be involved in a variety of hormone actions, many of which might be independent of phosphorylase activation. The actions of the catecholamines on a variety of tissues were amenable to investigation, and studies by HAYNES and his co-workers widened our horizons early when he showed that the steroidogenic effect of ACTH on the adrenal cortex was mediated by cyclic AMP (HAYNES et al., 1959; 1960). HAYNES initially felt that the action of cyclic AMP on steroidogenesis was expressed at the level of phosphorylase, but this has been questioned recently (FERGUSON, 1963). However, this does not detract from the very significant contribution made by HAYNES in showing that cyclic AMP did mediate the steroidogenic effect of ACTH and this has been confirmed repeatedly (FERGUSON, 1963; IMURA et al., 1965; KARABOYAS and KORITZ, 1965), even though the site of action of the cyclic nucleotide remains unclear as of this writing.

In general, three approaches have been used in studying the possible relationship of cyclic AMP to the action of a hormone. First, the effects of the hormone and blocking agents (where available) were examined on broken cell preparations of adenyl cyclase from the tissue under study. Qualitative and quantitative aspects of hormone specificity can be evaluated with this approach which is very useful because the particulate preparations can be washed, thus removing many small molecules and enzymes. Therefore, we have a more direct measurement since only ATP and Mg^{++} are added to the cyclase system in addition to the hormones, and there is less chance of indirect actions or displacements which might be misleading.

The second approach involves the measurement of cyclic AMP levels in the intact tissue under a variety of conditions (e.g., with and without hormones, blockers, etc.). The physiological state of the tissue (i.e., the response of the tissue to the hormone) is also monitored, and the changes at the level of cyclic AMP are compared with the ultimate physiological response. Qualitative, quantitative and temporal correlation (or lack of it) can be evaluated with this approach.

The third approach involves testing the ability of cyclic AMP or derivatives to reproduce the actions of the hormone on the tissue — i.e., to produce the "physiological" effect of the hormone. Although this approach has been used to good advantage in tissues like liver, fat, steroidogenic tissues and the parotid gland, it has not always been successful. Even in certain tissues where the data obtained by the first and second approaches were totally compatible with the hypothesis that cyclic AMP did mediate the action of the hormone, the third approach was negative. This was found to be related at least in part to the difficulty with which cyclic AMP enters cells and to the rapid inactivation of the compound by the phosphodiesterase (DAVOREN and SUTHERLAND, 1963a).

Cyclic AMP in the liver. All three approaches have been applied to the glycogenolytic effects of the catecholamines and glucagon on the liver. When the relative potencies of the catecholamines on cyclic AMP accumulation in highly washed particulate fractions from dog liver homogenates were tested (first approach), the activities of isoproterenol, 1-epinephrine, and 1-norepinephrine were of the order 4:1:1 respectively, in good agreement with values found by others measuring glycogenolysis in intact dog liver slices. In addition, dihydroergotamine inhibited the action of catecholamines but not of glucagon. The β-adrenergic blocking agents (e.g., pronethalol and DCI) antagonize the catecholamines but not glucagon, as also would be expected from the physiological results obtained by others.

Glucagon was found to be much more potent than the catecholamines in stimulating cyclic AMP production. Not only is the half-maximal dose much lower for glucagon, but the net increase in cyclic AMP by far exceeds that which can be approached by epinephrine. One explanation for this might be that the liver contains two adenyl cyclase systems, one which is activated by catecholamines and the other by glucagon. However, this has not been supported experimentally — liver preparations maximally stimulated by glucagon did not show an additive response when supramaximal concentrations of epinephrine were added (MAKMAN and

SUTHERLAND, 1964). Glucagon and epinephrine have also been compared in the isolated perfused rat liver (ROBISON et al., 1967 b), and glucagon was found to be much more effective than the catecholamine. The fact that different hormones can increase the intracellular concentration of cyclic AMP to different levels may be of importance when considering hormones with more than one action on a given tissue. It seems probable that different systems within the tissue may have different thresholds for stimulation by cyclic AMP. Glucagon, for example, might be expected to have certain effects in the liver not shared by epinephrine.

Finally, it was shown that exogenous cyclic AMP was capable of stimulating glucose release and glycogen disappearance. The addition of 1.5×10^{-5} M cyclic AMP or epinephrine to liver slices resulted in almost identical stimulation of glucose output and glycogen disappearance (SUTHERLAND, 1962).

It would be well to review the mechanism of the glycogenolytic action of epinephrine and glucagon on liver. First, phosphorylase activity was shown to be rate-limiting in the steps from glycogen to glucose. This activity was variable, the balance of active versus inactive phosphorylase being regulated by two enzyme systems, a phosphatase inactivating phosphorylase, and a Mg^{++} and ATP dependent kinase. The activity of the kinase system is increased by cyclic AMP, the production of which is under hormonal control at the level of the adenyl cyclase system. The qualitative, quantitative, and temporal requirements imposed by the criteria established earlier are satisfied by this mechanism at the level of phosphorylase, adenyl cyclase, and intracellular cyclic AMP.

The adenyl cyclase system, as the initial site of interaction of the catecholamines, has properties identical to those of the so-called adrenergic receptor in dog liver, with respect to both blockade and stimulation. In addition, the hormone tissue specificity has also been established e.g., glucagon is ineffective as a stimulant of the adenyl cyclase of skeletal muscle although the catecholamines are very active (RALL and SUTHERLAND, 1958).

Certain other actions of the catecholamines and glucagon on the liver have been shown to be mediated by cyclic AMP. These include the stimulatory effects of the hormones on gluconeogenesis (EXTON and PARK, 1966), ketogenesis, and lipolysis (BEWSHER and ASHMORE, 1966). Other effects of cyclic AMP have been reported but have not been studied in detail. These include K+ release by the liver (Tsujimoto et al., 1965), decreased incorporation of acetate into fatty acids and cholesterol (Berthet, 1960), and inhibition of the incorporation of amino acids into protein (Pryor and Berthet, 1960).

Cyclic AMP in adipose tissue. The lipolytic response of the rat epididymal fat pad to epinephrine was an attractive hormone action to study for a number of reasons. First, it seemed highly unlikely that increased FFA release could be related to phosphorylase activation, and thus it would provide an example of cyclic AMP mediating a hormone action not directly involving phosphorylase. Secondly, the relative lack of specificity of the response was fascinating — i.e., in addition to epinephrine, ACTH, TSH, glucagon, vasopressin and other hormones had been shown to activate lipolysis.

Some indirect evidence bearing on the lipolytic action of epinephrine was available. First, adipose tissue was found to be equipped with an epinephrine sensitive cyclase (Klainer et al., 1962). Also, Martha Vaughan (1960) had shown that lipolysis and phosphorylase activation were both stimulated by ACTH, glucagon and TSH as well as by epinephrine. She noted in her initial reports that serotonin stimulated phosphorylase without an effect on lipolysis, but it has since been found that serotonin does activate lipolysis (Bieck et al., 1966). In addition, Vaughan and Steinberg (1963) reported that caffeine, an inhibitor of the phosphodiesterase, acted synergistically with epinephrine in stimulating lipolysis. This suggested cyclic AMP mediation.

We started with the second approach — i.e., a correlation of intracellular cyclic AMP levels and lipolysis in collaboration with Drs. R. J. Ho and H. C. Meng (Butcher et al., 1965).

The results of these early experiments are illustrated in Tab. 1. The addition of epinephrine to the medium caused increased rates of lipolysis and cyclic AMP accumulation. The addition of 1 mM caffeine only mildly affected lipolysis and was without effect on cyclic AMP levels, but when caffeine was combined with 0.1 μg/ml epinephrine both lipolysis and cyclic AMP levels were markedly stimulated. This synergistic action of the two compounds served to illustrate that the activities of both adenyl cyclase and the phosphodiesterase can affect the intracellular levels of cyclic AMP and that the compound was turning over at high rates when adenyl cyclase was stimulated. Finally, the β-adrenergic blocking agent DCI partially inhibited the actions of epinephrine on both FFA release and cyclic AMP levels.

The increased cyclic AMP levels engendered by epinephrine were also found to occur rapidly enough to correlate temporally with the lipolytic action. Cyclic AMP levels were increased 3 fold in perfused

Table 1. *Effects of epinephrine, caffeine, and dichloroisopropylarterenol on cyclic 3′; 5′-AMP levels and FFA release in epididymal fat pads in 20 min incubations, from* BUTCHER *et al. (1965)*

Additions	Cyclic AMP m moles X 10^7/g, wet wt	p versus control	FFA μ moles X 10^2/g/min	p versus control
NaCl (control)[29]*	1.8 ± 0.08		1.26 ± 0.08	
Epinephrine, 10.0 μg/ml[6]	3.8 ± 0.3	0.01	4.55 ± 0.25	0.01
Epinephrine, 1.0 μg/ml[6]	2.9 ± 0.3	0.01	4.05 ± 0.25	0.01
Epinephrine, 0.1 μg/ml[6]	2.2 ± 0.2	0.05	2.26 ± 0.24	0.01
Caffeine, 1.0 mM[5]	2.1 ± 0.2	0.1	1.97 ± 0.19	0.01
Caffeine, 1.0 mM, plus epinephrine, 0.1 μg/ml[5]	4.3 ± 1.2	0.01	4.58 ± 0.6	0.01
Dichloroisopropyl-arterenol, 0.1 mM[3]	2.0 ± 0.06	0.1	1.81 ± 0.54	NS+
Dichloroisopropyl-arterenol, 0.1 mM, plus epinephrine, 1.0 μg/ml[3]	2.0 ± 0.07	0.2	2.73 ± 0.12	0.01

* Numbers in parentheses represent number of experiments.
+ NS, not significant.

fat pads within 30 sec after epinephrine was introduced. In incubated fat pads the levels of the cyclic compound were maximal at five minutes, the earliest time tested. More recently, glucagon and ACTH were found to increase cyclic AMP in fat pads (BUTCHER et al., 1966).

And now for the third approach. VAUGHAN (1960) found that exogenous cyclic AMP depressed rather than stimulated lipolysis in the whole fat pad and we also saw this. However, the problems involved in getting cyclic AMP into cells were known and in collaboration with Dr. THEO POSTERNAK of Geneva, some very useful derivatives of cyclic AMP were developed (Posternak et al., 1962). These compounds are acylated on the 6 amino-group (in an amide bond) and as an ester on the 2' hydroxyl. A large series was prepared (up to N^6-2'-0 distearyl cyclic AMP), and to date, the most successful compound has been N^6-2'-0 dibutyryl cyclic AMP. The reasons for the success of this compound as opposed to free cyclic AMP are unclear. They were synthesized to be more lipid soluble and hence to perhaps enter cells more freely, but when the dibutyryl compound was tested for better penetration, it could not be demonstrated. An added benefit of the substitution is that these compounds are resistant to the phosphodiesterase, and this may in large part explain their potency.

When whole fat pads were incubated with dibutyryl cyclic AMP lipolysis was markedly stimulated. In addition, cyclic AMP itself was found to be mildly stimulatory on isolated fat cells prepared by the method of Rodbell (1964) and the dibutyryl compound was about an order of magnitude more effective (Butcher et al., 1965).

The dibutyryl derivative was effective (where cyclic adenylic was not) in the parotid gland. This was reported by Bdolah and Schramm (1965) who found that the dibutyryl derivative mimicked the effect of epinephrine in causing release of amylase from the rat parotid gland. The dibutyrate has also been effective in the thyroid gland (Pastan, 1966).

Thus the role of cyclic AMP as a second messenger in the actions of lipolytic hormones on rat adipose tissue was established by all three approaches. In addition, cyclic AMP has recently been implicated in the anti-lipolytic actions of insulin and the prostaglandins.

The antilipolytic effect of insulin on fat *in vitro* in the absence of glucose was first described by Jungas and Ball (1963) who suggested that insulin might be acting to decrease the level of cyclic AMP. When fat pads were incubated under similar conditions but for shorter times by our group (in collaboration with Drs. J. G. T. Sneyd and C. R. Park), it was found that insulin decreased intracellular cyclic AMP markedly (Butcher et al., 1966). The exact mechanism by which insulin causes this is as yet unclear. Jungas (1966) has shown that homogenates of adipose tissue which were exposed to insulin prior to homogenization were less active in producing cyclic AMP than were controls, suggesting that adenyl cyclase was in some way inhibited. However, like us, he was unable to elicit an effect by adding insulin to homogenates.

The prostaglandins, especially prostaglandin E_1, are also potent antilipolytic agents (STEINBERG et al., 1964). Like insulin, PGE_1 acts to decrease intracellular cyclic AMP levels at very low concentrations and, as with insulin, the mechanism by which PGE_1 effects this is unknown since to date we have been unable to elicit the effect in broken cell preparations (BUTCHER et al., 1967).

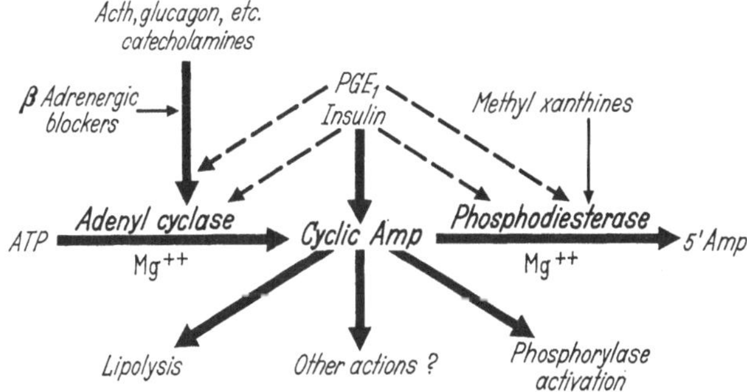

Fig. 8. The interrelationships between lipolytic and antilipolytic hormones and other agents on cyclic AMP levels in the rat epididymal fat pad (from BUTCHER et al., 1967)

Fig. 8 summarizes the current state of knowledge about the effects of hormones on cyclic AMP levels in adipose tissue. To recapitulate, the lipolytic hormones act by increasing the activity of adenyl cyclase and the methyl xanthines, which are also capable of stimulating lipolysis, inhibit the phosphodiesterase. $β$-adrenergic blocking agents antagonize the effects of the lipolytic hormones on adenyl cyclase, and PGE_1 and insulin act to lower cyclic AMP by a mechanism which is not clear as yet. Finally, thyroid hormone has been reported to increase the total amount of adenyl cyclase in adipose tissue by one group (BRODIE et al., 1966) and triiodothyronine to inhibit the phosphodiesterase by another (MANDEL and KUEHL, 1967). Thus, adipose tissue represents a fascinating crossroads of hormone action at the molecular level, and provides an excellent model system for future studies.

Cyclic AMP in the heart. The first evidence that cyclic AMP might be involved in the positive inotropic effect of the catecholamines came from studies on phosphorylase activation in the perfused heart (KUKOVETZ et al., 1959; HAUGAARD and HESS, 1965). In this preparation a good correlation was obtained between the inotropic response and the activation of phosphorylase when different sympathomimetic amines and adrenergic blocking agents were compared, suggesting that both responses were mediated by adrenergic β-receptors. It was later shown, however, that the two responses could be dissociated to at least some extent. Small doses of epinephrine were found to be capable of causing substantial increases in contractile force with little or no change in phosphorylase activity (MAYER et al., 1963; DRUMMOND et al., 1964). Also, when time courses were studied, the inotropic response was found to precede the change in phosphorylase activity (ROBISON et al., 1965; CHEUNG and WILLIAMSON, 1965; OYE, 1965; VALADARES and FRIESEN, 1966). These data suggested that although the two responses might have a common mediator, or second messenger, the activation of phosphorylase was probably not responsible for the increase in contractile force.

More direct evidence that cyclic AMP was involved in these responses came from studies on adenyl cyclase in broken cell preparations from dog heart (MURAD et al., 1962). The relative potencies of a series of catecholamines in stimulating adenyl cyclase were found to be similar to their relative potencies as inotropic agents *in vivo*, and these effects were blocked by DCI (MURAD et al., 1962) and pronetholol (ROBISON et al., 1967a). Isopropylmethoxamine, which is less potent than pronetholol in preventing the inotropic response to isoproterenol (BURNS et al., 1964; BLINKS, 1967), was also less effective than pronetholol in preventing the stimulation of adenyl cyclase *in vitro* (ROBISON et al., 1967a).

Tissue levels of cyclic AMP have been measured in the isolated perfused heart of the rat (ROBISON et al., 1965; CHEUNG and WILLIAMSON, 1965; DRUMMOND et al., 1966) and the rabbit (HAMMERMEISTER et al., 1965) and were found to be elevated in both tissues following the administration of epinephrine. The studies in the rat heart revealed that the activation of adenyl cyclase was an extremely rapid process, preceding both the inotropic response and the activation of phosphorylase. It was suggested earlier (SUTHER-

LAND and ROBISON, 1966; ROBISON et al., 1967a) that the threshold for stimulation by cyclic AMP might be higher for phosphorylase *b* kinase than for the system responsible for the inotropic effect. This hypothesis was not supported by the data of DRUMMOND et al. (1966), who found that the activity of phosphorylase *b* kinase in the heart was closely related to the intracellular level of cyclic AMP. It has not been possible to dissociate the rise in cyclic AMP levels from the inotropic response on the basis of dose-response studies (ROBISON et al., 1967a) and DRUMMOND et al., (1966) were likewise unable to dissociate the inotropic response from the increase in phosphorylase *b* kinase activity. The reason for the dissociation between the inotropic effect and phosphorylase activation (DRUMMOND et al., 1964) is thus not clear. In a recent abstract, NAMM and MAYER (1967) have reported that they were unable to detect changes in intracellular cyclic AMP levels in dog hearts exposed to epinephrine *in situ*, despite their finding that phosphorylase *b* kinase and phosphorylase in these hearts were both activated. It seems possible that under these *in vivo* conditions a factor or factors may be produced which interfere with the assay for cyclic AMP. On the other hand, graded increases in the level of cyclic AMP in the rat adrenal cortex have been measured in response to ACTH in hypophysectomized rats *in vivo* (GRAHAM-SMITH et al., 1967).

Other evidence supporting the hypothesis that the inotropic response to the catecholamines is mediated by cyclic AMP has been obtained from studies with adrenergic blocking agents (ROBISON et al., 1965; DRUMMOND et al., 1966) and also theophylline (RALL and WEST, 1963). Theophylline, which under some conditions potentiates the inotropic response to the catecholamines, enhances the accumulation of cyclic AMP by inhibiting the phosphodiesterase. It should be recognized, however, that theophylline has other actions, and therefore may not potentiate the effects of hormones which stimulate adenyl cyclase under all conditions. It is interesting to note that theophylline has recently been found to be an inhibitor of protein synthesis (HALKERSTON et al., 1966), just as puromycin is now known to be capable of inhibiting the phosphodiesterase (APPLEMAN et al., 1966).

It has so far not been possible to demonstrate an effect of exogenous cyclic AMP on either contractility or phosphorylase activity in the perfused rat heart (ROBISON et al., 1967a), presumably because of the permeability barrier. LEVINE and VOGEL (1966) have reported that in dogs and also in man the injection of cyclic AMP was followed by definite increases in heart rate and cardiac output. These may have been indirect effects, or they may possibly indicate that species differences exist with respect to the ability of cyclic AMP to penetrate myocardial cell membranes.

Our conclusion in this section on the heart is that the available evidence supports the hypothesis that the positive inotropic effect of the catecholamines is mediated by cyclic AMP, although more work remains to be done before a final conclusion can be reached. It may be especially interesting to measure the cyclic AMP response in certain molluscan hearts, in which contractility is increased not by the catecholamines but by serotonin (WELSH, 1957). It might be predicted that in these hearts serotonin will have an effect on cyclic AMP levels. As for the target system which might be affected by cyclic AMP in the heart, this is completely unknown, although it seems clear that the calcium ion must in some way be involved in this response (REITER, 1964).

The action of catecholamines on cyclic AMP levels in other tissues

The catecholamines are capable of stimulating cyclic AMP formation in a wide variety of tissues (Tab. 2). Although in a few cases the mechanism of action of catecholamines is fairly clear, in

Table 2. *Tissues in which formation of cyclic 3′, 5′-AMP is stimulated by catecholamines*

Tissue	Tissue
Liver	Lung
Heart	Spleen
Skeletal muscle	Brain (chiefly cerebellum)
Adipose tissue	Parotid gland
Intestinal smooth muscle	Avian erythrocytes
Uterus	Frog skin

several tissues (e.g., lung, spleen) the physiological significance of increased cyclic AMP is obscure, and often even the cell type(s) being stimulated are unknown.

The actions of other hormones or neurohumors on cyclic AMP levels

The other hormones or neurohormones which may have an action mediated through cyclic AMP are listed with the tissues which they stimulate in Tab. 3. In several cases the evidence supporting the relation of the hormone to cyclic AMP is well documented, while in other cases the evidence is less complete.

Table 3. *Hormones and neurohumors for which there is evidence that one or more actions may be mediated by cyclic AMP*

Hormone	Tissues
Catecholamines	Various tissues
ACTH	Adrenal cortex
	Adipose tissue
Vasopressin	Toad bladder
	Kidney
	Frog Skin
Luteinizing hormone (LH)	Corpus luteum,
	Ovary, testis
Insulin	Adipose tissue, liver
Thyroid stimulating hormone (TSH)	Thyroid, adipose tissue
Glucagon	Liver
	Adipose tissue
	Pancreas (β cells)
Melanocyte stimulating hormone (MSH)	Dorsal frog skin
Histamine	Brain
	Gastric Mucosa
Prostaglandin E_1	Adipose tissue
Serotonin	Fasciola hepatica
Acetylcholine	Heart muscle

We will consider briefly the actions of two of these hormones, vasopressin and luteinizing hormone. ORLOFF and HANDLER (1962) had shown that cyclic adenylic acid mimicked the effect of vasopressin on toad bladder i.e., the permeability to sodium and water was increased. In collaboration with our group it has been shown that vasopressin increases the accumulation of cyclic AMP in this tissue. This effect is specific for vasopressin and occurs rapidly enough to account for the changes in permeability. The action of vasopressin on accumulation of cyclic AMP is potentiated by theophylline (HANDLER et al., 1965).

In the other case Marsh and Savard (1966) have shown that cyclic AMP increases progesterone synthesis when incubated with corpus luteal slices. In collaborative experiments it was shown that L H increased the accumulation of cyclic AMP in this tissue. This effect was specific for L H and was rapid enough to account for changes in steroid production (Marsh et al., 1966).

Adenyl cyclase is found throughout the animal kingdom and is present in microorganisms (Makman and Sutherland, 1965). In

Table 4. *Cellular processes known to be influenced by cyclic AMP*

Enzyme or process affected	Change in activity or rate
Phosphorylase	Increased
Glycogen synthetase	Decreased
Phosphofructokinase	Increased
Fructose 1,6-diphosphatase	Decreased
Steroidogenesis	Increased
Lipolysis	Increased
Ketogenesis	Increased
Incorporation of amino acids into liver protein	Decreased
Acetate to liver fatty acids and cholesterol	Decreased
Gluconeogenesis	Increased
Release of amylase (rat parotid)	Increased
Permeability (toad bladder)	Increased
Pigmentation (dorsal frog skin)	Increased
HCl secretion (gastric mucosa)	Increased
DPNH oscillations (yeast extract)	Increased
Release of insulin	Increased

microorganisms the control agents are not hormones, at least as we know and define hormones. *E. coli* makes large amounts of cyclic AMP when deprived of a carbon source and addition of glucose or numerous other substrates brought about almost immediate cessation of this formation. The cyclic AMP content fell rapidly to low levels and appeared in the medium, and most likely was pumped out by the cells as in the case of avian erythrocytes (Davoren and Sutherland, 1963a). Hirata and Hayaishi (1965) have found a soluble cyclase system in *Brevibacterium liquefaciens*, the only soluble cyclase known to date. This enzyme is essentially inactive unless pyruvate, or related compounds are added. Thus, rapidly acting mechanisms exist in simple unicellular forms which are not dependent on new synthesis of protein (MAKMAN and SUTHERLAND, 1965).

A variety of enzymes or processes are affected by cyclic AMP (Tab. 4). Activation of phosphorylase was the first action described and is of some importance in certain tissues like liver and skeletal muscle. However, a relation of phosphorylase activation to most of the other events listed here is not obvious and in most cases is probably unimportant. It is our belief that the relation of cyclic AMP to phosphorylase has been overemphasized. Probably several dozen enzymes or structures will ultimately be found to be affected by cyclic AMP.

Adenyl cyclase probably varies in molecular configuration from tissue to tissue, which is not very surprising since there are numerous examples of enzyme variation from tissue to tissue and even within a single tissue. Such variations could account for the specificity of the hormone response. Also enzyme patterns vary in different tissues, thus allowing for variations in physiological responses when a single component such as cyclic AMP is altered. For example, increased cyclic AMP levels in the bovine corpus luteum result in mainly progesterone synthesis, while in the adrenal cortex glucocorticoids are made because the adrenal is endowed with enzymes which can catalyze the further transformations.

The production or secretion of steroids, probably thyroxin, insulin, and perhaps other hormones is stimulated by cyclic AMP. Such hormones can be viewed as third messengers which exit from their cells to influence other tissues. These hormones have basic actions of their own, but may at times collaborate with cyclic AMP. For example, the steroids and thyroxin may at times stimulate events so that the accumulation or action of cyclic AMP is enhanced. The reverse situation may occur also, for the cyclic nucleotide may assist directly or indirectly in the changes brought about by the third messengers.

The possibility that other cyclic nucleotides might have biological functions has recently come under investigation. Primary interest to date has been focused on cyclic GMP. This compound has been synthesized in several laboratories and was first found in nature in rat urine by ASHMAN et al. (1963) after injecting ^{32}P to label the nucleotide pool of rats.

An enzymatic method for measuring cyclic GMP (which can be applied to the measurement of other cyclic nucleotides) has been developed and applied to a study of the excretion of this nucleotide

$$\text{cyclic nucleotide} \xrightarrow{\text{phosphodiesterase}} \text{nucleoside monophosphate} \tag{1}$$

$$\text{nucleoside monophosphate} \xrightarrow[\text{ATP}]{\text{nucleoside monophosphate kinase}} \text{nucleoside diphosphate} \tag{2}$$

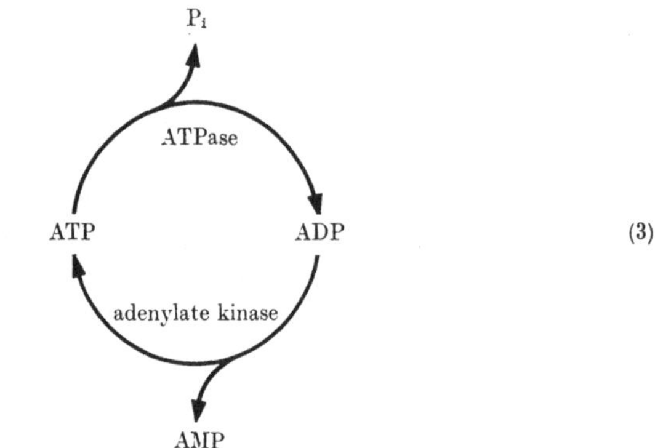

$$\tag{3}$$

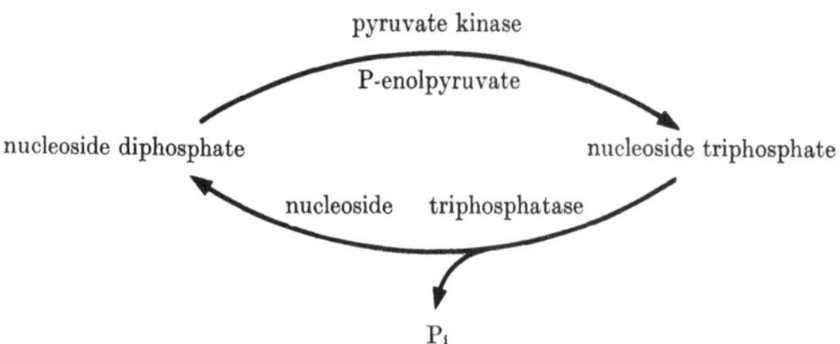

Fig. 9. A summary of the reaction sequence involved in assaying cyclic GMP (from Hardman et al., 1966)

in the urine of rats in various hormonal states (HARDMAN et al., 1966). Fig. 9 summarizes the series of reactions involved in this assay. Reaction 3 is necessary to remove ATP and ADP which otherwise would give a very high assay blank. The use of a cycling reaction (4) in which orthophosphate is generated makes possible the measurement of very small amounts of cyclic GMP.

A relatively large amount of cyclic GMP is excreted in the urine. The values range from about 40 to 75% of the amount of cyclic AMP excreted by rats. On removal of the pituitary, the excretion

Table 5. *Excretion of cyclic GMP and cyclic AMP in urine from normal and hypophysectomized rats*

Rats	No. of groups	Cyclic GMP	Cyclic AMP	Urine volume
		n moles/100 g rat		ml/100 g rat
Normal	17	40 ± 3.6	53 ± 3.3	36 ± 1.9
Hypophysectomized	11	17 ± 2.5	52 ± 5.7	12 ± 1.6
Hypophysectomized plus hydrocortisone	17	17 ± 2.1	53 ± 7.8	30 ± 2.7

Urine was collected for 24 h from groups of two or more female Sprague-Dawley rats. The animals were fed only a 10% dextrose solution during periods of urine collection. Operated and non-operated controls of the same weight range (80 to 100 g) were received from Hormone Assay Laboratories 1 or 2 days after surgery and urine was collected 3 to 5 days later. Hydrocortisone phosphate (MERCK, SHARP and DOHME), 0.25 mg/100 g, was injected subcutaneously twice a day for 24 to 72 h before and three time during the periods of urine collection. Values refer to total excretion in 24 h $\pm$ standard error (from HARDMAN et al., 1966).

of cyclic GMP falls to very low levels while the excretion of cyclic AMP is not altered. The administration of hydrocortisone restores the urine volume to near normal but does not affect cyclic GMP excretion (Tab. 5). Combined treatment with hydrocortisone and thyroxine causes a partial and, at times, complete restoration of the excretion. In addition to these two hormones, other pituitary factors, aging and some unknown factors appear to be involved in regulating either directly or indirectly the excretion of cyclic GMP. Considerably more work remains to be done before the physiological significance of the changes in excretion of cyclic GMP in altered hormonal states can be evaluated. Work in this area is in an early and very active phase.

Summary

Taken together, the data reviewed or mentioned here provide very strong evidence for a generalized mechanism of action for many hormones. That is, that nature often uses a two messenger system in the expression of hormone action. To date cyclic AMP represents the only clearly defined second messenger, but evidence for a similar role of cyclic GMP is accumulating and it seems not unreasonable that other second messengers will also be discovered. Among the many responses of cells to increased levels of cyclic AMP are the release of certain hormones (e.g., adrenal and gonadal steroids, thyroid hormone, and insulin), which may be considered third messengers after they are released. These third messengers may then go on to act on their target tissues: 1. as first messengers, thus changing second messenger levels in the target cell or; 2. by influencing the levels or activity of target cell constituents which are controlled by the second messenger; 3. by acting at a point independent of the second messenger.

What appear to be superficially identical second messenger systems in various tissues can act with the retention of hormone specificity because the adenyl cyclase systems of various tissues, although catalyzing the same reaction, discriminate between hormones with great specificity. In addition, cells of different tissues react in different ways to increased cyclic AMP levels because they have various enzymatic profiles — i.e., cells will do what they are equipped to do.

Finally, the location of the adenyl cyclase system in the cell membrane of at least three types of cells supports the time honored concept that many interactions between hormones and cells occur at the cell surface.

Abbreviations

Abbreviations used are: Cyclic AMP, adenosine $3', 5'$-monophosphate; LH, luteinizing hormone; PGE_1, prostaglandin E_1; DCI, 1 (3,4-dichlorophenyl)-2-isopropylaminoethanol hydrochloride.

For more extensive reviews on cyclic AMP the reader is referred to the following:

a) For a thorough review of the work up to 1960, SUTHERLAND, E. W., and T. W. RALL: Pharmacol. Rev. **12**, 265 (1960).

b) For recent reviews of the biological role of cyclic AMP; SUTHERLAND, E. W., and G. A. ROBISON: Pharmacol. Rev. **18**, 145 (1966).

ROBISON, G. A., R. W. BUTCHER, and E. W. SUTHERLAND: Proc. N.Y. Acad. Sci. **139**, 703 (1967).

BUTCHER, R. W., and E. W. SUTHERLAND: Proc. N.Y. Acad. Sci. **139**, 849 (1967).

References

APPLEMAN, M. M., R. G. KEMP, J. PERRY, and R. SORENSON: Biochem. biophys. Res. Commun. **24**, 564 (1966).

ASHMAN, D. F., R. LIPTON, M. M. MELICOW, and T. D. PRICE: Biochem. biophys. Res. Commun. **11**, 330 (1963).

BERTHET, J., E. W. SUTHERLNAD, and T. W. RALL: J. biol. Chem. **229**, 351 (1957).

BERTHET, J.: Proc. 4th Int. Congr. Biochem. **17**, 107 (1960).

BEWSHER, P. D., and J. ASHMORE: Biochem. biophys. Res. Commun. **24**, 431 (1966).

BIECK, P., K. STOCK, and E. WESTERMANN: Life Sciences **5**, 2157 (1966).

BLINKS, J. R.: Ann. N.Y. Acad. Sci. **139**, 673 (1967).

BRODIE, B. B., J. I. DAVIES, S. HYNIE, G. KRISHNA, and B. WEISS: Pharmacol. Rev. **18**, 273 (1966).

BURNS, J. J., K. I. COLVILLE, L. A. LINDSAY, and R. A. SALVADOR: J. Pharmacol. exp. Ther. **144**, 163 (1964).

BUTCHER, R. W., and E. W. SUTHERLAND: J. biol. Chem. **237**, 1224 (1962).

—, J. E. PIKE, and E. W. SUTHERLAND: Proc.: Nobel Symposium II. In press 1967.

—, R. J. HO, H. C. MENG, and E. W. SUTHERLAND: J. biol. Chem. **240**, 4515 (1965).

—, J. G. T. SNEYD, C. R. PARK, and E. W. SUTHERLAND: J. biol. Chem. **241**, 1652 (1966)

CHEUNG, W. Y., and J. R. WILLIAMSON: Nature (Lond.) **207**, 979 (1965).

DAVOREN, P. R., and E. W. SUTHERLAND: J. biol. Chem. **238**, 3009 (1963).

— — J. biol. Chem. **238**, 3016 (1963 b).

DRUMMOND, G. I., J. R. E. VALADARES, and L. DUNCAN: Proc. Soc. exp. Biol. (N.Y.) **117**, 307 (1964).

—, L. DUNCAN, and E. HERTZMAN: J. biol. Chem. **241**, 5899 (1966).

EXTON, J. H., and C. R. PARK: Pharmacol. Rev. **18**, 181 (1966).

FERGUSON, J. J.: J. biol. Chem. **238**, 2754 (1963).

GRAHAM-SMITH, D. G., R. W. BUTCHER, R. L. NEY, and E. W. SUTHERLAND: Clin. Res. In press 1967.

HALKERSTON, I. D. K., M. FEINSTEIN, and O. HECHTER: Proc. Soc. exp. Biol. (N.Y.) **122**, 896 (1966).

HAMMERMEISTER, K. E., A. A. YUNIS, and E. G. KREBS: J. biol. Chem. **240**, 986 (1965).

HANDLER, J. S., R. W. BUTCHER, E. W. SUTHERLAND, and J. ORLOFF: J. biol. Chem. **240**, 4524 (1965).

HARDMAN, J. G., J. W. DAVIS, and E. W. SUTHERLAND: J. biol. Chem. **241**, 4812 (1966).

HAUGAARD, N., and M. E. HESS: Pharmacol. Rev. **17**, 27 (1965).

Haynes, R. C., S. B. Koritz, and F. G. Peron: J. biol. Chem. **234**, 1421 (1959).

—, E. W. Sutherland, and T. W. Rall: Recent Progr. Hormone Res. **16**, 121 (1960).

Hirata, M., and O. Hayaishi: Biochem. biophys. Res. Commun. **21**, 361 (1965).

Imura, H., S. Matsukura, H. Matsuyama, T. Setsuda, and T. Miyake: Endocrinology **76**, 933 (1965).

Jungas, R. L.: Proc. nat. Acad. Sci. (Wash.) **56**, 757 (1966).

—, and E. G. Ball: Biochemistry **2**, 383 (1963).

Karaboyas, G. C., and S. B. Koritz: Biochemistry **4**, 462 (1965).

Klainer, L. M., Y. M. Chi, S. L. Friedberg, T. W. Rall, and E. W. Sutherland: J. biol. Chem. **237**, 1239 (1962).

Kukovetz, W. R., M. E. Hess, J. Shanfeld, and N. Haugaard: J. Pharmacol. exp. Ther. **127**, 122 (1959).

Larner, J., C. Villar-Palasi, and D. S. Richman: Ann. N.Y. Acad. Sci. **82**, 345 (1959).

Leloir, L. F., and C. E. Cardini: J. Amer. chem. Soc. **79**, 6340 (1957)

Levine, R. A., and J. A. Vogel: J. Pharmacol. exp. Ther. **151**, 262 (1966).

Lipkin, D., W. H. Cook, and R. Markham: Amer. Chem. Soc. **81**, 6075 (1959).

Makman, M. H., and E. W. Sutherland: Endocrinology **75**, 127 (1964).

Makman, R. S., and E. W. Sutherland: J. biol. Chem. **240**, 1309 (1965).

Mandel, L. R., and F. A. Kuehl: Fed. Proc. **26**, 810 (1967).

Marsh, J. M., and K. Savard: Steroids 8, 133 (1966).

—, R. W. Butcher, K. Savard, and E. W. Sutherland: J. biol. Chem. **241**, 5436 (1966).

Mayer, S. E., de V. M. Cotten, and N. C. Moran: J. Pharmacol, exp. Ther. **139**, 275 (1963).

Murad, F., Y. M. Chi, T. W. Rall, and E. W. Sutherland: J. biol. Chem. **237**, 1233 (1962).

Namm, D. H., and S. E. Mayer: Fed. Proc. **26**, 351 (1967).

Orloff, J., and J. S. Handler: J. clin. Invest. **41**, 702 (1962).

Oye, I.: Acta physiol. scand. **65**, 251 (1965).

—, and E. W. Sutherland: Biochim. biophys. Acta (Amst.) **127**, 347 (1966).

Pastan, I.: Biochem. biophys. Res. Commun. **25**, 14 (1966).

Posternak, Th., E. W. Sutherland, and W. F. Henion: Biochem. biophys. Acta (Amst.) **65**, 558 (1962).

Pryor, J., and J. Berthet: Biochim. biophys. Acta (Amst.) **43**, 556 (1960).

Rall, T. W., E. W. Sutherland, and W. D. Wosilait: J. biol. Chem. **218**, 483 (1956).

— —, and J. Berthet: J. biol. Chem. **224**, 463 (1957).

— — J. biol. Chem. **232**, 1065 (1958).

— — J. biol. Chem. **237**, 1228 (1962).

—, and T. C. West: J. Pharmacol. exp. Ther. **139**, 269 (1963).

Reiter, M.: 1964, In Pharmacology of Cardiac Function, p. 25. O. Krayer, ed. New York: Pergamon Press 1964.

ROBISON, G. A., R. W. BUTCHER, and E. W. SUTHERLAND: Ann. N.Y. Acad. Sci. **139**, 703 (1967a).

— —, I. ØYE, H. E. MORGAN, and E. W. SUTHERLAND: Molec. Pharmacol. **1**, 168 (1965).

—, J. H. EXTON, C. R. PARK, and E. W. SUTHERLAND: Fed. Proc. **26**, 257 (1967b).

RODBELL, M.: J. biol. Chem. **239**, 375 (1964).

SCHULTZ, G., G. SENFT und K. MUNSKE: Naturwissenschaften **53**, 529 (1966).

STEINBERG, D., M. VAUGHAN, P. J. NESTEL, O. STRAND, and S. BERGSTROM: J. clin. Invest. **43**, 1533 (1964).

SUSSMAN, K. E., and G. D. VAUGHAN: Diabetes, In press 1967.

SUTHERLAND, E. W.: Ann. N.Y. Acad. Sci. **54**, 693 (1951).

— The Harvey Lectures, Series 57, p. 17. New York: Academic Press 1962.

— Proc. of the 3rd International Congr. of Biochem., p. 318. New York: Academic Press 1956.

—, and T. W. RALL: J. biol. Chem. **232**, 1077 (1958).

— — Pharmacol. Rev. **12**, 265 (1960).

— —, and T. MENON: J. biol. Chem. **237**, 1220 (1962).

—, I. ØYE, and R. W. BUTCHER: Recent Progr. Hormone Res. **21**, 623 (1965).

—, and G. A. ROBISON: Pharmacol. Rev. **18**, 145 (1966).

TSUJIMOTO, A., S. TANINO, T. NISHIUE, and Y. KUROGOCHI: Jap. J. Pharmacol. **15**, 441 (1965).

TURTLE, J. R., G. K. LITTLETON, and D. M. KIPNIS: Nature (Lond.) **213**, 727 (1967).

VALADARES, J. R. E., and A. J. D. FRIESEN: Life Sciences **5**, 521 (1966).

VAUGHAN, M.: J. biol. Chem. **235**, 3049 (1960).

—, and D. STEINBERG: J. Lipid Res. **4**, 193 (1963).

WELSH, J. H.: Ann. N.Y. Acad. Sci. **66**, 618 (1957).

WOSILAIT, W. D., and E. W. SUTHERLAND: J. biol. Chem. **218**, 469 (1956).

Diskussion

DIRSCHERL (Bonn): Ich danke Herrn SUTHERLAND und eröffne die Diskussion.

CLAUSER (Paris): Dr. SUTHERLAND, does one know which enzyme systems respond to cyclic 3′-5′-AMP in E. coli?

SUTHERLAND (Nashville/USA): I have no idea. I think that nothing is known about the function of cyclic AMP in microorganisms.

KUKOVETZ (Graz): Dr. SUTHERLAND, you stated that IMA did not inhibit the stimulatory effect of epinephrine on 3′,5′-AMP production. In experiments on isolated guinea pig hearts we tried to dissociate with IMA the effects of catecholamines on phosphorylase activity and contractility. We found that IMA reduced both types of effects, but failed to dissociate them. We, therefore, believe that IMA is just a weak beta blocking agent. Did you use the L-isomer or the racemate? We used both, but found the L-form more active.

SUTHERLAND: We got the preparation from Dr. BYRN's group, and I can't recall at the moment whether it was the racemate or the L-form. — I would like to comment a little further on this. This is indeed a beta blocking agent, and if you use those species that can stand a large amount, e.g. the turtle, then you can show that it is a beta blocking agent even on the heart.

SENFT (Berlin): I would like to ask about the effect of cyclic AMP on insulin secretion you mentioned first. There is more and more evidence that cyclic AMP stimulates insulin secretion. But how is this compatible with the finding of RANDLE et al. that epinephrine decreases insulin secretion?

SUTHERLAND: This is indeed a very interesting point. In the first place, I think it fits teleologically, although we should not argue this way, I suppose. If a person becomes frightened and wants to flee, he must mobilize glucose for the activity of his brain; he wants the glucose to be used, not to be stored somewhere. — We have a theory about the two types of epinephrine receptors, the alpha and the beta, and there is a special case where they can work in opposite directions. In this case, you could obtain the results you just mentioned. If isopropyl nor arterenol is used instead of epinephrine insulin secretion is stimulated.

SCHREIBER (Freiburg): In relation to the increase by 3′,5′-cyclic AMP of the secretion of amylase in the rat parotis I would like to ask: what are the ultrastructural changes in the endoplasmic reticulum and the golgi apparatus brought about by cyclic AMP?

SUTHERLAND: Buried in the literature, there is a report by PORTER et al. from the Rockefeller Institute. After injection of glucagon he found some structures in the liver cell which are similar to lysosomes. They look like the sheath of lysosomes. There is apparently a rapid formation of some phosphorlipid membrane in the cell. In adrenal and thyroid glands similar phenomena are described. But how this is related to the known structures is unclear; it may have something to do with the lysosomes.

CHANDRA (Frankfurt): Would you suggest from your results that the hormonal stimulation of cyclic adenylic acid is organ specific? In one of your slides you showed that thyroid stimulating hormone (TSH) stimulates cyclic adenylic acid in the thyroid gland. — Did you check up other organs too?

SUTHERLAND: We have not worked with TSH recently; this is the work of other laboratories. As far as I know, there is a response not only in the thyroid, but also in the fat pad which responds to everything. Otherwise, I don't know of any breaks in the specificity.

KUKOVETZ: I would like to briefly comment on our results with the dibutyryl-3′,5′-AMP. We did experiments in isolated perfused guinea pig hearts, and we found that rather high doses of 5 and 10 μM produced definite increases in rate, contractility and coronary flow, but not initially in phosphorylase activity of the isolated heart. In samples, taken six minutes after administration of the dibutyryl derivative, phosphorylase activity was also increased. The inotropic and chronotropic effects disappeared and reoccurred repeatedly over a period of 10 to 15 min. When dibutyryl-3',5'-AMP was given during infusions of 0.5 to 1 mg/min theophylline, its effects on contractility were more pronounced, fluctuated less and there was also an immediate rise in phosphorylase activity.

SUTHERLAND: This seems very interesting. It is something which we tried to do without success using rat hearts.

BÜCHER (München): You mentioned that in adipose tissue, a small increase in cyclic AMP gives rise to a large secretion of fatty acids. Would you think that this is due to compartmentation? You get a production of about one nano-mole of cyclic AMP which produces this effect as compared to the two nano-moles already present in the tissue.

SUTHERLAND: I don't know. I think that is a possibility. Another possibility is that the cyclic AMP already present is firmly bound to some structures and therefore not active. The small increase above this is sufficient to activate lipolysis.

CLAUSER: Does one know to what extent the activity of adenyl cyclase in membrane fragments is dependant on the integrity of lipoprotein structures? What are the effects of phospholipases?

SUTHERLAND: There are studies by IVAR OYE from Norway that after destructions of membranes by pressure homogenization, there is still cyclase activity, but the hormone effect is gone. Thus, the structural integrity seems to be important for the hormone effect. However, after addition of phospholipase C, both hormone and cyclase effects are retained, though the membrane is damaged. However, the membranes may not be really dispersed in these experiments. If we take a particulate fraction which responds to epinephrine and really disperse it with a 2% solution of triton X 100, we usually have some cyclase left, but our hormone effects are gone.

MOSEBACH (Bonn): You have demonstrated that in brain histamine affects cyclic AMP. Can you say which parts of the brain?

SUTHERLAND: Drs. KAKIUCHI and RALL have done this work, and there is a report on this in the press.

SENFT: I should like to comment on the topic "action of 3',5'-AMP as a mediator of the glycogenolytic effect of epinephrine". We have been interested to learn if hormones may regulate glycogen metabolism not only by stimulating adenyl cyclase but also by varying the rate of breakdown of 3',5'-AMP by influencing 3',5'-AMP phosphodiesterase activity.

Insulin deficiency caused by administration of alloxan or by withdrawal of food has been found to cause a decrease in activity of this enzyme in liver, skeletal muscle and adipose tissue but not in kidney. Administration of nsulin or stimulation of its secretion by feeding the animals results in an increase in 3′,5′-AMP phosphodiesterase activity within a relatively short period of time. Maximal activity is measured half an hour after intraveneous injection of 0.5 U insulin/kg b.w. The hormonal effect is prevented by pretreating the animals with actinomycin D [SCHULTZ, G., G. SENFT und K. MUNSKE: Naturwissenschaften **53**, 529 (1966); Naunyn-Schmiedebergs Arch. Pharmak. exp. Path. **257**, 62 (1967), and unpublished data]. Opposite effects have been observed with glucocorticoids. 3′,5′-AMP phosphodiesterase activity in lever, adipose tissue and kidney is increased in adrenalectomized animals. Administration of 6-methylprednisolone (1 mg/kg and day) has been found to depress the activity of the enzyme to control values (unpublished data).

SUTHERLAND: I think these are quite interesting developements; they may be a basis for a more unified theory of insulin action. You might extent it to an enzyme in a membrane, which will give you transport changes.

SEIF (Tübingen): You have presented results indicating a decrease in [cyclic-3′,5′-GMP] in rat urine after hypophysectomy, which increases to normal levels after thyroid hormone administration. Have you got any idea about the lag period until the onset of increase of cyclic-GMP in urine?

SUTHERLAND: We havn't studied this in great detail. I have the impression that it is a matter of one or several days. A glucocorticoid as well as thyroxin is required.

WOHNLICH (Paris): We have found a glyconeogenetic effect of epinephrine in liver homogenate. Could this be a cyclase effect?

SUTHERLAND: It could be. You can find cyclase activity in the supernatant of a homogenate; this depends on the method of homogenization. You could check this if you add cyclic AMP to the homogenate and see if it works. You may have to add it several times, because it is destroyed quite rapidly.

Hormonwirkungen an isoliert durchströmten Organen

Von H. Schimassek

Physiolog.-chem. Institut der Philipps-Universität, Marburg/Lahn

Mit 3 Abbildungen

Die Spiegel der Substrate im Blut sind erstaunlich konstant. Die Analysen dieser Substratgrößen dienen daher in vieler Hinsicht sowohl der theoretischen als auch der praktischen Biologie als Indikatoren für Änderungen und Störungen im Stoffwechsel bestimmter Organe. Die Halbwertszeit der einzelnen Substrate im Blut ist unterschiedlich; sie kann Tage betragen (z. B. für Cholesterin[1]; sie beträgt für Glucose nur 50 min[2], für Pyruvat 22 min[3] und für Fettsäuren sind noch kürzere Zeiten gemessen worden. Die Leistung des Organismus, die Spiegel der so rasch umgesetzten Substrate konstant zu halten, ist also beträchtlich. Die Spiegeleinstellung ist oft das Resultat von Substratabgabe des einen und Substrataufnahme des anderen Organs. Regulation von Abgabe und Aufnahme der Substrate und Koordination des Stoffwechsels einzelner Gewebe untereinander ist u. a. eine der Aufgaben der Stoffwechselhormone und damit zugleich ein gutes Beispiel für die Wirkung der Hormone.

Um die Zusammenhänge dieser Regulation und Koordination analysieren zu können, müssen wir die Wirkungen des einzelnen Hormons in den verschiedenen Organen kennen. Diese Effekte zu analysieren, ist nur in vitro möglich; denn in vivo können wir den Einzeleffekt eines Hormons nicht von Parallel- oder Gegenregulationen anderer Hormone oder von verschiedensten Substrateffekten differenzieren. Welche der „in vitro-Techniken" für derartige Untersuchungen gewählt wird, ist allein von der Fragestellung abhängig. Je komplexer die Fragestellung, je mehr sie auf die Analyse von bestimmten Stoffwechselwegen und auf das Zusammenspiel von Stoffwechselketten hinausläuft — und das Studium der Hormonwirkungen wird in diesen Bereich führen —, desto aufwendiger wird auch die Versuchsanordnung werden. Dieser Weg führt zu Versuchen mit isoliert durchströmten Organen, da nur diese Methode es ermöglicht, Zellverband und Zellstrukturen zu erhalten.

Krebs, der viele und wesentliche Entdeckungen auf dem Sektor des Intermediärstoffwechsels in Homogenaten und an Gewebeschnitten gemacht hat, hat in einer neueren Arbeit auf den Wert von Perfusionsexperimenten hingewiesen[4].

Auf Technik und Methodik der Perfusionsexperimente soll an dieser Stelle nicht weiter eingegangen werden*. Eine Diskussion kann eventuelle Unklarheiten rascher und gezielter klären. Allgemein sei hierzu gesagt: Die Fragestellung allein entscheidet über die anzuwendende Technik. Die Versuchsanordnung ihrerseits limitiert die Möglichkeit der Interpretation der Versuchsergebnisse. Wesentlicher als das Thema der Methodik ist die Frage, inwieweit ein isoliert durchströmtes Organ uns auf so komplexe Fragen wie die der Wirkung von Hormonen in bestimmten Stoffwechselwegen noch Antwort geben kann, d. h. mit welchen Kriterien wir seine Funktion sinnvoll nachweisen können. Es ist sicher ungenügend, das histologische Bild oder die Gallesekretion als einzigen Nachweis der Funktion des isolierten Organs zu verwenden. Für die isoliert perfundierte Leber können wir unsere Daten in Stichworten nennen, die zeigen, in welch hohem Maße das isolierte Organ funktionstüchtig bleibt[6]. Im durchströmten Organ bleiben Enzymmuster und Enzymaktivitäten unverändert, so daß die Möglichkeit der Stoffumsetzungen gegeben ist. Es treten keine Enzyme der Leber in das Außenmedium; das bedeutet, daß Zelle und Zellmembran intakt bleiben. Isolierung von Mikrosomen und — in Zusammenarbeit mit Klingenberg — von Mitochondrien, belegen, daß auch subcelluläre Partikel ihre Funktion während der Perfusion behalten. Als integrierende Größe haben wir den O_2-Verbrauch des isolierten Organs gemessen[7]. Er beträgt 2.2 μ Mole/g/min und liegt damit im Bereich vergleichbarer in vivo-Werte. — Die Technik der Organperfusion ermöglicht — wie sich aus diesen Kriterien entnehmen läßt — das gleichzeitige Studium vieler Fragen, angefangen von Tauschprozessen über die Membran, von Stoff-

---

* Allen, die diese Technik übernehmen wollen, sei das Studium eines grundlegenden Artikels der Chirurgen über extracorporalen Kreislauf empfohlen[5].

wechselvorgängen im Organ, von Analysen subcellulärer Strukturen
bis hin zu dem molekularen Bereich.

Für das Studium der Hormone ist ferner wesentlich zu wissen,
daß ein isoliertes Organ den in vivo hormonal vorgestellten Status
behält und darüber hinaus auf Zusätze von Hormonen ansprechbar
bleibt. Die Tab. 1 belegt an Hand von Substratumsätzen und Stoff-
wechselgrößen, daß sowohl der in vivo vorgestellte Status während

Tabelle 1. *Einfluß verschiedener Hormone auf Umsatzgrößen und den Redox-
Quotienten Lactat/Pyruvat in der isoliert perfundierten Rattenleber (hormonaler
Status in vivo vorgestellt bzw. durch Hormoninfusion in vitro eingestellt)*

	Lactat-umsatz (μ Mole/g/ Std)	$Q\,\dfrac{\text{Lactat}}{\text{Pyruvat}}$	O_2-Verbrauch (μ Mole/g/ min)	
unbehandelt	20	10	2,2	
Hyperthyreose	60	10	>4	in vivo
Hypothyreose	3	17	1,8	vorbehandelt
Prednisolon (70 μ Mole/Std)	5	18	1,9	in vitro infundiert
Adrenalin (190 n Mole/Std)	50	9	2,6	
Glucagon (0,5 n Mole/Std)	60	6	2,2	

der Perfusion erhalten bleibt als auch das isolierte Organ auf Hormonzusatz noch reagiert. —

Wenn ein isoliert durchströmtes Organ auf Grund so vieler empfindlicher Kriterien voll funktionsfähig bleibt, so wird es auch entsprechend der Funktion, die es in vivo zu erfüllen hat, in vitro seine
Aufgaben widerspiegeln.

Die Abb. 1 zeigt Abgabe und Aufnahme verschiedener Substrate
durch die isoliert perfundierte Rattenleber. Dem Medium waren
keine Substrate gesondert zugesetzt. Wir finden, daß das isolierte
Organ Lactat, Pyruvat und Glucose an das Außenmedium abgibt,
freie Fettsäuren dagegen aufnimmt. Für Lactat und Pyruvat stellt
das isolierte Organ konstante Spiegel ein, die im Bereich physiologischer Werte liegen. Die Glucosespiegel dagegen steigen im Verlauf der Perfusion kontinuierlich an. Sie liegen nicht im Bereich

physiologischer Werte, sondern sind 2- bis 2,5mal so hoch wie im Rattenblut. Umgekehrt ist das isolierte Organ bestrebt, freie Fettsäuren möglichst rasch aus dem Außenmedium aufzunehmen.

Für Lactat und Pyruvat werden nicht nur konstante Spiegel eingestellt, sondern jedes Über- oder Unterangebot wird wieder auf den Ausgangswert korrigiert[8]. Das gleiche gilt auch für das komplette Spektrum der Aminosäuren und beleuchtet damit eindrucksvoll die Leistung des isolierten Organs für die Regulation der Sub-

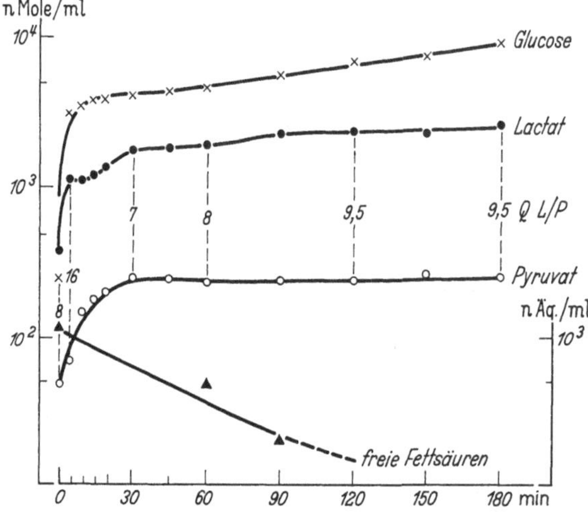

Abb. 1. Funktion der Leber in Hinblick auf Spiegeleinstellung bzw. Abgabe und Aufnahme verschiedener Blutsubstrate

stratspiegel[9]. Auch hier werden für alle Aminosäuren konstante Spiegel eingestellt, von denen der größte Teil im Bereich physiologischer Grenzen liegt. — Die gleichen Gehaltsunterschiede, die zwischen Leberzelle und Außenmedium in vivo für die einzelnen Aminosäuren bestehen, werden auch in vitro eingestellt[9].

Mit diesen Daten dokumentiert uns das isolierte Organ eine ganze Reihe von Funktionen, die es im Zusammenspiel mit anderen Organen zu erfüllen hat und die für uns die Basis für die zu besprechenden Hormonwirkungen im Leberstoffwechsel bilden. Sie lauten: Verantwortliche Einstellung der Substratspiegel von Lactat, Pyruvat und den gesamten Aminosäuren, Aufnahme von freien Fettsäuren und Umwandlung in Fettsäureester sowie Produktion von Glucose für den gesamten Organismus.

Glucose ist ein Produkt der Leber und kein Substrat für die Leber. Sie nimmt auch bei zusätzlicher Glucoseinfusion keine deutlich meßbaren Mengen an Glucose aus dem Außenmedium auf, auch dann nicht, wenn wir eine glykogenfreie Leber eines Hungertieres durchströmen[10].

Eine letzte notwendige Basis für das Studium der Stoffwechselregulation und damit auch der Wirkung der Hormone ist die Charakterisierung des metabolischen Status des isolierten Organs. Er wird repräsentiert durch die Substratspiegel. An Hand von Substratgehalten des Embden-Meyerhof-Weges und des Citratcyclus ist belegt worden, daß der metabolische Status des isolierten Organs für 3—4 Std mit dem Status in vivo erstaunlich gut übereinstimmt, obwohl der Leberstoffwechsel während der O_2-freien Operationszeit einem Extremzustand unterworfen war[6]. — Generell liegen die in vitro gemessenen Spiegel etwas zu hoch. Das ist verständlich, da die Peripherie als Donator und vor allem als Akzeptor für den Substrattausch wegfällt und das Organ sich in einem Selbstkreislauf befindet.

Bei Untersuchungen, die Substratumsetzungen unter Hormonwirkungen betreffen, ist eine gewisse Sorgfalt am Platze. Denn WOOL beschreibt, daß bei Herzperfusionen mit einfachen Salzlösungen das durchströmte Herz beispielsweise erhebliche Mengen an Aminosäuren an das Außenmedium verliert[11]. Dieser Verlust ist nur dadurch zu kompensieren, daß zusätzlich zur Zugabe aller Aminosäuren auch Albumin im Medium enthalten ist. Ferner beschreibt WILLIAMSON bei Herzperfusionen mit einfachen Salzlösungen, daß eine ganze Reihe von intracellulär zuständigen Substraten wie die Hexosephosphate im Außenmedium umgesetzt werden, wodurch besonders bei Versuchen mit radioaktiven Verbindungen grobe Fehler zustande kommen können[12].

In beiden Fällen blieben unter den experimentellen Bedingungen die Gewebszellen nicht vollständig intakt, so daß sowohl Substrate als auch Enzyme in das Außenmedium gelangten.

Mit der Tab. 2 soll gezeigt werden, auf welch weitem und verzweigtem Feld auf dem Hormongebiet mit isoliert durchströmten Organen gearbeitet wird. Die kleinsten Objekte, die wir kennen, sind die Perfusionen von isolierten Nebennieren der Ratte[13] und die Perfusion der isolierten Retina des Frosches[14]. Sie seien hier als Anregung und Ermutigung genannt. Es muß nicht immer eine

Rinderleber sein — auch dafür gibt es Beispiele[15]. Literatur für
Tab. 2 ([16–30]).

Im Rahmen dieses Colloquium dürften die Arbeiten besondere
Beachtung verdienen, die sich mit Proteinsynthese bzw. Enzyminduktion beschäftigen. Es erhebt sich nämlich die Frage, ob ein
isoliertes Organ, das über Stunden ohne kontinuierliche Hormon-
kontrolle ist, gleich große Proteinsynthese zu leisten vermag. Die
Antwort auf diese Frage fällt unterschiedlich aus. Nach den Arbei-

Tabelle 2. *Beispiele für Untersuchungen von Hormonwirkungen an isoliert
perfundierten Organen*

Fragestellung	Organ	Befund
Hormonsynthese	Herz	Tyrosin-^{14}C → Adrenalin, Acetat-^{14}C → NNR-Hormone
Hormonfreisetzung	Pankreas	Glucosespiegel ↑ Insulin ↓ Glucagon
Hormonabbau	Leber	NNR-Hormone, Insulin, Glucagon
Substratsynthese, Enzymaktivierung	Herz Leber	Adrenalin Glucagon $> 3'5'$ AMP → Phosphorylase
Proteinsynthese	Herz Leber Milz	Insulin Thyroxin ⎱ Einbau NNR-Hormone ⎰ ^{14}C-markierter Aminosäuren
Enzyminduktion	Leber	NNR-Hormone: Tryptophanpyrrolase Tyrosin-Ketoglutarat- Transaminase Thyroxin: Hydroxymethylglutaryl- Reduktase

ten von Miller[31] ist die Albuminsynthese der isoliert durchströmten Leber gleich groß wie die der Leber in vivo. Gordon und
Mutschler[32] haben die Synthese zweier spezifischer Plasmapro-
teine, Transferrin und Fibrinogen, untersucht. Sie finden, daß die
Leistung des isolierten Organs für Transferrin 85%, für Fibrinogen
jedoch nur 15% des in vivo-Wertes beträgt. Fest steht, daß bisher
alle je untersuchten Eiweißkörper auch im isolierten System synthetisiert werden konnten[33, 34]. Fest steht ferner, daß die Eiweiß-
synthese isoliert durchströmter Organe über Stunden wesentlich
höher liegt als die von Organschnitten.

Ein weites und interessantes Feld ergibt sich im Hinblick auf die Induktion von Enzymen im isolierten System, wie es erstmals von KNOX[28] für die Tryptophanpyrrolase unter Cortisol nachgewiesen wurde. WIELAND[30] konnte in Versuchen mit durchströmten Rattenlebern unter Zusatz von Trijodthyronin die Hydroxymethylglutaryl-reduktase teilweise um den Faktor 10 in ihrer Aktivität steigern. BARNABEI und SERENI[29] erreichten unter Cortisol ebenfalls an isoliert durchströmten Rattenlebern eine Aktivitätssteigerung der Tyrosin-α-Ketoglutarat-transaminase.

Der Aktivitätsanstieg war verhindert sowohl durch Hemmstoffe der Proteinsynthese als auch der Nucleinsäuresynthese. Cortisol stimulierte darüber hinaus auch die Markierung der Ribonucleinsäure der Kernfraktion und in geringem Maße auch die der cytoplasmatischen Ribonucleinsäure nach Zugabe von [6-^{14}C]-Orotsäure zum Perfusionsmedium. Der Hormoneffekt war auch hierbei durch Actinomycin D und Mitomycin C zu unterbinden. Die Autoren fanden ferner, daß Lebern, die mit Cortisol perfundiert waren, gegenüber Kontrollen ein höhere Aktivität der Ribonucleinsäure-Polymerase besaßen. Die Tatsache, daß die Synthese von Ribonucleinsäuren um 30 min früher einsetzte als der Anstieg der Enzymaktivität und die Tatsache, daß auch die Polymerase durch das Hormon aktiviert wurde, sehen die Verfasser im Zusammenhang und deuten die Versuche dahin, daß unter Cortisol die Synthese von Messenger-Ribonucleinsäure-Molekülen für die Synthese des Enzyms, also für die Neusynthese verantwortlich ist.

Diese Arbeitsrichtung, nämlich das Studium der Proteinsynthese bis hin zu den Nucleinsäuren des Kernes, bedient sich der Perfusionstechnik, um die primären Angriffspunkte („action") einer bestimmten Gruppe von Hormonen im Stoffwechsel fassen zu können. Die mehr indirekt gestellte Frage nach der Hormonwirkung betrifft die der Hormoneffekte, d. h. die distinkten Wirkungen der Hormone in den einzelnen Stoffwechselketten. Auf diesem indirekten Wege lassen sich Hormonwirkungen durch eine Fülle von „Effekten" beschreiben. Ziel der Untersuchungen muß es aber sein, das Feld der vielfältigen Symptome einzuschränken auf — im Idealfalle — einen einzigen, steuernden Effekt. Dieser kann dann für die Molekularbiologen der Ansatzpunkt zur Klärung der Frage nach der „action of hormones" sein.

Es wurde eingangs erwähnt, daß eine der Aufgaben der Hormone darin besteht, nicht nur in einem Zellverband den Stoffwechsel zu regulieren, sondern die Koordination der Einzelgewebe untereinander zu erreichen. Verbindendes Glied zwischen den Organen ist das Blut. Wir haben uns daher eingangs eingehend mit den Aufgaben der Leber im Hinblick auf die Blutsubstrate beschäftigt. Aufgabe der Leber ist es, Glucose an den Organismus abzugeben und Lactat aus der Peripherie aufzunehmen. Am Cori-Cyclus hat sich nach unserer Meinung im Hinblick auf den Substrattausch nichts geändert.

Tabelle 3. *Glucoseabgabe und Harnstoffproduktion der isoliert perfundierten Leber gefütterter und hungernder Ratten unter verschiedenen Konzentrationen von Glucagon*

	Glucagon (n Mole/100 ml/ Std infundiert)	Glucoseabgabe (μ Mole/10 g/Std)	Harnstoff- produktion (μ Mole/10 g/Std)
gefüttert:	0,01	500	30
	0,1	1200	100
	3,0	1300	200
14 bis 18 Std	0,1	100 bis 150	300
Nahrungsentzug:	3,0	bis 200	300

Akzeptor für die Glucose ist in erster Linie die Muskulatur. Starke Muskelarbeit verbraucht vermehrt Glucose und liefert über das Blut vermehrt Lactat zur Leber. Wie erfolgt die Koordination des Leberstoffwechsels für diese Situation? Verminderung des Blutzuckerspiegels durch höheren Verbrauch setzt Glucagon in Freiheit. Wir hatten in der Tab. 1 bereits gesehen, daß Glucagon den Lactatstoffwechsel der Leber verdreifacht — und damit bereits eine Anpassung erreicht. Glucagon ist auch für die Glucoseproduktion der Leber verantwortlich.

Infundieren wir in Versuchen mit isoliert perfundierten Lebern Glucagon in das Perfusionsmedium, so kommt es unmittelbar zu einer starken Glucoseausschüttung aus der Leber in das Außenmedium[35]. Der Effekt ist in engen Grenzen von der Hormonkonzentration abhängig. Der Anstieg erfolgt linear und beträgt bei einer Konzentration von 0,01 n Mol/100 ml/h 500 μ Mole Glucose/Std/ 10 g Leber und bei Konzentration von 0,1 n Mol an aufwärts etwa

1 m Mol/Std (Tab. 3). Die bisher übliche Erklärung für die Glucoseproduktion unter Glucagon war, daß Glucagon zur Bildung von
3′5′ AMP führt, daß dadurch das Phosphorylasesystem aktiviert,
Glykogen abgebaut, G 6 P gebildet und über die Glucose-6-Phos-
phatase Glucose ins Außenmedium abgegeben wird[36]. Stellt man
aber Bilanzen auf über Glucoseabgabe und Glykogenverbrauch, so
gehen diese niemals auf. Es wird immer mehr Glucose abgegeben
als der Glykogenverminderung entspricht. Die Differenz wird noch
krasser, wenn man innerhalb eines Versuches die Bilanzen über die
Zeit aufstellt. Die Tab. 4 gibt einen Versuch wieder, in dem Glu-

Tabelle 4. *Glucose und Glykogenbilanzen unter Glucagon*

Perfusionszeit	unbehandelt	Glucagoninfusion (0,3 n Mole/100 ml/Std)
	0 bis 120 min	120 bis 180 min
Δ Glykogen (μ Mole Glucose)	640	280
Glucoseabgabe (μ Mole)	532	875
Glykogenanteil an Glucoseabgabe	100%	32%

Wistar-Ratte ♂ 218 g
Lebergewicht: 8,5 g; Glykogengehalt: 323 μ Mole/gf.

coseausschüttung und Glykogenabnahme über 120 min ohne Hor-
monzugabe verglichen werden. Die Glucoseabgabe entspricht in
diesem Zeitraum der Glykogendifferenz. Nach 120 min wurde Glucagon zum Außenmedium infundiert und nach 1 Std wiederum
Glucoseausschüttung und Glykogendifferenz bestimmt. Unter Glucagon kann die in das Außenmedium abgegebene Glucose niemals
allein aus dem Glykogen stammen.

Unsere ausgedehnten Basisuntersuchungen gaben uns den Hinweis für einen wesentlichen Effekt des Glucagons. Es hatte sich
gezeigt, daß sich die Gehalte von G 6 P und F 6 P zum FDP im
Verlauf der Perfusion gegensinnig verändern. Der Gehalt an G 6 P
sinkt im Verlauf der Perfusion ab und der von FDP steigt an.
G 6 P und F 6 P sind auch in vitro über die Hexoseisomerase stets
in einem Gleichgewicht von 5:1. Wir können daher im folgenden
der Einfachheit halber stets von G 6 P und FDP sprechen. Um

die Änderungen in den Substratspiegeln in diesem Bereich deutlicher zu machen, haben wir den Q G6P/FDP gebildet. Dieser Q beträgt in vivo 17, in perfundierten Lebern dagegen 4 (Tab. 5). Auf Grund der vollen Funktion, die das isolierte Organ in vitro aufweist, konnten diese Abweichungen nicht durch Schädigung des Organs bedingt sein, sondern gaben einen Hinweis dafür, daß hier steuernde Faktoren fehlten. Substitution von Glucagon gleicht die Spiegel von G 6 P und FDP sofort denen in vivo wieder an, wobei wir uns mit den Hormonkonzentrationen absolut im physiologischen Bereich bewegen. Wir können sogar sagen, daß wir damit

Tabelle 5. *Substratgehalte (n Mol/gf) und Quotienten von G 6 P, F 6 P und FDP in der isoliert perfundierten Rattenleber unter Glucagon- und Lactatinfusionen im Vergleich zu Kontrollen in vivo und in vitro*

	in vivo	in vitro	Glucagon (3 n Mol/h)	Lactat (500 μ Mol/h)	Lactat (500 μ Mol/h) Glucagon (3 n Mol/h)
G 6 P	370	130	390	327	1000
F 6 P	74	26	78	65	197
FDP	22	31	19	90	9
G 6 P/FDP	17	4,5	21	3,6	110
F 6 P/G 6 P	0,2	0,2	0,2	0,2	0,2

im Bereich molekularer Dimensionen sind; denn Infusion von z. B. 0,1 n Mol Glucagon/100 ml/h ergibt eine Konzentration von $1,7 \times 10^{-13}$ Molen Glucagon/min/Leber. Die Rechnung/Zeiteinheit ist erlaubt, da Glucagon durch die in der Leber befindliche Glucagonase sofort inaktiviert wird. Diese Hormonkonzentration ist so klein, daß die Umrechnung in Moleküle sinnvoll wird. Sie entspricht 1×10^{11} Molekülen/Leber. Eine Rattenleber von 10 g enthält 2×10^{9} Zellen. Damit kommen auf eine Leberzelle nur 50 Hormonmoleküle. Eine solche Bilanz ist ermutigend, über die Topographie der Hormonmoleküle in der Zelle nachzudenken.

Kommen wir zurück zu unserer Tabelle. Den Ausgleich des G6P-Spiegels können wir auch durch einfache Lactatinfusion erreichen. Ist diese Infusion so groß, daß die Kapazität der Leber zum Lactatumsatz überschritten wird, so steigen die Lactatspiegel innen und außen an. Der Lactatanstieg innerhalb der Zelle führt zu einem generellen Anstieg der Substrate. Dabei erfahren G6P und FDP eine gleichsinnige Erhöhung und der Q G6P/FDP bleibt unverändert 4.

Infundieren wir zum Lactat zusätzlich Glucagon, so kommt es zu einer extremen Erhöhung des Spiegels an G6P und zu einer Verminderung des Spiegels von FDP (Tab. 5). Auf derartige gegensätzliche Substratbewegungen haben wohl erstmals LYNEN und CORI hingewiesen. Die Spiegeländerungen dieser Art wurden für

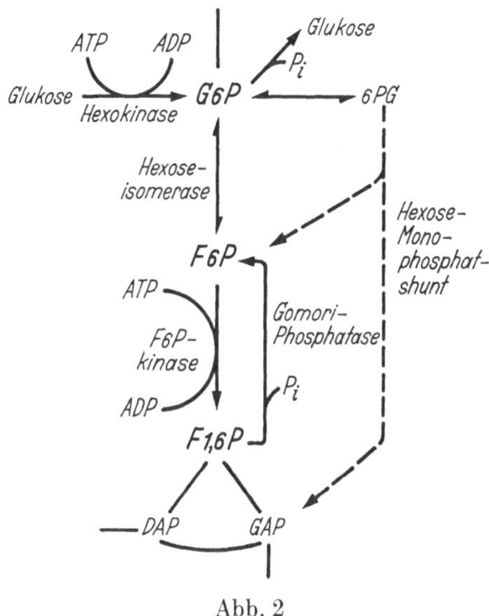

Abb. 2

den Muskelstoffwechsel dahin gedeutet, daß Anstieg an G6P und Abfall an FDP mit einer Hemmung der Phosphofruktokinase-Reaktion gleichgesetzt wurden. Im umgekehrten Falle sollte eine Aktivierung des Enzyms vorliegen. Über die Richtigkeit dieser Aussage brauchen wir nicht zu diskutieren, da die Verhältnisse im Leberstoffwechsel völlig anders sind. Wir bewegen uns im Gebiet der Substrate G6P bis FDP in einem stark verzweigten Abschnitt der Embden-Meyerhof-Kette (Abb. 2). Allein von und zum G6P sind die Wege fünffach gegeben. Auf diesen interessanten Abschnitt der Glykolysekette als vermutliche Steuerstelle hatten bereits HASTINGS[37], LEUTHARDT[38] und BÜCHER[39] aufmerksam gemacht. Die wesentliche Differenz beruht darauf, daß wir im Leberstoffwechsel

zwei zusätzliche Enzyme haben, die Glucose-6-Phosphatase und die
Fruktosediphosphatase. Im Hinblick auf die hier vorgezeichneten
Stoffwechselwege können wir unsere beschriebenen Glucagoneffekte
nur dahin deuten, daß die Spiegeländerung der Gehalte G6P und
FDP im reziproken Verhältnis Umkehr der Flußrichtung bedeutet.
Es wird Lactat nach Glucose umgewandelt, wir haben Umkehr der
Glykolysekette.

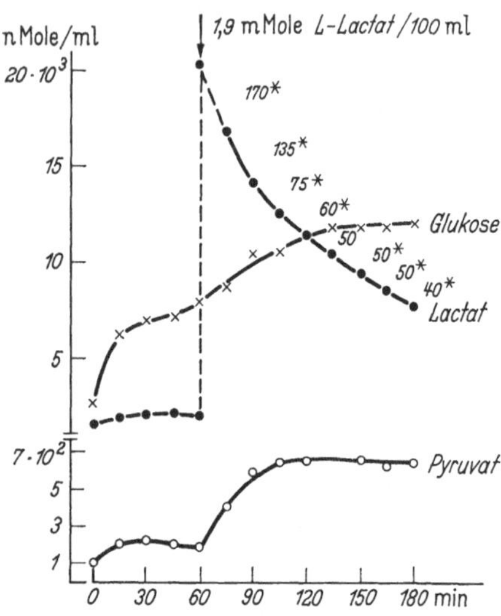

Abb. 3. Lactataufnahme der isoliert perfundierten Rattenleber bei gleichbleibendem Blutdurchfluß unter extrem hohem Lactatangebot (Ausgangskonzentration L-Lactat: 20 m M). * Lactataufnahme aus dem 15 min-Intervall berechnet auf μ Mole/g Leber/Std

Die Arbeitsgruppe von Park[40] in Tennessee hat diese Befunde
aufgegriffen und zunächst die von uns beschriebenen Substratänderungen bestätigt. Sie haben ferner unter Einsatz von ^{14}C-markiertem Lactat nachgewiesen, daß unter der Wirkung von Glucagon
Lactat zu 90% in Glucose umgewandelt wird. In ihren Versuchen
haben sie einmalig Lactat zum Außenmedium zugegeben, jedoch
in sehr hohen Konzentrationen. Sie haben den Lactatgehalt auf
20 mM erhöht (das ist mehr als das zehnfache des physiologischen
Wertes). Unter diesem wirklich gewaltigen Substratdruck ändert
sich vieles, was noch der Klärung bedarf. Die Abb. 3 zeigt, daß
Umsatzraten an Lactat bei derart überhöhten Spiegeln konzentra-

tionsabhängig werden und die Bedingungen unübersichtlich gestalten. Die Umsatzraten an Lactat schwanken unter diesen Bedingungen zwischen 170 und 40 μ Molen Lactat/g/h. Infundiert man dagegen kontinuierlich 500 μ Mole L-Lactat/h, so beträgt die Umsatzrate nur 20 μ Mole Lactat/g/h[8]. Da die Spiegel im Außenmedium in den Infusionsversuchen linear ansteigen, muß das System so weit gesättigt sein, daß der Wert von 20 μ Molen einen Sättigungswert darstellt. — Unter Zusatz von Glucagon wird der

Tabelle 6. *Einfluß von Glucagon auf den Stoffwechsel von $^{14}C_2$-Lactat in isoliert perfundierten Rattenlebern. Einsatz: 20 μC $^{14}C_2$-DL-Lactat/100 ml Medium. Wiedergefunden: 65% der eingesetzten Aktivität (Daten berechnet auf 10 g Leber)*

	Kontrolle	Glucagon 0,1 n Mole/Std infundiert
Einbau in Glucose (L-Anteil)		
nach 30 min	5,4%	62%
nach 120 min	15%	85%
Spez. Aktivität nach 120 min (dpm/μ Mol Glucose)	$3,7 \times 10^3$	$9,2 \times 10^3$
Glucoseanstieg im Außenmedium (μ Mole/Std)	+ 12	+ 822
Einbau in Leberprotein (L-Anteil)	5,5%	3,7%
$^{14}CO_2$-Produktion (DL-Anteil)	12%	10%
Einbau in Glykogen nach 120 min (L-Anteil)	0,7%	0,04%

Lactatumsatz — wie bereits erwähnt — verdreifacht. Wir haben unter Einhaltung kleinerer Lactatkonzentrationen (bis zum Doppelten des physiologischen Wertes) den Einbau von $^{14}C_2$-markiertem Lactat in Glucose (d. h. die Umkehr der Glykolyse) mit und ohne Glucagonzusatz verfolgt. Die Tab. 6 zeigt, daß in 60 min ohne Hormonzusatz immerhin 15% des Lactats nach Glucose umgewandelt werden. Dieser Anteil schnellt unter Glucagon auf 85% herauf. Da das System physiologisch durch Glucagon laufend gesteuert wird, wird sowohl der L-Lactatumsatz als auch der Fluß in der rückläufigen Glykolysekette zwischen diesen beiden Extrem-werten — den in vitro-Kontrollen und den Glucagonwerten — zu finden sein.

Der Glucagoneffekt ist, wie in diesem Falle zu erwarten, rein cytoplasmatisch. Das läßt sich unabhängig von Sauerstoffmessungen an einem Parallelsubstrat darstellen. In der Leber wird nicht nur L-Lactat, sondern auch D-Lactat in ganz erheblichem Maße metabolisiert. Das der Lactatdehydrogenase entsprechende Enzym ist für D-Lactat die D-Lactatoxydase, ein Flavoprotein, das mitochondrial lokalisiert ist[41]. Der Umsatz an D-Lactat liegt bei etwa 15 μ Mol/g Leber/h. Infusion von Glucagon verdreifacht sofort den Umsatz von L-Lactat und auch von Pyruvat, so daß diese Spiegel

Tabelle 7. *Wirkung von Insulin auf die Substratgehalte (n Mol/gf) und Quotienten von G 6 P, F 6 P und FDP mit und ohne Lactatbelastung in der isoliert perfundierten Rattenleber im Vergleich zu Kontrollen in vitro und unter Glucagoneinfluß*

	in vitro	Glucagon (3 n Mol/h)	Insulin (3 IE/h)	Lactat (500 μ Mol/h)	Lactat (500 μ Mol/h) Insulin (3 IE/h)
G 6 P	130	390	160	327	146
F 6 P	26	78	30	65	30
FDP	31	19	60	90	36
G 6 P/FDP	4,5	21	2	3,6	4
F 6 P/G 6 P	0,2	0,2	0,2	0,2	0,2

im Außenmedium sofort absinken. Der Glucagoneffekt ist außerdem an der Glucoseabgabe ablesbar, die unmittelbar einsetzt.

Am Umsatz von D-Lactat dagegen ändert sich nichts. Infusion von Adrenalin, das den O_2-Verbrauch der Leber stimuliert, erhöht auch sofort den Umsatz von D-Lactat[8].

Als Antagonist zum Glucagon gilt das Insulin und dieser Antagonismus läßt sich auch an der Schaltstelle zwischen G6P und FDP nachweisen. In der Tab. 7 wird nochmals die Einstellung der Fließgleichgewichte von G6P und FDP verglichen. Der Effekt von Insulin wird wiederum am deutlichsten im Vergleich des Q G6P/ FDP, der in Kontrollen 4, unter Glucagon 20 und unter Insulin nur 2 beträgt. Der Insulineffekt wird unter gleichzeitiger Lactatinfusion noch deutlicher. Im Gegensatz zur reinen Lactatinfusion steigt unter gleichzeitiger Insulingabe der Gehalt an G6P nicht an.

Ein direkter Antagonismus zwischen Insulin und Glucagon ist in den Blutsubstraten schwerer demonstrierbar. MILLER[42] fand eine Abschwächung der durch Glucagon bewirkten Harnstoffproduktion durch Insulin. Die Befunde anderer Autoren, die starke Insulineffekte auf die Glucoseaufnahme der Leber beschreiben, können wir nicht bestätigen[43–46].

Die zwei Tabellen über die Bewegungen von G6P und FDP zeigten als bemerkenswertes Phänomen, daß sich die Spiegel dieser Substrate reziprok verhalten. Wir haben daraufhin aus den Daten in vivo und dem daraus errechneten Mittelwert eine Hyperbel in ein Koordinatensystem eingezeichnet mit den Koordinaten G6P auf der Ordinate und FDP auf der Abszisse. In dieses Diagramm sind Einzeldaten aus Versuchen mit Glucagon, Insulin, Lactat oder Adrenalin eingetragen. Trotz der nicht geringen Streuung der Einzeldaten ergibt sich dennoch eine recht gute Zuordnung zur Hyperbel. Wir finden die Insulinversuche in einem Extrem der Hyperbel, die der Glucagonversuche auf dem anderen Schenkel der Hyperbel. Wir erhalten aus dieser Darstellung wesentliche Aussagen. Die Umkehr der Glykolyse spiegelt sich in der strengen Zuordnung der Substrate G6P und FDP wieder, und zwar nur unter den Hormonen Insulin und Glucagon. Die Werte der Kontrollperfusionen liegen zwischen den Daten der in vivo-Kontrollen (die bereits durch Glucagonsteuerung bedingt sind) und den Insulinwerten, jedoch stark zur Seite der Insulinwerte verschoben.

Das isolierte Organ steht daher unter dem Einfluß von Insulin. Die kontinuierliche Anwesenheit von Insulin ist dazu nicht erforderlich. Insulin gehört zu den Hormonen, die die Proteinsynthese fördern. Es kann induktiv über Enzymspiegel einen Basiswert im Stoffwechsel einstellen, dessen Fixierung von den Resyntheseprozessen bestimmt ist. Glucagon als sofort wirksames Hormon wirkt als Gegenregulans. — Wir können uns die Funktion beider Hormone am besten am Beispiel eines Thermostaten verdeutlichen. Um ein Wasserbad bei konstanter Temperatur zu halten, nehmen wir den Basiswert einer konstant fließenden Wasserkühlung und heizen mit einer starken Heizung dagegen. Der Wasserkühlung entspricht das Insulin, der Heizung das Glucagon.

Unseres Wissens nach ist mit diesen Versuchen erstmals ein direkter Antagonismus zweier Hormone in ein und derselben

Stoffwechselkette und zusätzlich an einem eng umschriebenen Gebiet dieser Stoffwechselkette beschrieben. Ob der Effekt der Hormone im Bereich G6P und FDP der einzige Angriffspunkt innerhalb der Glykolysekette ist, wissen wir noch nicht; wir halten ihn jedenfalls für sehr wesentlich. Über diese Schaltstelle und durch die Umschaltung der Flußrichtung in der Embden-Meyerhof-Kette im Gebiet zwischen G6P und FDP sind jedoch viele Glucagoneffekte erklärbar. Mit der Umschaltung der Flußrichtung erklärt sich die erhöhte Aufnahme und der erhöhte Umsatz an Lactat in der Leber. Mit dem erhöhten Abfluß nicht nur an Lactat, sondern auch an Pyruvat wird ein Partner aus der Reaktion der Transaminierung von Alanin und Pyruvat fortwährend aus dem Gleichgewicht entfernt; damit stellt sich auch ein stärkerer Fluß von Alanin nach Pyruvat ein, der zu stärkerem Eiweißabbau führen muß. Damit erklärt sich zugleich der Effekt des Glucagon auf die erhöhte Harnstoffproduktion.

Miller[47] fand, daß unter der Wirkung von Glucagon die Oxydation von Glucose gehemmt ist und stellte die Frage, wie sich dieser Effekt durch unsere Befunde erklären ließe. Wenn die Glykolysekette auf Gluconeogenese umgestellt ist, wird auch zwangsläufig die Oxydation der Glucose eingeschränkt sein. Viel interessanter wäre die Frage, ob die Umschaltung der Glykolysekette an dieser Schaltstelle auch eine Mobilisierung des Glykogens bewirken kann und wie groß der Anteil des Glykogens an der Glucoseproduktion ist. Aus Isotopenversuchen mit markiertem Lactat und Aminosäuren errechneten wir einen Anteil der beiden genannten Quellen von mehr als 50%. Es ist uns leider bisher nicht gelungen, eine Leber in den Versuch nehmen zu können, die ^{14}C-markiertes Glykogen enthält. Ein solcher Versuch wäre von großem Interesse im Hinblick auf die Frage nach den Wegen und den Umschaltstufen innerhalb der Embden-Meyerhof-Kette. —

Das Hormonpaar Insulin-Glucagon verdient noch aus anderen Gründen einiges Interesse. Beide Hormone werden in derselben innersekretorischen Drüse gebildet. Die Ausschüttung beider Hormone erfolgt über ein und dasselbe Substrat, nämlich über die Konzentration der Glucose im Blut. Für beide Hormone ist die Leber das unmittelbare und erste Erfolgsorgan im Stoffwechsel. Glucagon wird normalerweise in der Leber sofort durch die Glucagonase inaktiviert und erscheint daher kaum in der Peripherie. —

Die Insulinase ist nicht ganz so aktiv wie die Glucagonase, so daß Insulin zu einem Teil stets in die Peripherie gelangt.

Für die Klärung der letzten Wirkung dieser Hormone im Stoffwechsel reichen unsere Daten bisher nicht aus. Wir haben jedoch das Feld auf den umschriebenen Bezirk zwischen G6P und FDP eingeschränkt. Die Wirkung der beiden Hormone dürfte in erster Linie die Phosphatasen G6Pase und FDPase betreffen. LEUTHARDT u. Mitarb.[38] fanden, daß im Diabetes sowohl die G6Pase (auch von anderen beschrieben) als auch die FDPase in ihrer Aktivität erhöht sind, durch Insulininjektionen aber wieder auf die Grundwerte eingestellt werden können.

Es wäre interessant zu prüfen, ob es noch andere derartige Hormonpaare gibt, die in einem anderen Gewebe als der Leber im gleichen Sinne wie Insulin-Glucagon wirken. Das naheliegendste wäre, an Adrenalin und die Nebennierenrindensteroide zu denken. Sie werden auch in ein und derselben innersekretorischen Drüse gebildet und könnten im peripheren Gewebe einen ähnlichen, gezielten Steuereffekt und Antagonismus wie Insulin-Glucagon übernehmen. Fehlregulation in einem Hormonpaar (Fehlen von Insulin z. B.) würde kompensatorisch Fehlsteuerung durch Gegenregulation des anderen Hormonpaares auslösen und auf Zeit einen pathologischen Zustand fixieren. Ein derartiger Aspekt könnte auch für die praktische Medizin von Interesse sein.

Die Untersuchungen wurden von der Deutschen Forschungsgemeinschaft unterstützt.

Literatur

[1] LONDON, J. M., and D. RITTENBERG: J. biol. Chem. **184**, 687 (1950).
[2] FELLER, D. D., E. H. STRISOVER, and J. L. CHAIKOFF: J. biol. Chem. **187**, 57 (1950).
[3] MACHO, L.: Experientia (Basel) **18**, 73 (1962).
[4] ROSS, B. D., R. HEMS, and H. A. KREBS: Biochem. J. **102**, 942 (1967).
[5] CLOWES, jr. G. H. A.: Physiol. Rev. **40**, 826 (1960).
[6] SCHIMASSEK, H.: Biochem. Z. **336**, 460 (1963).
[7] — Life Sciences **11**, 629 (1962).
[8] — Ann. N.Y. Acad. Sci. **119**, 1013 (1965).
[9] —, u. W. GEROK: Biochem. Z. **343**, 407 (1965).
[10] — Unveröffentlicht.
[11] SCHARFF, R., and J. G. WOOL: Biochem. J. **97**, 257 (1965).
[12] WILLIAMSON, J. R., and D. L. DI PIETRO: Biochem. J. **95**, 226 (1965).
[13] CESSION-FOSSION, A.: Rev. belge Path. **30**, 349 (1964).
[14] SICKEL, W., and E. P. MAC NICHOL: Fed. Proc. **23**, 517 (1964).

15 Condon, R. E., L. M. Nyhus, and H. N. Harkins: Arch. Surg. 89, 602 (1964).
16 Spector, S., A. Sjöerdsma, P. Zaltzman-Nirenberg, M. Levitt, and S. Udenfriend: Science 139, 1299 (1963).
17 Hechter, O., M. M. Soloman, A. Zaffaroni, and G. Pincus: Arch. Biochem. 46, 201 (1953).
18 Anderson, E., and J. A. Long: Recent Progr. Hormone Res. 2, 209 (1948).
19 Grodsky, G. M., and P. H. Forsham: J. clin. Invest. 39, 1070 (1960).
20 Grodsky, G. M., L. L. Benett, A. Batts, N. McWilliams, and C. Vcella: Fed. Proc. 21, 202 (1962).
21 Robinson, G. A., R. W. Butcher, J. Øye, H. E. Morgan, and E. W. Sutherland: Molec. Pharmacol. 1, 168 (1965).
22 Sokal, J. E., E. J. Sarcione, and A. M. Henderson: Endocrinology 74, 930 (1964).
23 Levine, R. A.: Amer. J. Physiol. 208, 317 (1965).
24 Northop, G., and R. E. Parks: J. Pharmacol. exp. Ther. 145, 135 (1964).
25 Scharff, R., and J. G. Wool: Biochem. J. 97, 257, 272 (1965).
26 Wool, J. G., and K. L. Manchester: Nature (Lond.) 193, 345 (1965).
27 Pryor, J., and J. Berthat: Arch. int. Physiol. 68, 227 (1960).
28 Knox, W. E., and E. J. Behrman: Amer. Rev. Biochem. 28, 223 (1959).
29 Barnabei, O., and F. Sereni: Biochim. biophys. Acta (Amst.) 91, 239 (1964).
30 Gries, F. A., F. Matschinsky, and O. Wieland: Biochim. biophys. Acta (Amst.) 56, 615 (1962).
31 Miller, L. L., C. G. Bly, M. L. Watson, and W. F. Bale: J. exp. Med. 94, 431 (1951).
32 Gordon, A. H., and L. E. Mutschler: Protedes Biol. Fluids 12, 475 (1964).
33 Mattii, R., J. L. Ambrus, J. E. Sokal, and J. Mink: Proc. Soc. exp. Biol. (N.Y.) 116, 69 (1964).
34 Krauss, St., and F. J. Sarcione: Biochim. biophys. Acta (Amst.) 90, 301 (1964).
35 Schimassek, H., u. H. J. Mitzkat: Biochem. Z. 337, 510 (1963).
36 Sutherland, E. W., and G. A. Robison: Pharmacol. Rev. 18, 145 (1966).
37 Ashmore, J., G. F. Gahill jr., A. B. Hastings, and S. Zottu: J. biol. Chem. 224, 225 (1957).
38 Leuthardt, F.: Helv. physiol. pharmacol. Acta 19, 234 (1961).
39 Bücher, Th.: 7. Symp. dtsch. Ges. Endocrinol., S. 129. Homburg (Saar) 1960.
40 Exton, J. H., and C. R. Park: Pharmacol. Rev. 18, 181 (1966).
41 Tubbs, P. K.: Ann. N.Y. Acad. Sci. 119, 920 (1965).
42 Miller, L. L.: Recent Progr. Hormone Res. 17, 539 (1961).
43 Gordon, E. R.: Canad. J. Physiol. Pharmacol. 43, 617 (1965).
44 Sokal, J. E., L. L. Miller, and F. J. Sarcione: Amer. J. Physiol. 195, 295 (1958).
45 Mortimore, G. E.: Amer. J. Physiol. 204, 699 (1963).
46 — Amer. J. Physiol. 200, 1315 (1961).
47 Miller, L. L.: Fed. Proc. 24, 737 (1965).

Diskussion

DIRSCHERL (Bonn): Ich danke Herrn Kollegen SCHIMASSEK für seinen fundierten Vortrag. Darf ich vielleicht selbst gleich eine Diskussionsfrage stellen. Wenn ich Sie recht verstanden habe, Herr SCHIMASSEK, so ist das, was Sie hier als action bezeichnen, der Wirkungsmechanismus der Hormone? Dann würde ich es besser finden, keinen neuen Ausdruck einzuführen. — Und noch eine zweite Frage: Wenn ich Ihre Ausführungen recht verstanden habe, sollte es das Ziel sein, den einheitlichen Wirkungsmechanismus für ein gegebenes Hormon schließlich aufzuklären. Ich glaube nicht, daß auch nur ein einziges Hormon alle seine Wirkungen über ein und denselben Mechanismus entfaltet. Dazu sind die Wirkungen zu verschieden, z. B. dle Advenalinwirkung auf die Leber und den Kreislauf.

SCHIMASSEK (Marburg): Ich habe nicht durch den Gebrauch von action und Wirkungsmechanismus nebeneinander Verwirrung stiften wollen; mir war nur daran gelegen, die Differenz zwischen ,action' und ,effect' mit einem Wort auszudrücken. — Zur zweiten Frage: Ich stimme Ihnen völlig zu. Ich glaube nicht, daß alle Hormone gleich wirken; ich würde zumindest zwischen sofort wirkenden Hormonen und langsam wirkenden Hormonen unterscheiden, wobei ich das Glucagon z. B. unter die akut wirkenden Hormone zählen würde, das Insulin als Zeithormon bezeichnen würde. Ob nun das einzelne Hormon über nur einen Wirkungsmechanismus alle seine verschiedenen Effekte entfaltet, können wir heute wohl noch nicht beantworten.

KUKOVETZ (Graz): Ich wollte zur Frage des Mechanismus der Kreislaufwirkung von Katecholaminen, die vom Herrn Vorsitzenden angeschnitten worden ist, kurz Stellung nehmen. Die Ergebnisse von Herrn SUTHERLAND und auch unsere eigenen Versuche weisen darauf hin, daß die Adrenalinwirkung auf das Herz über das $3',5'$-Cyclo-AMP geht. Das gleiche dürfte für die Wirkung auf die glatte Muskulatur der Gefäße, soweit sie über eine Erregung adrenerger Beta-Receptoren zustandekommt, zutreffen. Die Erregung der Alpha-Receptoren ist dagegen in ihrem Mechanismus noch strittig.

CLAUSER (Paris): Wir haben vor einigen Wochen mit Dr. FR. MEYER an unserem Laboratorium Experimente durchgeführt, die sich mit dem Wirkungsmechanismus der hypoglykämischen Biguanide befassen. Diese Ergebnisse wurden nun veröffentlicht: die Biguanide wirken auf die Leber und den Nierencortex anscheinend ausschließlich als spezifische Inhibitoren der gluconeogenetischen Enzymkette. Außerdem scheinen die verschiedenen Biguanide substratspezifisch zu sein. Es wäre also möglich, den Effekt des Glucagon auf die Gluconeogenese mit Hilfe dieser Inhibitoren zu testen und vor allem die verschobenen Gleichgewichte von G6P/FDP, falls sie der Gluconeogenese zuzuschreiben sind, rückgängig zu machen.

SCHIMASSEK: Wir haben selber nie mit diesen Biguaniden gearbeitet, ich kann Ihre Frage also nicht direkt beantworten. Allerdings kann man sagen, daß es nicht nur Hormone sein müssen, die in dieses Wechselspiel eingreifen; wir haben auch Verbindungen in die Hand bekommen, die die Umkehr der

Flußrichtung bewirken, die also vermehrte Lactataufnahme und erhöhte Glucoseabgabe zeigen. Man wird wahrscheinlich mit vielen synthetischen Substanzen ähnliche Effekte erreichen; das Ausmaß der Steuerung dürfte aber in diesen Fällen kleiner sein als mit den natürlichen Hormonen.

WIELAND (München): Ich habe zwei Fragen, Herr SCHIMASSEK: 1. Sie sagten am Anfang Ihres Vortrages, daß die Leber keine Glucose aufnimmt. Ich weiß nicht, ob man das so allgemein sagen darf. Wir haben in vielen Experimenten Glykogenansatz in der perfundierten Leber erzielen können durch Einstellung sehr hoher Glucosekonzentrationen. — Die zweite Frage betrifft die Kontrollstelle der Gluconeogenese. Wir haben uns auch mit der Wirkung des Glucagon auf die Gluconeogenese beschäftigt und verschiedene Substrate als Vorstufen für die Glucose angeboten. Dabei hat sich herausgestellt, daß das Glucagon den Glucoseaufbau stimuliert, sofern man Pyruvat oder Lactat anbietet, nicht aber, wenn man Glycerin als Glucosevorstufe anbietet. Für die Steuerung des Gesamtprozesses ist vielleicht noch wichtig, daß man mit langkettigen Fettsäuren die Bildung von Glucose aus Lactat oder Pyruvat in ganz ähnlicher Weise stimulieren kann wie mit Glucagon.

SCHIMASSEK: Ich möchte zunächst auf die Frage der Glucoseaufnahme eingehen. Ich bezweifle nicht, daß man mit hohen Substratdrucken viel erreichen kann, auch eine Glucoseaufnahme und Glykogenbildung. Das ist u. a. auch von dem Arbeitskreis um KREBS gefunden worden.

Zu Ihrer Frage mit dem Glycerin: Ich möchte vorschlagen, die einzelnen Hormone ganz bestimmten Schaltstellen in den einzelnen Ketten zuzuordnen. Es würde sehr viel mehr Klarheit in die Gesamtzahl der Befunde bringen. — Der Glycerineffekt ist interessant; allerdings hat KREBS gefunden, daß das Glycerin unter Glucagon gleichfalls einen Glucoseanstieg macht, aber einen wesentlich geringeren als Lactat. — Zum Effekt der Fettsäuren: das zielt ja letztlich auf die Frage hin, wo der Wasserstoff für die Gluconeogenese herkommt, wenn wir Pyruvat nehmen. Es ist natürlich theoretisch denkbar, daß er von den Fettsäuren kommen könnte. KREBS ist nicht der Meinung; er nimmt an, daß der Wasserstoff gleichfalls vom Lactat geliefert wird. Wir möchten vorläufig der Meinung von Prof. KREBS den Vorzug geben, aber die Frage ist wohl wirklich noch ungeklärt.

KARLSON (Marburg): Darf ich zum Glycerineffekt noch eine Frage an Herrn WIELAND richten: Wieweit kommt hierbei die Kompartmentierung der Zelle ins Spiel? Wenn wir Glycerin als Substrat geben, dann wird es ja über Glycerophosphat zum Dihydroxyacetonphosphat dehydriert. Das geschieht zum Teil auch in den Mitochondrien. Frage: Wie bleibt das Dihydroxyacetonphosphat dann in den Mitochondrien drin, wird es dort eventuell weiter umgesetzt oder geht es wieder ins Cytoplasma? Hat man einmal mit markierten Glycerin nachgeguckt, wo die Markierung nachher auftritt?

WIELAND: Ich weiß nicht, ob jemand einmal mit markiertem Glycerin nachgesehen hat, welche Produkte daraus entstehen. Die Phosphorylierung des Glycerins zum Alpha-Glycerophosphat dürfte jedenfalls im Cytoplasma erfolgen; die Glycerinkinase ist ein Enzym, das vornehmlich im Cytoplasma oder — wie man neuerdings sagt — im Cytosol lokalisiert ist. α-GP kann durch das Baranowski-Enzym dehydriert werden zum Dihydroxyacetonphosphat; damit ist der Anschluß an die Glykolyse hergestellt. — In der Leber ist die Aktivität der mitochondrialen Alpha-Glycerophosphatoxydase wohl vergleichsweise gering.

SCHIMASSEK: Darf ich hierzu noch eines ergänzen. Wir haben bei einigen Versuchen gleichfalls Glycerin infundiert, um das Glycerophosphatsystem zu überladen. Glycerophosphat selbst geht als phosphorylierter Stoff ja nicht in die Leberzelle hinein. Tatsächlich bekommt man in der Leber sehr schön die Umwandlung zu Glycerophosphat, aber dann passieren merkwürdige Dinge. Der Quotient Glycerophosphat zu Dihydroxyacetonphosphat, der normalerweise etwa 7 beträgt, steigt wirklich ins Überdimensionale, er steigt um zwei Größenordnungen (700). Alpha-Glycerophosphat bleibt also in der Zelle liegen; Dihydroxyacetonphosphat reagiert kaum, es sinkt sogar eher noch ab. — In Leberschnitten ist der Effekt geringer, und im Homogenat, wo „jeder jedem Guten Tag sagen kann", ist der Koeffizient 2. Das zeigt doch, daß hier eine gewisse Kompartmentierung oder Kettenzugehörigkeit eine Rolle spielen mag, die nur im intakten System zu beobachten ist.

STAIB (Düsseldorf): Ich möchte Sie fragen, Herr SCHIMASSEK, ob Sie den Glucagoneffekt einmal mit entfettetem Albumin geprüft haben, und zweitens wie sich Cortisol auf den Glucagoneffekt auswirkt.

SCHIMASSEK: Wir haben uns nicht besonderes fettfreies Albumin hergestellt. Wir haben ein sog. „Fatty acid poor albumin" von ARMOUR verwendet und damit keinen Unterschied gegenüber Normalalbumin gesehen. — Die zweite Frage kann ich Ihnen nicht beantworten. Das Gegenspiel Glucagon/Cortisol haben wir nicht geprüft.

THORN (Hamburg): Herr Kollege SCHIMASSEK hat uns eine sehr übersichtliche Darstellung seiner am isolierten, durchströmten Organ erhobenen Befunde gegeben. Man kann ihn zum Referat und zu den schönen Befunden beglückwünschen.

Der metabolische Status eines Organs und die Relation der einzelnen Metabolitgehalte zueinander sind keine unabhängigen Größen, sie unterliegen der ständigen hormonalen und nervösen Kontrolle. Es ist verdienstvoll, am isolierten Organ den Einfluß einzelner Faktoren aufzuklären. Herr SCHIMASSEK hat einen Teil seiner Ausführungen dem Gehalt an Fructose-1,6-Diphosphat gewidmet, dabei die Bedeutung der Fructokinase erwähnt und darauf hingewiesen, daß das FDP auf sehr unterschiedlichen Wegen entstehen kann.

Wegen der Bedeutung der nervösen und hormonalen Einflüsse auf die Organfunktion haben wir den größten Teil unserer Untersuchungen an Organen in situ durchgeführt und finden in der Leber bei hypoxischen und anaeroben Belastungen, bei Leberlappenunterbindungen usw. einen recht gleichmäßigen, niedrigen Gehalt an FDP. Der Leber stehen mehrere Möglichkeiten offen, ihren Substrat- und Energiebedarf zu decken. Fehlt die Glucose als Substrat, ergänzt sie ihren Glykogengehalt durch Gluconeogenese. Im Gehirn, im Herz- und Skeletmuskel findet man hingegen bei Störungen des oxydativen Stoffwechsels stets einen erhöhten Gehalt an FDP. Man kann daraus und aus anderen Befunden schließen, daß die glykolytischen Reaktionen in Nerven- und Muskelzellen eine größere Bedeutung haben als etwa in der Leber. Man kann aus dem Anstieg des Gehalts an FDP ebenfalls folgern, daß für die Umstellung von aeroben Stoffwechsel auf anaerobe Glykolyse die Fructokinasereaktion in Nerven- und Muskelzellen nicht zum limitierenden Faktor wird. Der FDP-Gehalt erreicht erst seinen Ausgangswert, wenn die Geschwindigkeit der Milchsäurebildung sich verringert. Die Leber entwickelt trotz hohen Glykogengehalts nur eine mäßige anaerobe Glykolyse, die anaerobe Stoffwechselsituation der Leber ist durch einen sehr hohen Gehalt an freier Glucose gekennzeichnet.

Die von Herrn SCHIMASSEK vorgetragenen Befunde über die Glucoseaufnahme durch eine an Glykogen verarmten Leber habe ich nicht verstanden. Konnte der Gehalt an Glykogen des perfundierten Organs bei Versuchsbeginn und Versuchsende direkt kontrolliert werden? Die Organe wurden, wenn ich das richtig verstanden habe, mit albuminhaltigem Perfusat durchströmt. Wie steht es mit der Aufnahme von Aminosäuren? Kann eine Gluconeogenese etwa aus dem Abbau aufgenommener Aminosäuren für die vorgetragenen Untersuchungen ausgeschlossen werden?

SCHIMASSEK: Glucose ist ein Produkt der Leber und kein Substrat für die Leber. Selbst die Leber einer Ratte nach 16 Std Nahrungsentzug, die kein Glykogen enthält (Kontrolle der Glykogenwerte zu Beginn und Ende der Perfusion) nimmt praktisch keine Glucose aus dem Außenmedium auf, sondern gibt noch Glucose an das Außenmedium ab. Die an das Außenmedium abgegebene Glucose stammt aus der Gluconeogenese. Ausgangsprodukte dafür sind Lactat, Pyruvat (die aus dem Außenmedium vermehrt aufgenommen werden) und natürlich auch Aminosäuren. Die für die Gluconeogenese notwendigen Aminosäuren kann die Leber auch selber durch Eiweißabbau bilden.

SCHREIBER (Freiburg): In Versuchen über die Leberschädigung durch sehr kleine Mengen CCl_4 fanden SMUCKLER und BENDITT als empfindliche und sehr früh eintretende Veränderungen eine Abnahme der Synthese von Fibrinogen und eine Auflösung und Neuordnung des endoplasmatischen Reticulums. Eine ähnliche Veränderung findet sich auch bereits sehr früh, etwa 30 min, nach partieller Hepatektomie. Da Sie eingangs erwähnten, daß in Versuchen mit Perfusion der Rattenleber *in vitro* ebenfalls die Synthese von Fibrinogen herabgesetzt ist, liegt es nahe zu fragen, ob — und wenn ja — welche Veränderungen in der ultramikroskopischen Feinstruktur zwischen normaler Leberzelle und den Zellen einer perfundierten Leber bestehen.

SCHIMASSEK: Wir persönlich haben keine elektronenoptischen Untersuchungen über die Feinstruktur der perfundierten Leber durchgeführt.

Aus Untersuchungen anderer Arbeitskreise wissen wir, daß unter den von uns verwendeten Techniken keine Abweichungen bestehen. An sich sollte man aber erwarten, daß die von uns durchgeführten Nachweise der Leberfunktion (und zwar mit empfindlichen biochemischen Kriterien) empfindlicher sind als das submikroskopische Bild. — Sie haben aber indirekt die Frage gestellt, ob die verminderte Fibrinogensynthese nicht doch ein Zeichen einer Leberschädigung ist. Es ist sehr wahrscheinlich nicht der Fall (bei aller Einschränkung dieser Möglichkeit bei einem *in-vitro*-System). Die Leber synthetisiert Antikörper und jeden bisher untersuchten Eiweißkörper, jedoch verschieden schnell. Es ist daher eher die Frage zu stellen, was dem isolierten System zur Synthese, z. B. des Fibrinogens fehlt als die nach einer möglichen Schädigung.

STAIB: Ich möchte nochmals auf die Glucoseaufnahme zurückkommen. Es gibt Arbeiten von GORDON, der über längere Zeit mit verschiedenen Glucosekonzentrationen infundiert hat und schon bei vergleichsweise niedrigen Konzentrationen ein steady state bekommt.

SCHIMASSEK: Soweit ich die Arbeiten von GORDON kenne, hat er eigentlich keine Nullwerte gemacht, sondern nur die Perfusionen mit und ohne Insulin verglichen. Mich würde als erstes interessieren, wie seine normalen Spiegel aussehen, ob sie nicht von allein absinken, und dann würde ich vermuten, auf Grund eigener schlechter Erfahrungen, daß dabei Bakterien eine Rolle spielen. — Vielleicht kann ich noch anfügen, daß die Bedingungen der Perfusion nicht bei allen Autoren gleich sind. Das ist sehr bedauerlich. KREBS hat es seinerzeit erreicht, daß die Schnittechnik vereinheitlicht wurde und nicht die verschiedensten Untersucher durch verschiedene Salzkonzentrationen und dergleichen Bedingungen geschaffen haben, bei denen alle möglichen Werte zu messen sind. Es wäre zu begrüßen, wenn auch auf dem Gebiet der Leberperfusion eine derartige Vereinheitlichung der äußeren Bedingungen durchgeführt würde.

REINAUER (Düsseldorf): Es wäre interessant, neben den Metabolitgehalten der Glykolyse die Enzymaktivitäten des Glucoseabbaus zu betrachten. Die Leber hat durch das Vorkommen einer spezifischen Glucokinase eine Sonderstellung beim Glucoseabbau. Die Aktivität dieses Enzyms ist vermindert im Hunger und Diabetes und steigt nach Wiederfütterung bzw. Insulingabe an. Falls eine Glucoseutilisation in der Leber nicht erfolgen soll, welche Bedeutung soll dann diesem Enzym zukommen?

SCHIMASSEK: Die Steuerung der Glucokinase ist in der Leber ebenso abhängig vom Glucose-6-Phosphatspiegel wie im Muskel. Wir haben die Aktivitäten der Glucokinase allerdings nicht selbst gemessen; ich halte aber eine solche Steuerung doch für möglich.

REINAUER: Das besondere an der Glucokinase ist ja, daß ihre Aktivität durch Glucose-6-Phosphat nicht gesteuert wird.

SCHIMASSEK: Die Bedeutung des Enzyms kann aus bisherigen *in-vitro*-Untersuchungen nicht näher definiert werden. Die während der Diskussion gegebenen Daten mit $^{14}C_u$-markierter Glucose — mit und ohne Insulin — zeigen, daß der Insulineffekt auf den Glucosestoffwechsel der Leber jedoch sehr gering ist.

HASSELBLATT (Göttingen): Leberschnitte, die *in-vitro* inkubiert wurden, bilden Ketonkörper aus Fettsäuren, die aus Fettsäureestern des Gewebes freigesetzt werden. Die Versorgung mit unveresterten Fettsäuren, die *in-vivo* die Leber mit dem Blut erreichen, ist bei Inkubation *in-vitro* abgeschnitten. Ein Zusatz von Glucagon zu Leberschnitten *in-vitro* steigert die Ketonkörperbildung durch das Gewebe. Durch ASHMORE u. Mitarb. (Fed. Proc. **25**, 720 (1966)] ist darauf hingewiesen worden, daß Glucagon diesen Effekt auslöst, weil es eine hepatische Lipase aktiviert und dadurch die Abspaltung freier Fettsäuren aus den Fettsäureestern der Leber stimuliert. Dadurch sollen vermehrt freie Fettsäuren im Gewebe für die Ketonkörperbildung zur Verfügung stehen.

Es wäre doch möglich, daß auch an der perfundierten Leber Glucagon die hepatische Lipolyse aktiviert und so in den Leberzellen Fettsäuren freisetzt, die in den Stoffwechsel eingehen. Ein Anstieg der Ketonkörper könnte, ähnlich wie an Leberschnitten *in-vitro,* in diesem Sinne sprechen.

SCHIMASSEK: Unter hohem Glucagonspiegel steigen die Ketonkörper an. Es ist allerdings die Frage, ob alle Ketonkörper nur aus den Fettsäuren kommen. Wir haben schließlich unter Glucagon auch erhöhten Eiweißabbau, und eine ganze Reihe von Aminosäuren laufen über den Fettsäureweg, besonders diejenigen, mit denen die Leber ohnehin schon schwer fertig wird. So hat GEROK gefunden, daß unter Glucagon die Spiegel von Valin, Leucin und Isoleucin ungewöhnlich stark ansteigen.

WIELAND: Wir hatten zunächst an die Möglichkeit gedacht, daß das Glucagon eine hepatische Lipase aktiviert und damit zu einer Anhäufung der freien Fettsäuren in der Zelle führt, und daß dann diese freien Fettsäuren die zweiten Messenger sind, die die Gluconeogenese stimulieren. Wir haben uns sehr viel Mühe gegeben, diese Stimulation nachzuweisen, haben sie aber nicht finden können. Die Untersuchungen von ASHMORE, die Sie angeführt haben, zeigen zwar kleine Effekte, wirken aber nicht sehr überzeugend. Ich glaube nicht, daß man heute schon sagen kann, daß das Glucagon eine Leberlipase aktiviert, in Analogie zu den Verhältnissen im Fettgewebe.

WEISS (München): Herr SCHIMASSEK, Sie haben uns ein Bild über die Glucagonwirkung aus einer Arbeit von EXTON und PARK gezeigt. Dann sagten Sie uns, Sie konnten den Effekt des Glucagons auf das Substratpaar G-6-P und FDP beschränken. In der Arbeit von EXTON und PARK ist eine Erhöhung des PEP-Spiegels von Glucagon angegeben. Haben Sie nun den Effekt beschränken können oder haben Sie Ihre Untersuchungen beschränkt?

SCHIMASSEK: Wir haben sinnvollerweise die Konzentration bzw. Infusion an Lactat beschränkt. Ein hoher Lactatdruck erhöht auch die Intermediate innerhalb der Glykolysekette, damit auch den Gehalt an PEP — und diesen Effekt haben Sie auch ohne Glucagonzugabe. Unnötig überhöhte Substratspiegel komplizieren die an sich schon recht komplexen Verhältnisse im Leberstoffwechsel.

STAIB: Konnte der Effekt von Insulin auf die Glykogensynthetase auch an der isolierten Rattenleber nachgewiesen werden?

SCHIMASSEK: Die Frage kann ich nicht beantworten.

DIRSCHERL: Da keine weiteren Wortmeldungen vorliegen, darf ich nochmals Herrn SCHIMASSEK und den Diskussionsrednern danken und schließe die Vormittagssitzung.

Über Hormon-„Receptoren"
Die oestrogenbindenden Prinzipien
der Erfolgsorgane*

Von P. W. Jungblut, I. Hätzel, E. R. DeSombre
und E. V. Jensen

Ben-May-Laboratory for Cancer Research, University of Chicago, USA

Mit 14 Abbildungen

Auf die Frage nach einem besonderen Merkmal der Endokrinologie ließe sich mit der Diskrepanz zwischen definierter chemischer
Struktur der Hormone und der Komplexität ihrer Wirkungen in
Organismen antworten. Bei der Vielfalt der von einem Hormon
gesteuerten Vorgänge liegt es nahe, nur für eine Reaktion einen
direkten Einfluß des Hormons anzunehmen und die anderen als
Folgereaktionen aufzufassen. Diese Vorstellung findet ihren Ausdruck in der im Schrifttum ausführlich besprochenen „Hormon-
Receptor-Hypothese"[1], die als Primärreaktion die Bindung des
Hormons an einen spezifischen Receptor fordert. Für die Prüfung
der Hypothese stehen im wesentlichen zwei Wege offen. Der am
meisten begangene ist die Suche nach der Primärreaktion durch
Verfolgung des zeitlichen Ablaufs der Hormonwirkung. Der zweite
Weg besteht in dem direkten Versuch, die hypothetischen Hormonreceptoren nachzuweisen und zu isolieren. Dabei muß vorausgesetzt werden, daß die Spezifität des Receptors einer besonders
hohen Affinität für das Hormon gleichzusetzen ist. Die zur Zeit
vorhandenen experimentellen Möglichkeiten und Grenzen einer
Receptorisolierung sollen hier am Beispiel der spezifischen Bindungsfaktoren für Oestrogene diskutiert werden.

* These investigations were supported by a research grant (GA 02897)
and a research contract (PH-43-65-596) from the National Institutes of
Health, United States Public Health Service.

1. Nachweis der spezifischen Bindung
von Oestrogenen in Erfolgsorganen

Vor 9 Jahren synthetisierten E. V. JENSEN und H. I. JACOBSON
$6,7,{}^{3}$H-Oestradiol-17β durch katalytische Reduktion von 6,7-De-
hydrooestradiol in trägerfreiem Tritiumgas[2]. Das erhaltene Produkt
hatte eine theoretische spezifische Aktivität von 50 Curie/mMol und
war deshalb besonders geeignet zur Beantwortung der von den
Autoren zunächst gestellten Frage: Was macht das Gewebe mit
dem Hormon ? Die Antwort war überraschend und klar[3]. Injektion
von markiertem Oestradiol in physiologischen Dosen führt bei infantilen und ovarektomierten Sprague-Dawley-Ratten zu einer
Akkumulation und Retention von Radioaktivität in den Erfolgsorganen Uterus, Vagina und Hypophysenvorderlappen, während
der Gehalt an radioaktiven Verbindungen in der Leber (nach kurz-
fristigem initialen Anstieg), in Nieren, Nebennieren, Ileum, Knochen,
Lunge, Großhirn und quergestreiftem Muskel der Konzentration
im Blut parallel verläuft. Die Erfolgsorgane enthalten als einzige,
nicht covalent gebundene, radioaktive Substanz in allen Zeitpunk-
ten nach der Injektion unverändertes Oestradiol. In der Leber wird
das Hormon sehr schnell metabolisiert. Die Analyse der anderen
Gewebe zeigt Oestradiol und Metaboliten in den gleichen Propor-
tionen, wie sie im Blut gefunden werden.

Mit prinzipiell der gleichen Versuchsanordnung wurde eine Reihe
weiterer wichtiger Ergebnisse erzielt. Die Retentionsfähigkeit des
Rattenuterus für Oestradiol strebt einer Sättigung bei etwa
2×10^{-8} Mol/kg zu. Nach Injektion eines Gemisches von 6,7 tritiiertem und 17 tritiiertem Oestradiol enthält der Uterus die beiden
Steroide im unveränderten Verhältnis. Es erfolgt also kein Wasser-
stoffaustausch über den von TALALAY und WILLIAMS-ASHMAN[4]
vorgeschlagenen Transhydrogenasemechanismus. Oestron wird
offenbar nicht als solches vom Uterus retiniert, sondern muß zunächst — wahrscheinlich in der Leber — zu Oestradiol reduziert
werden. Die Bindungskurve des synthetischen Oestrogens Hex-
oestrol gleicht der von Oestradiol. Hemmung der uterotrophen
Wirkung des Hormons durch Aktinomycin D und Puromycin hat
keinen Einfluß auf die Bindung von Oestradiol, Antioestrogene wie
Upjohn U 11 100* konkurrieren dagegen. Als weiteres Erfolgsorgan

---

* 1-{2-[p-(3,4-dihydro-6-methoxy-2-phenyl-1-naphthyl) phenoxy]äthyl}
Pyrrolidin Hydrochlorid.

für Oestradiol hat sich das mit 7,12 Dimethylbenzanthrazen (DMBA) bei Sprague-Dawley-Ratten induzierte Adenocarcinom der Brustdrüse herausgestellt[5, 6, 7].

Wir haben inzwischen diese *in-vivo* erzielten Ergebnisse mit überlebenden Rattenuteri, DMBA-Tumor- und Kalbsendometriumschnitten *in-vitro* reproduzieren und erweitern können[7, 8]. Eine einfache Perfusionskammer wurde für die ersten Versuche mit Rattenuteri benutzt. Vor allem für die schlecht zu fixierenden Tumorschnitte empfiehlt sich das jetzt von uns angewendete diskontinuierliche Verfahren, bei dem die Gewebe in adäquat

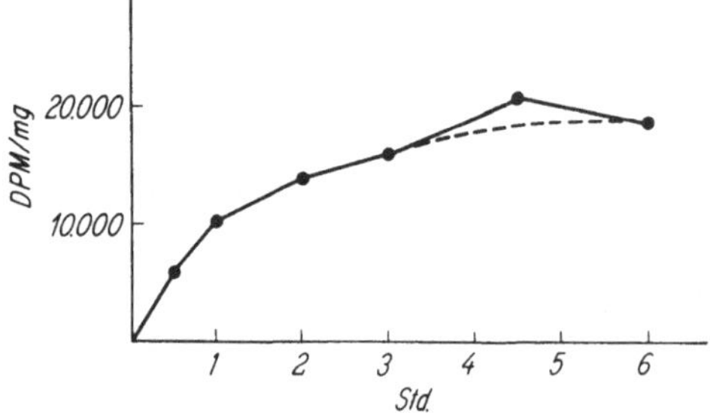

Abb. 1. Aufnahme von 6,7³H-Oestradiol durch Rattenuteri in vitro. Uteri von ovariektomierten, nicht vorbehandelten Tieren. Inkubationsmedium: Krebs-Ringer-Henseleit-Lösung mit 1 g Glucose/L und 9×10^{-10}M/L 6,7 ³H-Oestradiol. Temperatur: 38 °C. Gefäßvolumen: 10 ml. Flußgeschwindigkeit: 0,24 ml/min. 20000 DPM/mg Trockengewicht entsprechen etwa $1,5 \times 10^{-8}$M Oestradiol/kg Feuchtgewicht

dimensionierten Bechergläsern langsam agitiert und in vorgegebenen Zeitabständen in frische Lösungen übertragen werden. Abb. 1 und 2 zeigen als Beispiele der Resultate die Sättigung von Rattenuteri mit Oestradiol, die Diskriminierung in der Retentionsfähigkeit für Oestradiol und Oestron nach kurzer Aufnahmeperiode und die begrenzte Aufnahme und mangelnde Retention von Oestradiol durch Zwerchfellstreifen. Die Bindung von radioaktivem Oestradiol durch den Uterus ist reversibel. Unmarkiertes Oestradiol wäscht die radioaktive Verbindung schnell aus. Ein neben ihrer Einfachheit besonderer Vorteil der *in-vitro*-Anordnung besteht darin, die Durchführung von Versuchen zu erlauben, die mit dem intakten Tier nicht möglich sind. Sättigung der Inkubationslösung mit N_2 anstelle von O_2 beeinträchtigt die Bindung von Oestradiol durch Rattenuteri nicht. Anwesenheit von

10^{-3} m CN⁻ ist ebenfalls ohne wesentlichen Einfluß, Sulfhydrylreagentien wie p-Hydroxymercuribenzoat, N-Äthylmaleinamid
und Jodacetamid in 10^{-3} molarer Konzentration verringern dagegen die Aufnahme von Oestradiol durch Uteri und heben ihre
Retentionsfähigkeit auf. Wir werden später sehen, daß es sich dabei
offenbar um einen direkten Einfluß auf die oestrogenbindenden
Faktoren des Uterus handelt.

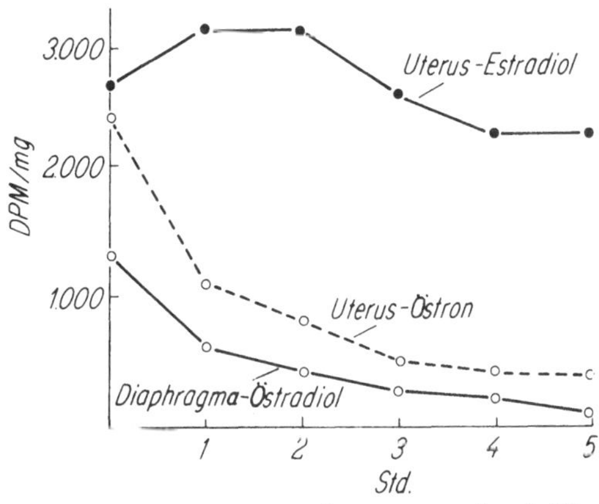

Abb. 2. Vergleich der Retention von Oestrogenen durch Uterus und Diaphragma. Gewebe: Uteri von ovariektomierten, nicht vorbehandelten
Ratten; etwa gleich schwere Zwerchfellstreifen aus der pars costalis. Inkuba-
tionsperioden im Durchflußgefäß: 1. 30 min, KRH, 2,8 ml/min; 2. 15 min,
KRH, mit 5,2 × 10^{-10}M/l, 6,7 ³H-Oestradiol bzw. Oestron, 0,24 ml/min;
3. 60 bis 300 min, KRH, 2,8 ml/min

Dieser kurze und unvollständige Überblick von Ergebnissen
unseres Labors, die von anderen Arbeitsgruppen mehrfach bestätigt
wurden, legt den Schluß nahe, daß eine spezifische Bindung von
Oestradiol in den Erfolgsorganen tatsächlich die Primärreaktion
der Hormonwirkung darstellt.

2. Intracelluläre Lokalisation von 6,7, ³H Oestradiol

Wenn man voraussetzt, daß die nicht kovalente Bindung von
Oestradiol an den hypothetischen Receptor relativ stabil ist, d. h.
bestimmte Aufarbeitungsmethoden überlebt, sollte es möglich sein,
aus der intracellulären Verteilung von Oestradiol auf die Lokalisation des Receptors zu schließen. Die sich anbietenden Methoden
sind die Fraktionierung von Zellelementen durch differentielles
Zentrifugieren und die Autoradiographie. Eines oder beide Ver-

fahren sind verschiedentlich u. a. von Noteboom und Gorski[9], King u. Mitarb.[10, 11], Maurer und Chalkley[12] sowie von uns mit prinzipiell gleichen Resultaten angewendet worden. Ich möchte an dieser Stelle bemerken, daß Uterus ein außerordentlich schwierig zu homogenisierendes Gewebe ist. Wir haben uns deshalb für die Fraktionierungsversuche auf Endometrium beschränkt. Oestradiol wird auch von Myometrium gebunden, allerdings in nur etwa halber Konzentration, bezogen auf das Trockengewicht.

Die Verteilung von radioaktivem Oestradiol in den Zellelementen des Endometriums und des ebenfalls untersuchten DMBA-Tumors ist unabhängig von der Zeit nach der Injektion bzw. der Länge der Inkubation. Tab. 1 zeigt die Oestradiolverteilung 1 Std nach der

Tabelle 1. *Prozentuale Verteilung von 6,7 ^{3}H-Oestradiol in Erfolgsorgan-Homogenaten*

	Endometrium			DMBA-Tumor
	Ratte	Kaninchen	Kalb	Ratte
„Kerne"	60,3	53,6	62,8	78,3
„Mitochondrien"	8,4	[1] 8,1	5,0	2,2
„Mikrosomen"	8,9	4,5	4,8	3,1
Überstand	22,3	[1]33,8	27,4	16,3

[1] Fraktionen enthalten bis zu 30% Oestron.

Injektion physiologischer Dosen und im Falle von Kalbsendometrium nach einer 30minutigen Inkubation mit 10^{-9} molarer Oestradiollösung in Krebs-Ringer-Henseleit-Puffer. Die Werte wurden durch ein Differenzverfahren ermittelt, bei dem die Radioaktivität in den ursprünglichen 0,44 m Saccharosehomogenaten (weniger als 0,5% intakte Zellen) und in den Überständen nach Abzentrifugieren der Kerne, der Kerne und mittelgroßen Partikel und schließlich aller partikularen Elemente gemessen wurde. Man erhält damit nicht nur eine optisch günstige Addition zu 100%, sondern ist vor allem vor zusätzlichen Homogenisierungsartefakten geschützt, die die Verteilung entscheidend beeinflussen können. Als prinzipieller Bindungsort für Oestradiol in allen untersuchten Erfolgsorganen hat sich der Zellkern herausgestellt. In dem aus methodischen Gründen verläßlichsten Homogenat von DMBA-Tumor befinden sich bis zu 80% des Oestradiols in der Kernfraktion. Bei der Radioaktivitätsverteilung in Kaninchenendometriumfraktionen muß eine Einschränkung gemacht werden. Im Gegensatz zum Ratten-

und Kalbsuterus findet im Kaninchenuterus eine Oxydation von Oestradiol statt. 1 Std nach der Injektion von $2\,\mu g$ ^{3}H-Oestradiol an 3 kg schwere, ovarektomierte und vorbehandelte Kaninchen liegen etwa 30% der aus dem Homogenatüberstand und der Mitochondrienfraktion isolierten Radioaktivität als Oestron vor.

Während bei den Zellfraktionierungsverfahren das Haupterfolgsorgan Uterus Schwierigkeiten bereitet, ist bei der Autoradiographie

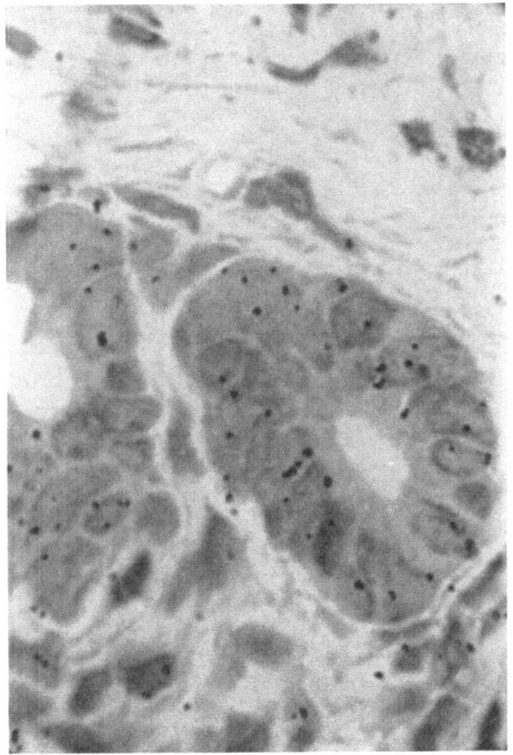

Abb. 3. Autoradiographie: Endometrium einer 290 g schweren, ovariektomierten Sprague-Dawley-Ratte, 2 Std nach s. c. Injektion von $2\mu g$ 6,7 ^{3}H-Oestradiol in 0,5 ml 0,9%iger NaCl-Lösung. Uterus nach Entnahme in flüssigem Propan gefroren, $1\,\mu$ Gefrierschnitt (Messertemperatur $-60\,^{\circ}\mathrm{C}$) lyophylisiert, auf vorgetrocknete Kodak NTB-3-Emulsion aufgepreßt ($-8\,^{\circ}\mathrm{C}$) und 18 Tage bei $-15\,^{\circ}\mathrm{C}$ exponiert. Entwicklung mit Kodak D-19, H-E-Färbung, Photomikroskop Zeiss, rotes Sperrfilter, Endvergrößerung 1300 $\times$. (W. E. Stumpf und L. J. Roth)

die gute Löslichkeit von Oestradiol in organischen Lösungsmitteln besonders zu beachten. Wir hatten das Glück, die Herren Dr. Stumpf und Dr. Roth vom Dept. of Pharmacology der Univ. of Chicago für unser Problem interessieren zu können. Dr. Stumpf und Dr. Roth haben ein Verfahren zur autoradiographischen Lokalisierung von nicht kovalent gebundenen Substanzen entwickelt, das den Kontakt mit Lösungsmitteln völlig vermeidet[13]. Das Ge-

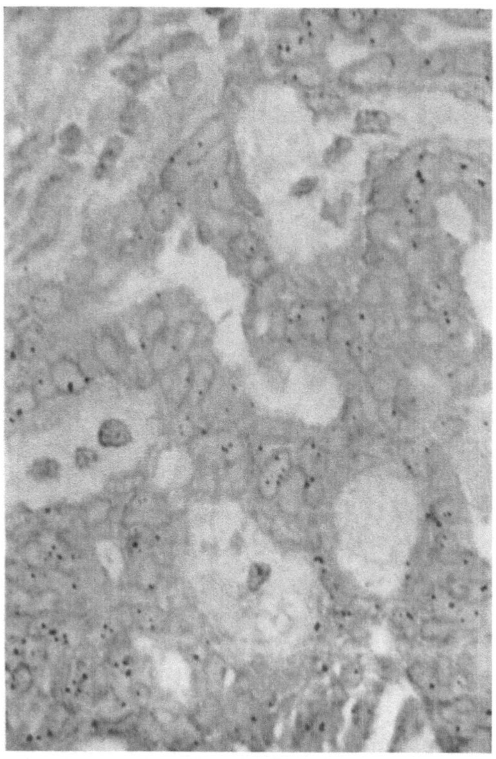

Abb. 4. Autoradiographie: DMBA-Mammacarcinom einer 250 g schweren Sprague-Dawley-Ratte, 6 Std nach s. c. Injektion von 0,3 µg 6,7 ³H-Oestradiol. Gleiche Aufarbeitung wie in Abb. 3, 62 Tage exponiert, Methylgrün-Pyronin-färbung, Endvergrößerung 900 ×. (W. E. Stumpf und L. J. Roth)

webe wird dabei sofort nach der Entnahme in flüssigem Propan gefroren, Gefrierdünnschnitte von 0,5 bis 1 µ werden lyophylisiert und mit Teflonspateln auf getrocknete Emulsion aufgepreßt.

Abb. 3 und 4 sind Beispiele ihrer Ergebnisse. Sowohl im Rattenuterus als auch im DMBA-Tumor finden sich Silberkörner zu allen Zeiten nach der Injektion von tritiiertem Oestradiol fast ausschließlich über den Zellkernen, vornehmlich in Nähe der Kernmembranen.

Beide Methoden, die Zellfraktionierung und die Radioautographie weisen also in guter Übereinstimmung den Zellkern als bevorzugten Bindungsort für Oestradiol und damit als wahrscheinliche Lokalisation des Receptors aus.

3. Demonstration und Eigenschaften
von spezifischen Oestradiol-Makromolekülkomplexen

In schonend hergestellten Extrakten aus oestradiolbehandelten Erfolgsorganen liegt das Hormon in Bindung an Makromoleküle vor. Mehrtätige Dialyse solcher Extrakte bei 0 bis 2° C führt zu keinem Verlust an Radioaktivität, und bei Passage über Sephadex wird Oestradiol fast ausschließlich mit der makromolekularen Fraktion eluiert. Als bestes Verfahren zur weiteren Charakterisierung der Oestradiol-Makromolekülkomplexe hat sich bis heute die Dichtegradientenzentrifugation erwiesen. Sie wurde meines Wissens zuerst von MAURER und CHALKLEY für einen Chromatinextrakt aus Kalbsendometrium angewendet[12]. Unter ihren Bedingungen, isopyknische Zentrifugation in einem $CsCl_2$-Gradienten bei 22° C, denaturierte eine Proteinfraktion mit einer Dichte von 1,24, die in einer scharfen Bande den Großteil des im Extrakt vorhandenen Oestradiols enthielt. Leider wird die Signifikanz dieses Ergebnisses durch die hohe Temperatur beeinträchtigt, bei der die spezifischen Komplexe (s. unten) nicht stabil sind. TOFT und GORSKI[14] homogenisierten Rattenuteri — nach *in-vivo*-Aufnahme von $6,7,^3$H-Oestradiol — mit verdünnten Pufferlösungen (am besten bewährte sich 0,01 m TRIS, 0,0015 m EDTA P_H 7,4) und zentrifugierten die lösliche Phase nach Überschichten auf einen 5- bis 20%igen Saccharosegradienten bei niederer Temperatur. Sie fanden Oestradiol bevorzugt in Bindung an eine mit etwa 9 S sedimentierende Komponente. In mit gepufferter KCl-Lösung (0,3 m KCl; 0,001 m NaN_3; 0,001 m $CaCl_2$; 0,01 m TRIS P_H 7,5) hergestellten Kernextrakten aus $6,7$ ^{3}H-Oestradiol enthaltenden Kalbsund Rattenuteri und DMBA-Tumoren sedimentiert die Radioaktivität in Bindung an eine 5 S-Fraktion. Es sei hier vorausgeschickt, daß die 9 S- und die 5 S-Komponente nicht einfache, lösungsmittelabhängige Varianten im Aggregationszustand ein und desselben Faktors sind. Austausch und Rücktausch der Lösungsmittel durch Dialyse oder Passage über Sephadex ändern das Sedimentationsverhalten nicht. Wir haben es demnach, nicht wie erwartet und gewünscht, mit einem „Receptor" zu tun, sondern mit zwei Bindungsprinzipien, deren Eigenschaften, Herkunft, Spezifität und mögliche Funktion nun zu erörtern sind.

a) Der 9 S-Faktor aus TRIS-EDTA-Extrakten. Der von Toft und Gorski entdeckte 9 S-Faktor hat eine experimentell besonders angenehme Eigenschaft: Er läßt sich auch aus unbehandelten Uteri leicht und vollständig extrahieren und bindet zum Extrakt zugesetztes Oestradiol. Seine Charakterisierung wird dadurch wesentlich erleichtert, und es erscheint zunächst nicht ausgeschlossen, ihn mit konventionellen Methoden isolieren zu können. Ein solches Vorhaben wird jedoch durch seine Wärmelabilität eingeschränkt. Erwärmen des Komplexes auf Zimmertemperatur führt nicht nur zu einer Dissoziation des gebundenen Oestradiols, sondern auch zum Verlust der Bindungsfähigkeit des Makromoleküls. Oestradiolfreie TRIS-EDTA-Extrakte können ohne wesentlichen Bindungsverlust der 9 S-Komponente gefroren aufbewahrt werden. Aufbewahrung bei 0 bis 2° C verringert innerhalb eines Monats die Bindungsfähigkeit im 9 S-Bereich um etwa 70%. Gefriertrocknung hat Verluste bis zu 90% zur Folge. Toft und Gorski haben durch Verdauungsversuche nachgewiesen, daß der 9 S-Faktor wahrscheinlich ein Protein ist oder zumindest einen für die Oestradiolbindung wichtigen Proteinanteil enthält. Proteasen setzen Oestradiol aus dem Komplex frei, Nucleasen lassen ihn intakt. P-Hydroxymercuribenzoat, N-Äthylmaleinamid und Jodacetamid (10^{-3} molar) zerstören in halbstündiger Inkubation bei 0 bis 2° C die Bindungsfähigkeit des 9 S-Faktors für Oestradiol[15]. Der Sedimentationskoeffizient des Faktors verschiebt sich in Abhängigkeit vom P_H von etwa 9 S bei P_H 7,4 bis zu etwa 7 S bei P_H 9,5. Dialyse oestradiolfreier Extrakte gegen P_H 6-Puffer führt zu einer massiven Fällung, wobei die 9 S-Komponente kopräcipitiert. Bei der Chromatographie eines mit markiertem Oestradiol versetzten TRIS-EDTA-Extrakts an Carboxymethylcellulose wird die Radioaktivität zwischen P_H 6,6 und P_H 7,0 eluiert (Abb. 5). Eine Retention des 9 S-Faktors ist auch an Diäthylaminoäthylcellulose nachzuweisen, allerdings ist das Bild dabei weniger klar, weil Oestradiol selbst von dem basischen Austauscher gebunden wird. Anreicherungsversuche aus unmarkierten Extrakten mit fraktionierter $(NH_4)_2SO_4$-Fällung waren unbefriedigend. Die von 0 bis 10% (Gew./Vol.) und von 10 bis 15% gefällten Fraktionen enthalten neben wenig bindender Substanz im 9 S-Bereich vor allem schneller sedimentierende Aggregate. Etwa 50% der Bindungskapazität fällt zwischen 15 und 30%

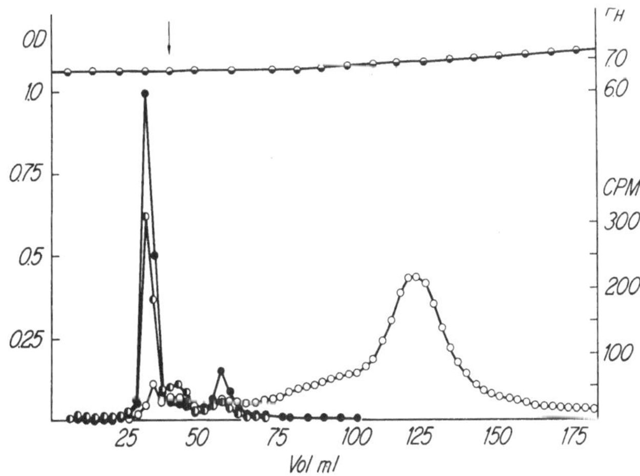

Abb. 5. Chromatographisches Verhalten von makromolekular-gebundenem Oestradiol in Kalbsuterus TRIS/EDTA-Extrakten. Extraktherstellung: Grob zerkleinerte Kalbsuteri (Durchschnittsgewicht 18 g) werden 3 × 30 min mit dem dreifachen Vol. gepufferter, 0,25 m Saccharoselösung gerührt, gesiebt, kurz mit TRIS/EDTA-Puffer (s. Text) gewaschen, im Fleischwolf fein zerkleinert und nach Zugabe von 1 Vol. TRIS/EDTA-Puffer mit einem Ultraturrax homogenisiert. Der Brei steht 12 bis 16 Std. Abtrennung der unlöslichen Bestandteile durch Zentrifugation; 15 min 15000 U/min Servall Ro SS, 5 Std 30000 U/min Ro 30 Spinco. Chromatographie: Carboxymethylcellulose, 0,8 maeq/g, 1,5 × 30 cm Säule; Auftrag: 1 ml Extrakt, dialysiert gegen Ausgangspuffer, 40000 cpm 6,7 ^{3}H-Oestradiol zugegeben; Gradientenelution (↓) mit 0,05 m Citronensäure/0,1 m Na_2HPO_4-Puffer, pH 6,8 bis 8,0. Alle Operationen bei 0 bis 2 °C. ●—● OD 280 mμ; ◖—◖ OD 260 mμ; o—o CPM/0,2 ml; ◓—◓ P_H im Elnat

$(NH_4)_2SO_4$. Die gelösten Niederschläge dieser Fraktionen binden Oestradiol nur in der 9 S-Zone. Alle diese Befunde sprechen mehr für eine komplexe Natur des 9 S-Faktors, als für ein wohldefiniertes und stabiles Proteinindividuum.

Der 9 S-Faktor kann in Lösung (d. h. nach Extraktion aus unbehandelten Erfolgsorganen) mit Oestradiol gesättigt werden. Bei dem in Abb. 6 dargestellten Versuch wurden steigende Mengen von 6,7 ³H-Oestradiol aliquoten Teilen eines TRIS/EDTA-Extraktes aus gewaschenen Kalbendometriumschnitten zugesetzt und die 9 S-Gipfel in der Dichtegradientenzentrifugation gemessen. Halbsättigung wurde bei etwa $1,3 \times 10^{-9}$ M/L Oestradiol erreicht. Ein weiteres Affinitätsmaß gibt der Vergleich der Oestradiolbindung an den 9 S-Faktor mit der an Albumin, dem Bindungsprinzip des

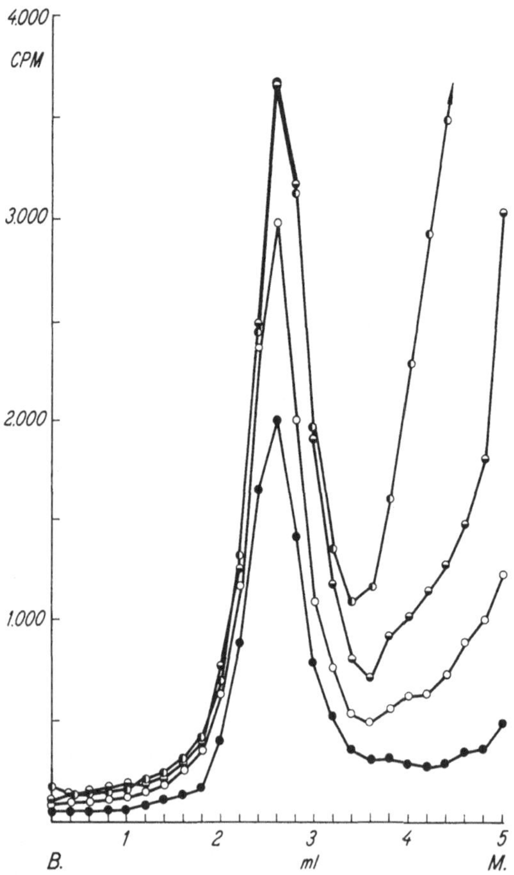

Abb. 6. Sättigung des 9-S-Bindungsfaktors mit Oestradiol. Aliquote Teile eines Kalbsendometrium TRIS/EDTA-Extraktes wurden 5 Std bei $0 \rightarrow 2$ °C mit 6,7 ³H-Oestradiol inkubiert und je 0,2 ml auf Dichtegradienten aufgetragen. $10 \rightarrow 30\%$ Saccharose, 10 Std, 50000 U/min. SWL 50, 9 °C. Oestradiolkonzentrationen in 10^{-9} M/L: ●—● 1,35; ○—○ 2,23; ◑—◑ 3,34; ◐—◐ 6,09.

Serums[16]. Extrakte aus grob zerkleinerten, weniger gut gewaschenen Kalbsuteri enthalten erhebliche Mengen an Serumproteinen. Bei einer Gesamtproteinkonzentration von 13 bis 15 mg Protein/ml in den Extrakten wird eine Albuminkonzentration von 3 bis 4 mg gemessen, d. h. die Extrakte enthalten etwa 40% Serumproteine. Aus der Oestradiolbindung im 9 S- und im 4- bis 5 S-Bereich solcher Extrakte und der bekannten Albuminkonzentration errechnet sich ein Affinitätsverhältnis von 10000 bis 50000:1.

Als Herkunftsort des 9 S-Faktors wird von seinen Entdeckern TOFT und GORSKI das Cytoplasma vermutet. Wir sind nicht uneingeschränkt dieser Meinung. Homogenisation mit hypotonen Medien führt sicher zu einer Zerstörung von Zellstrukturen und damit zu einer möglichen Extraktion von *in-vivo* strukturgebundenen Substanzen. Zur Klärung der Frage haben wir folgenden Versuch unternommen: Gleiche Teile eines Kalbsuterusbreies wurden nach Zugabe von je einem Volumen 0,25 m gepufferter Saccharoselösung bzw. TRIS/EDTA-Puffer unter gleichen Bedingungen homogenisiert. Die Überstände hatten identische Gesamtprotein- und Albuminkonzentrationen. Nach Zusatz von markiertem Oestradiol und Zentrifugation in 10 bis 30%igen Saccharosegradienten sedimentierte beim TRIS/EDTA-Extrakt fast alle Radioaktivität im 9 S-Bereich. Das Sedimentationsdiagramm des Saccharosehomogenatüberstandes zeigte dagegen um 60% weniger Oestradiol im 9 S-Gipfel, eine vermehrte Bindung im Albuminbereich und freies Oestradiol (Abb. 7). Dieses Ergebnis bekräftigt unseren Verdacht, daß der 9 S-Faktor keine in Lösung befindliche cytoplasmatische Substanz ist, sondern aus einer Zellstruktur stammt, die wahrscheinlich nicht der Zellkern ist. TRIS/EDTA-Extrakte aus Endometriumkernen enthalten den 9 S-Faktor nicht (Abb. 7).

Das Vorkommen des 9 S-Faktors scheint auf Erfolgsorgane, zu denen auch die Samenblase zählt, beschränkt zu sein. Abb. 8 zeigt Sedimentationsdiagramme von mit markiertem Oestradiol versetzten TRIS/EDTA-Extrakten aus Vagina, Samenblase, Leber und Thymus. Nur die beiden ersteren enthalten Aktivität im 9 S-Bereich. Tab. 2 gibt eine Zusammenfassung der Oestradiolbindung im 9 S-Bereich von bisher untersuchten Organextrakten. Danach ist die Bindungskapazität von Uterus-, Vagina- und Brustdrüsenextrakten etwa gleich groß, die von Samenblasenextrakten um mehr als die Hälfte niedriger und Hypophysenvorderlappenextrakte

enthalten nur etwa 10% der Bindungskapazität von Haupterfolgs-
organen. In Nieren ist wie in Thymus und Leber der 9 S-Faktor
nicht nachzuweisen.

Für die Hormonbindungsspezifität des 9 S-Faktors liegen bereits Ergebnisse von Toft und Gorski vor. Danach konkurriert

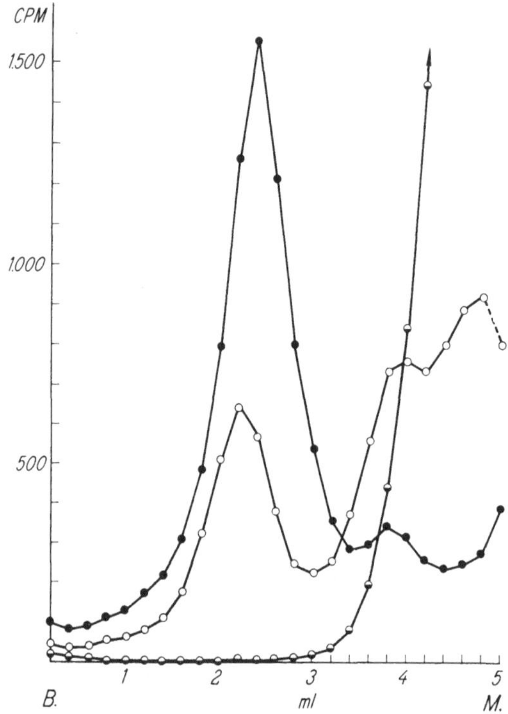

Abb. 7. Untersuchung der Herkunft des 9-S-Faktors. Dichtegradientenzentrifugation 5 Std nach Zugabe von 6,7 ³H-Oestradiol zu den Extrakten.
(s. Text) 10 → 30% Saccharose, 11 Std. 15 min, 50000 U/min, SWL 50, 9 °C
●—● TRIS/EDTA-Extrakt Uterus; ○—○ Saccharoseextrakt Uterus; ◐—◐
TRIS/EDTA-Extrakt aus Endometriumkernen

in-vivo verabreichtes Diäthylstilboestrol mit der Bindung von
Oestradiol. Einen geringen Einfluß hat ebenfalls 17-Alpha-Oestra-
diol. Wir fanden, daß auch ein 10⁴facher Überschuß des Oestrogen-
antagonisten U 11 100 in-vivo, in-vitro (Inkubation von Gewebe) und
in Lösung (Zusatz zum Extrakt) ein Kompetitor ist. Die relative
Spezifität des 9 S-Faktors für Oestradiol geht ferner aus Bindungs-

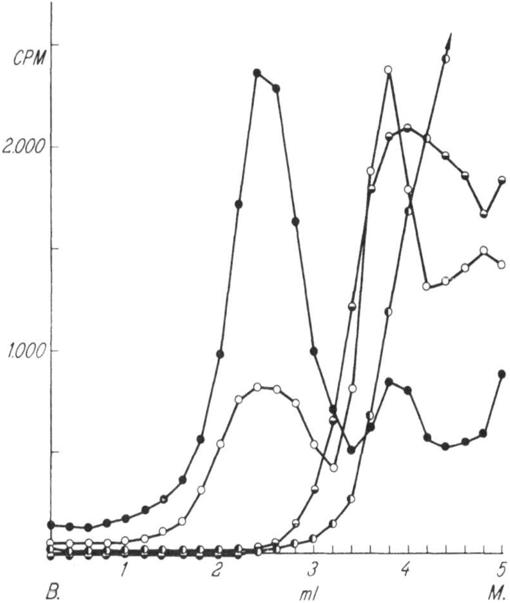

Abb. 8. Organspezifität des 9-S-Faktors. Dichtegradientenzentrifugation 5 Std nach Zugabe von 6,7 ³H-Oestradiol zu den TRIS/EDTA-Extrakten. 10 → 30% Saccharose, 10 Std, 50000 U/min, SWL 50, 9 °C. ●—● Vagina; ○—○ Samenblasen; ◐—◐ Leber; ◐—◐ Thymus

Tabelle 2. *Relative Konzentrationen von 9-S-Faktor in Kalbsgeweben*

(1:1 Extrakte mit TRIS/EDTA-Puffer, Dichtegradientenzentrifugation nach Zugabe von 2,2 × 10⁻⁹ M/L 6,7 ³H-Oestradiol.)

	mg/ml Protein (Folin)	mg/ml Albumin (Immuntest)	10⁻⁹ M/L Oestradiol im 9-S-Bereich
Endometrium	14,5	3,5	1,57
Vagina	17,2	> 5	1,58
Milchdrüse	14,0	3,4	1,32
Samenblase	20,1	3,3	0,62
Adenohypophyse	19,3	2,4	0,12
Leber	28,2	1,9	0
Niere	28,2	1,4	0
Thymus	22,4	1,5	0

und Austauschversuchen mit anderen Steroiden in Kalbsuterusextrakten hervor. Bei Zusatz etwa gleicher molarer Konzentrationen von radioaktiven Oestradiol, Oestron oder Testosteron verhalten sich die Steroidkonzentrationen im 9 S-Bereich wie 100:57:0,5. Vorher gebundenes radioaktives Oestradiol tauscht mit einem Überschuß der unmarkierten Verbindung nicht bis zum Gleichgewicht aus. Ein etwa um die Hälfte geringerer Austausch erfolgt sowohl mit Oestron als auch mit 17-Methoxyoestradiol. Die Ergebnisse der Bindungs- und Austauschversuche lassen zwei Schlüsse zu: 1. Die Bindung von Oestradiol an den 9 S-Faktor ist — wenigstens unter den gewählten Versuchsbedingungen — nicht voll reversibel und 2. die „Substratspezifität" des Faktors ist begrenzt. Oestradiol wird bevorzugt, aber nicht ausschließlich gebunden. Selbst das nicht aromatische Testosteron weist eine geringe Affinität auf.

Über die Funktion des 9 S-Faktors lassen sich in diesem Zeitpunkt nur Vermutungen anstellen. Man kann mit verschiedenen Methoden nachweisen, daß er nur etwa 20% der Oestradiolbindungskapazität des Uterus einnimmt. Am naheliegendsten erscheint mir ein Zusammenhang mit der Steroidaufnahme in die Erfolgsorganzellen.

b) Der 5 S-Faktor. Während der 9 S-Faktor für die Bindung von etwa 20% des bei Sättigung im Uterus vorhandenen Oestradiols verantwortlich ist, werden die restlichen 80% vom 5 S-Faktor gebunden. Seine Herkunft ist definierter als die des 9 S-Faktors. Nach Inkubation von Kalbsendometrium mit markiertem Oestradiol kann alle kerngebundene Aktivität durch wiederholte Extraktion mit gepufferter 0,3 m KCl-Lösung extrahiert werden und sedimentiert in der Dichtegradientenzentrifugation in Bindung an ein 5 S-Makromolekül. Das gleiche trifft für *in-vivo*- und Inkubationsversuche mit Rattenuteri und DMBA-Tumor zu. Vor der weiteren Beschreibung des Faktors muß zunächst ein naheliegender Verdacht ausgeräumt werden: Er ist nicht mit dem als Verunreinigung in den Extrakten enthaltenen Albumin identisch. Das ergibt sich nicht mit Sicherheit aus den nur wenig verschiedenen Sedimentationskoeffizienten, wohl aber durch den in Abb. 9 dargestellten Versuch. Fällung des Albumins aus einem Extrakt mit der Gamma-Globulinfraktion eines Antiserums führt zu keiner Ab-

nahme des 5 S-Gipfels, verglichen mit der Kontrolle. Das bei
niedrigerer Temperatur erhaltene Sedimentationsdiagramm des
frischen Ausgangsextraktes weist einen beträchtlich höheren 5 S-
Gipfel auf, was nicht allein aus der für die Immunfällung notwendigen Verdünnung erklärt werden kann. Der 5 S-Oestradiolkom-

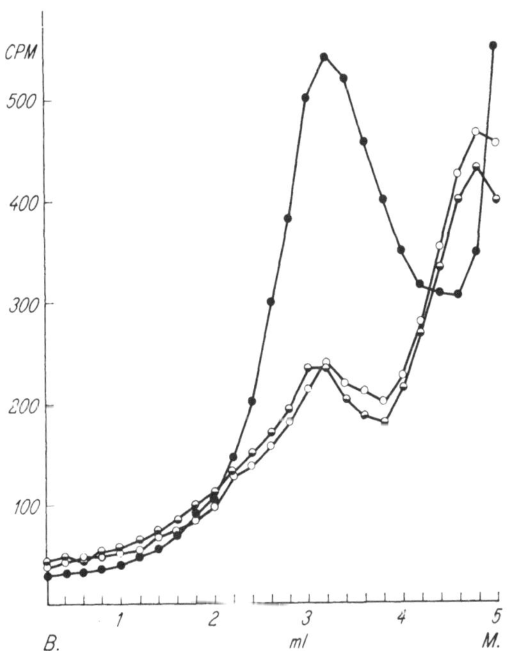

Abb. 9. Ausschluß der Identität von 5-S-Faktor und Serumalbumin. Albu-
minfällung mit der Gamma-Globulinfraktion eines Antiserums bei 0 → 30%
Saccharose. Frischer Extrakt: 10 Std, 55000 U/min, SWL 65, 2,5°C; albumin-
freier Extrakt und verdünnter Kontrollextrakt: 10 Std, 50000 U/min,
SWL 50, 9 °C

plex ist in noch größerem Maße wärmeempfindlich als der 9 S-
Komplex. Auch in der Kälte (0 bis 2° C) erfolgt der Zerfall rasch
und ist in etwa 2 Wochen vollständig. Frieren wird nur im begrenzten Maßstab überstanden, Gefriertrocknung führt zu völligem Verlust. Bei mehrtägiger Dialyse (0 bis 2° C) von KCl-Extrakten diffun-
diert kein Oestradiol aus dem Schlauch, jedoch nimmt die Höhe
des 5 S-Gipfels in der Dichtegradientenzentrifugation ab, wie es
auch nach der Dialyse von markierten TRIS-EDTA-Extrakten für

den 9 S-Gipfel beobachtet wird. Ob Oestradiol dabei von anderen, in den Extrakten vorhandenen Makromolekülen zurückgehalten oder an Fragmente der Bindungsfaktoren gebunden bleibt, ist ungeklärt. Die Oestradiolbindung an den 5 S-Faktor ist ebenfalls empfindlich gegen Sulfhydrylreagentien[15]. Nach Behandlung von KCl-Extrakten mit 10^{-3} m p-Hydroxymercuribenzoat (30 min bei 0 bis 2° C) erscheint Oestradiol am Meniskus des Sedimentationsdiagramms. Das Verfahren erlaubt eine einfache Unterscheidung des 5 S-Faktors von Albumin, dessen Bindungsfähigkeit für Oestradiol durch diese PHMB-Konzentration nicht beeinträchtigt wird.

Neben der Verschiedenheit in Herkunft, Sedimentationsverhalten und Anteil an der Oestradiolbindungskapazität gibt es noch weitere, wichtige Unterschiede zwischen den 9- und 5 S-Faktoren. Mit markiertem Oestradiol beladener 5 S-Faktor zeigt weder mit 10- noch mit 100fachem Überschuß von nicht radioaktivem Hormon einen Austausch. Besonders auffällig ist die unterschiedliche Extrahierbarkeit der beiden Faktoren aus unbehandelten Erfolgsorganen. Extrakte aus oestradiolfreien Uteri enthalten, gemessen an der Bindung von zugesetztem radioaktiven Oestradiol im fraglichen Bereich, nur wenig 5 S-Faktor. In Abb. 10 ist ein KCl-Kernextrakt aus mit Oestradiol inkubierten Kalbsendometriumschnitten, verglichen mit Extrakten aus Kernen von unbehandeltem Endometrium und Nieren, denen 6,7 ^{3}H Oestradiol zugesetzt wurde. Das unter vergleichsweise *in-vivo*-Bedingungen aufgenommene Oestradiol erscheint vornehmlich in der 5 S-Region, der in Lösung markierte Endometriumkernextrakt zeigt eine Schulter an der richtigen Stelle, der Extrakt aus Nierenkernen dagegen nur Oestradiol am Meniskus. Dieses Ergebnis demonstriert die Andeutung einer Organspezifität, stellt aber zugleich zwei Fragen: Ist ein *in-vivo*-Prozeß für das Auftreten des 5 S-Faktors aus Erfolgsorganzellkernen erforderlich, ist er eventuell eine neusynthetisierte Kernkomponente? oder macht die *in-vivo* erzielte, hohe Konzentration von Oestradiol im Zellkern den 5 S-Faktor erst löslich? Hemmung der Proteinsynthese mit 10^{-3} m Puromycin im Inkubationsmedium von Kalbsendometriumschnitten oder Einschränkung der RNS-Synthese durch 2,5 μg Aktinomycin D/ml haben weder einen Einfluß auf die Gesamtoestradiolaufnahme noch auf die Höhe des 5 S-Gipfels in den Kernextrakten. Auch die zweite Frage ist in den letzten Wochen beantwortet worden: Gewaschene albuminfreie Kalbsendometrium- und Milchdrüsenkerne wurden mit gepufferter KCl-Lösung in Anwesenheit von markiertem Oestradiol extrahiert, dessen Konzentration der *in-vivo* beobachteten Akkumulation im Kern entsprach. Nach Abtrennung von über-

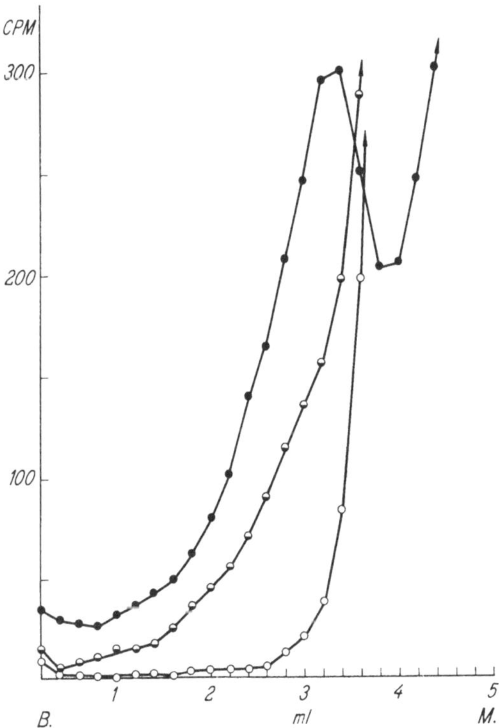

Abb. 10. Extrahierbarkeit und Organspezifität des 5-S-Faktors. KCl-Kernextrakt aus mit 6,7 ^{3}H-Oestradiol inkubierten Kalbsendometriumschnitten, verglichen mit Kernextrakten aus unbehandeltem Endometrium und Nieren, denen markiertes Oestradiol zugesetzt wurde. Dichtegradientenzentrifugation: 10 → 30% Saccharose, 10 Std, 50000 U/min, SWL 50, 9 °C

schüssigem Oestradiol an Sephadex G-75 zeigten die makromolekularen Fraktionen in der Dichtegradientenzentrifugation deutliche 5 S-Oestradiolgipfel, die nach Behandlung mit p-Hydroxymercuribenzoat verschwanden. Die ohne Anwesenheit von Oestradiol hergestellten Kontrollextrakte wiesen dagegen nur eine niedrige 5 S-Schulter auf.

Mit diesem Befund läßt sich eine vorsichtige Antwort auf die Frage nach der Funktion der 5 S-Komponente geben. Sie zeigt zumindest *eine* repressorähnliche Eigenschaft. Oestradiol erhöht ihre Extrahierbarkeit aus Zellkernen.

4. Versuche zur präparativen Isolierung von oestrogenbindenden Prinzipien

Das letzte Kapitel beschreibt Versuche, die einige Zeit vor der Entdeckung der 9 S- und 5 S-Faktoren begonnen wurden. Wir waren damals noch Unitarier und versuchten zunächst *den* Receptor mit konventionellen Methoden zu isolieren. Da wir etwa 1 bis 2 mg Receptor in 1 kg Uterus vermuteten, schied die Ratte aus verständlichen Gründen als Versuchstier aus. Wir wählten schließlich Kalbsuteri als Ausgangsmaterial, in denen genau wie in Rattenuteri und im Gegensatz zu Kaninchenuteri keine metabolische Veränderung von Oestradiol stattfindet. Die Verfolgung eines markierten Oestradiolreceptorkomplexes über viele Anreicherungsschritte erschien uns weder praktisch noch ist sie — wie wir heute wissen — erfolgversprechend. Wir gingen von der (für den 5 S-Faktor nicht korrekten) Annahme aus, daß unbeladener Receptor durch die gleichen Extraktionsverfahren freigesetzt werden kann, wie der *in vivo* gebildete Oestradiolreceptorkomplex. Als einziges Merkmal des Receptors war seine hohe Affinität für Oestradiol vorauszusetzen und demnach bot sich die Gleichgewichtsdialyse oder ein abgekürztes Sephadexverfahren[17] als Test an. Leider stellte es sich bald heraus, daß das phenolische Steroid in unterschiedlichem Ausmaß von jedem Biopolymeren gebunden wird, was den Wert der Gleichgewichtsdialyse als Anreicherungstest erheblich einschränkt. Abb. 11 zeigt die makromolekulare Bindung von Oestradiol durch Desoxycholatextrakte von Kalbsuteruskernen, -mitochondrien und -mikrosomen bei Passage über eine mit markiertem Oestradiol vorbeladene Sephadexsäule. Nach diesem Ergebnis wäre der Receptor in der Mikrosomenfraktion zu suchen, in der er nicht ist.

. Wir haben dann ein Verfahren ausgearbeitet, das in gewissem Sinn Affinitätstest und Isolierung in einem Arbeitsgang vereint[18]. Wenn die Bindung von Oestradiol an den Receptor keiner Aktivierungsenergie bedarf, sollte man — in Umkehrung des *in-vivo*-Vorganges — mit fixiertem Oestradiol den aus der Zellstruktur gelösten Receptor selektiv anreichern und so isolieren können. Wir

kuppelten gereinigte, diazotierte Paraaminobenzylcellulose mit
Oestradiol (Abb. 12). Die Kapazität der Adsorbentien wurde durch
Zusatz von tritiiertem Oestradiol zum Kupplungsgemisch und an-
schließende Verbrennungsanalyse bestimmt. Sie lag in der Größenordnung von 1 μMol Oestradiol/g Cellulose, was bei gleichmäßiger
Verteilung etwa einem Molekül Oestradiol auf 5000 Glucosemole-
küle entspräche. Die durch Reduktion der Azobindung entstehen-

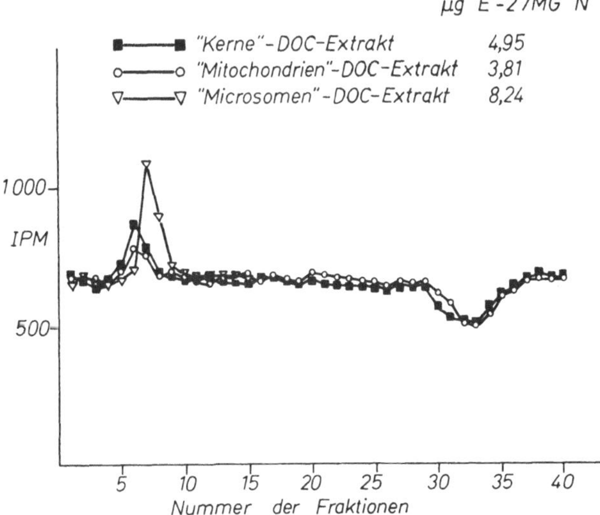

Abb. 11. *Untersuchung der markromolekularen Bindung von Oestradiol in
Desoxycholatextrakten von Kalbsendometriumkernen, -mitochondrien und -mi-
krosomen.* Sephadex G-75-Säule, 1×34 cm, äquilibriert mit 0,1 m Trispuffer
pH 7,5 enthaltend $1,1 \times 10^{-5}$ M/L Oestradiol (davon 17,5 μC 6,7 ^{3}H-Oestra-
diol); Auftrag von je 1 ml Extrakt, Elution in 1,5-ml-Fraktionen mit gleichem Puffer; Zimmertemperatur. Die Bindung im makromolekularen Gipfel
ist in μg Oestradiol/mg N der Extrakte angegeben

den Aminooestradiole können chromatographisch getrennt werden.
2-Aminooestradiol ist polarer als die 4-Verbindung, gibt einen
grünen Ninhydrin-Kupfer-Komplex und zeigt eine einzelne, aromatische CH-Schwingung bei 11,4 μ im Infrarot. Das Kupplungs-
verhältnis läßt sich aus der chromatographischen Analyse von
markiertem Aminooestradiolen bestimmen. Es beträgt 60:40:
für die 4- bzw. 2-Position.

Das Adsorbens ist für den gewünschten Zweck sicher nicht als
ideal anzusehen. Entweder die Bindung in 2- oder in 4- hindert

wahrscheinlich die Anlagerung des Receptors an das phenolische
Hydroxyl. Vor allem aber bewirkt die Wasserstoffbrückenbindung
des Phenols mit dem Azostickstoff eine Polarisierung, die das Adsorbens zu einem schwachen Kationenaustauscher macht. Wir
haben es trotz dieser Bedenken verwendet und wider Erwarten mit
Erfolg.

In einer im vorigen Frühjahr abgeschlossenen Versuchsserie
filtrierten wir — in der Kälte — KCl-Extrakte von je etwa 1 kg
gefrorener Kalbsuteri über 5 g Portionen des Adsorbens. Danach

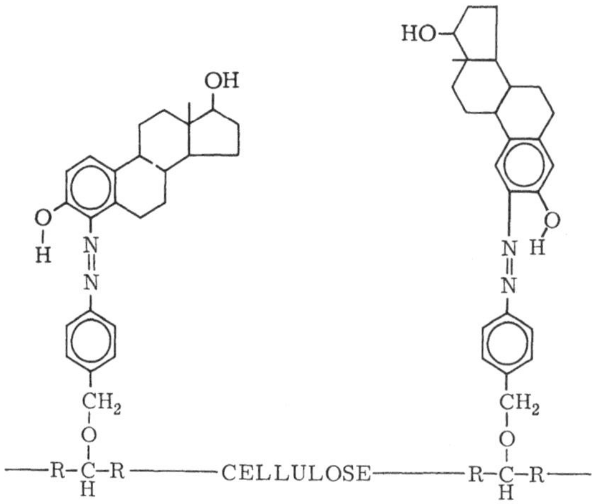

Abb. 12. Adsorbens zur Isolierung von Oestrogen-Bindungsfaktoren

wurde die Säule mit einem Puffer von P_H 9,0 gewaschen und anschließend das retinierte Material durch Zusatz von Desoxycholat
zum Puffer eluiert. (Die Verwendung von Oestradiol ist wegen
seiner geringen Löslichkeit in Wasser nicht möglich. Desoxycholat
hat Nachteile, setzt aber Oestradiol aus Uteri in ungebundener
Form frei. Ob es sich dabei um eine direkte Bindungskonkurrenz
oder eine Deformierung der bindenden Faktoren handelt, ist nicht
bekannt. Jedenfalls konnten wir nach der Desoxycholatelution und
anschließender Reduktion des Adsorbens keine zurückgebliebenen
makromolekularen Substanzen nachweisen.) Nach weitgehender
Entfernung von Desoxycholat aus den Eluaten blieben jeweils
weniger als 1 mg makromolekularer Substanz zurück. Kaninchen-
antikörper gegen diese Substanz fällten makromolekular gebunde-

nes Oestradiol aus KCl-Extrakten von Kalbsendometriumkernen quantitativ. Wir verwendeten die Gamma-Globulinfraktion der Antiseren und setzten in Kontrollserien für unspezifische Fällung den Extrakten Albumin und Antialbumin zu (Abb. 13).

Auf Grund unserer heutigen Kenntnisse über die Extrahierbar-keit der 9 S- und 5 S-Faktoren müssen wir annehmen, daß nur ein kleiner Teil des von dem Adsorbens gebundenen Materials 5 S-Faktor war, dessen Immunpräcipitation wir testeten. Der größere Teil entsprach sicher dem auch mit 0,3 m KCl aus unbehandelten

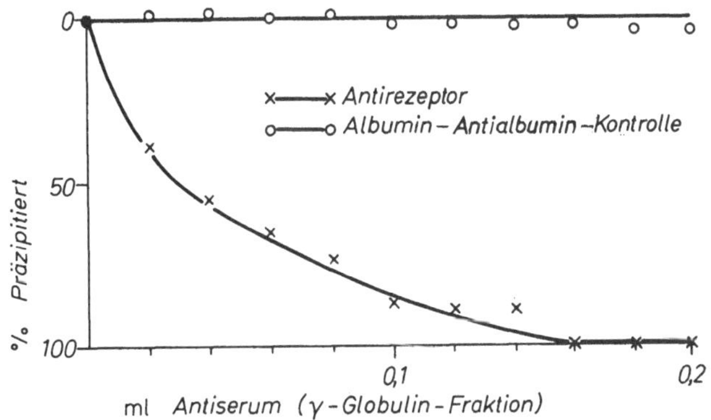

Abb. 13. *Immunfällung von makromolekular gebundenem Oestradiol aus KCl-Extrakt von Kalbsendometriumkernen.* Zu aliquoten Teilen Extrakt (Schnitte mit 6,7 ^{3}H-Oestradiol inkubiert) wurden steigende Mengen Antikörperlösung gegeben und mit Puffer auf gleiche Endvolumina aufgefüllt. Kontrollreihe: Extrakt + Albumin + Antialbumin. Messung der nicht gefällten Radioaktivität nach 20stündiger Inkubation bei 0 → 2 °C

Uteri besser zu extrahierenden 9 S-Faktor. Außerdem wurden einige μg Gamma-Globulin retiniert. Die durchgeführten Analysen sind deshalb nur insofern erwähnenswert, als sie neben Aminosäuren auch die Anwesenheit von Phosphor — teilweise als Lipid-P — und Ribose, jedoch nicht von Desoxyribose nachwiesen.

Wir haben inzwischen mit einer zweiten Versuchsserie im 10-kg-Maßstab begonnen, in der zunächst der 9 S-Faktor isoliert werden soll. Abb. 14 indiziert die Brauchbarkeit des Adsorbens und gibt zugleich eine weitere Information über die „Substrat"-Spezifität des 9 S-Faktors. Aliquote Teile eines Kalbsuterus-TRIS-EDTA-Extraktes wurden über Nacht in der Kältekammer mit Adsorben-

tien gleicher Kapazität gerührt und die Filtrate nach Zusatz von
³H Oestradiol in Saccharosegradienten zentrifugiert. Während das
Oestradiol enthaltende Adsorbens den 9 S-Faktor fast vollständig
adsorbierte, war der Austauscher mit 17-Methoxyoestradiol weniger wirksam, der mit Oestron am wenigsten. Die Gesamtproteinkonzentration des Extraktes, gemessen nach Folin, und die immu-

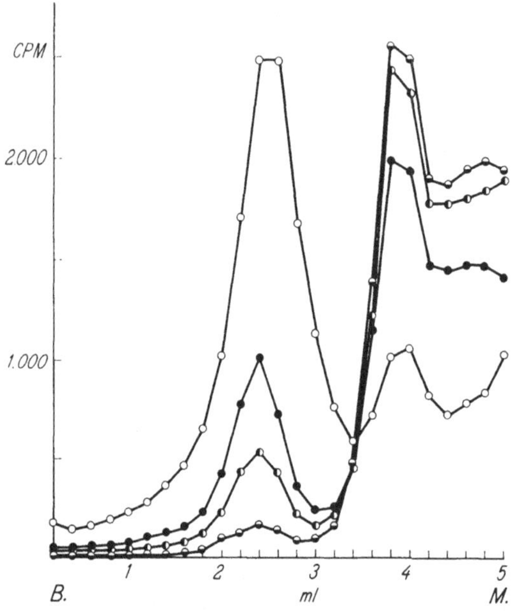

Abb. 14. Bindung von 9-S-Faktor aus Kalbsuterus TRIS/EDTA-Extrakt an
oestrogenenthaltende Adsorbentien. Kapazität der Adsorbentien: 1,1 μ Mol
Steroid/g Cellulose. Dichtegradientenzentrifugation der Filtrate nach 20stün-
diger Adsorption bei 0 → 2 °C und Zusatz von 6,7 ³H-Oestradiol: 10 → 30%
Saccharose, 10 Std, 50000 U/min, SWL 50, 9 °C. ◐—◐ Oestradioladsorbens;
◑—◑ 17-Methoxyoestradioladsorbens; ●—● Oestronadsorbens; ○—○ Kontrollextrakt

nologisch bestimmte Albuminkonzentration wurden während des
Adsorptionsvorganges nicht verändert.

Zur Isolierung des 5 S-Faktors arbeiten wir zur Zeit Extraktionsverfahren für unbehandelte Erfolgsorgane aus, die einen höheren
Ertrag bringen sollen.

Ausblick

An dem hier dargestellten und heute wohl am besten bekannten Beispiel ist zu erkennen, daß unsere Kenntnisse über Hormonreceptoren noch um einiges von Vollständigkeit entfernt sind. Dabei ist die Suche nach den oestrogenbindenden Faktoren durch die relativ stabile Bindung des phenolischen Steroids, die wahrscheinlich einen Ausnahmefall darstellt, sogar noch begünstigt. Für die beschriebenen Faktoren muß zunächst, trotz vieler ähnlicher Eigenschaften, eine Verschiedenheit angenommen werden. Die 9 S-Komponente ist offenbar extranuclearer, die 5 S-Komponente sicher nuclearer Herkunft. Es widerspräche der Theorie, beide als Receptoren zu bezeichnen. Diese Bezeichnung dürfte der 5 S-Komponenten mehr adäquat sein, die nicht nur 80% des Hormons bindet, sondern auch in ihrer durch Oestradiol veränderten Extrahierbarkeit eine „repressorähnliche" Eigenschaft aufweist. Ob diese Eigenschaft ein Ausdruck des von KARLSON[19] postulierten Wirkungsmechanismus der Steroidhormone ist, bleibt abzuwarten. Andere Vorstellungen, wie die hormonale Beeinflussung bestimmter Zellstrukturen, erscheinen nicht unrealistisch.

Unsere Versuche mit spezifischen Adsorbentien versprechen, wenn nicht bessere Analysen — der mg-Maßstab dürfte kaum überschritten werden —, so doch wenigstens die Gewinnung von Antiseren, die eine nähere Untersuchung der fraglichen Verwandtschaft beider Komponenten und ihrer Lokalisation in der Zelle erlauben. Eine genaue Lokalisation der Oestradiolbindungsfaktoren mit elektronenoptischen Methoden sollte bei der Aufklärung ihrer Funktion von Nutzen sein.

Der Gedanke, isolierte Hormonreceptoren in einem biologischen System zu testen, läßt sich vorläufig noch nicht verwirklichen.

Literatur

[1] HECHTER, O., and J. D. K. HALKERSTON: The Hormones, V, 697. New York: Academic Press 1964.

[2] JENSEN, E. V., and H. I. JACOBSON: Biological activities of steroids in relation to cancer, p. 129. New York: Academic Press 1960.

[3] — — Recent Progr. Hormone Res. 18, 387 (1962).

[4] TALALAY, P., and H. G. WILLIAMS-ASHMAN: Recent Progr. Hormone Res. 16, 1 (1960).

[5] HUGGINS, C., L. C. GRAND, and F. P. BRILLANTES: Lancet 1961, ii, 796.

[6] KING, R. J. B., D. M. COWAN, and D. R. INMAN: J. Endocr. 32, 83 (1965).

[7] JUNGBLUT, P. W., E. R. DeSOMBRE und E. V. JENSEN: UICC Panel Diskussion über Hormone in Genese und Therapie des Mamma-Carcinoms, Berlin 1965. Abh. d. Dtsch. Akad. d. Wissensch. (Im Druck.)

[9] —, R. J. MORROW, G. L. REEDER, and E. V. JENSEN: Endocrinology 76, 56 abstr. (1965).

[8] NOTEBOOM, W. D., and J. GORSKI: Arch. Biochem. 111, 559 (1965).

[10] KING, R. J. B., J. GORDON, and D. R. INMAN: J. Endocr. 32, 9 (1965).

[11] INMAN, D. R., R. E. W. BANFIELD, and R. J. B. KING: J. Endocr. 32, 17 (1965).

[12] MAURER, H. R., and G. R. CHALKLEY: J. Molecular Biol. (Im Druck).

[13] STUMPF, W. E., and L. J. ROTH: J. Histochem. Cytochem. 14, 274 (1966).

[14] TOFT, G., and J. GORSKI: Proc. nat. Acad. Sci. (Wash.) 55, 1574 (1966).

[15] DeSOMBRE, E. R., D. HURST, T. KAWASHIMA, P. W. JUNGBLUT, and E. V. JENSEN: Fed. Proc. 26, 536 (1967).

[16] WESTPHAL, U.: In mechanism of action of steroid hormones, p. 33. New York: Pergamon Press 1961.

[17] HUMMEL, J. P., and W. J. DREYER: Biochim. Biophys. Acta 63, 530 (1962).

[18] JUNGBLUT, P. W., and E. V. JENSEN: Endocrinology 78, 30 abstr. (1966).

[19] KARLSON, P.: Perspect. Biol. Med. 6, 203 (1963).

Diskussion

EBEL (Straßburg): Ich danke Herrn JUNGBLUT für seinen Vortrag; die Diskussion ist eröffnet.

MOSEBACH (Bonn): Ich hätte eine Menge Fragen, weil wir im gleichen Umfang wie die Chicagoer Gruppe über ein ähnliches Problem arbeiten, nämlich die Verteilung und Bindung von Testosteron im männlichen Organismus. Wir beobachten 10 min nach Injektion in den Erfolgsorganen meist keine ausgeprägte Lokalisation im Kern: in den *accessorischen Sexualorganen* im basalen und apikalen Bereich des Epithels — die Kerne erscheinen hierbei ausgespart (Abb. 1a) — sowie im Sekret; im *Hypothalamus* (Tuber cinerium) dicht gepackt im ganzen Zellbereich der Gliazellen (Abb. 1b); im *Muskel* (Adduktoren) im Sarkolemm; nur in den *Spermatocyten* 1. Ordnung deutlich im Zellkern (Abb. 1c). In Leber und Niere beobachten wir nach enzymatischem Abbau von Kern- *und* Cytoplasmaproteinen sehr feste Bindungen von ^{14}C-Metaboliten des Testosterons (wahrscheinlich hydroxyliertes Testosteron) an basische Peptide [Nature 214, 917 (1967)].

Meine grundsätzliche Frage lautet: Haben Sie keinerlei Hinweise dafür, daß auch in der Cytoplasmafraktion primäre Angriffspunkte des Oestradiols zu suchen sind?

JUNGBLUT (Chicago): Es ist schade, daß Herr STUMPF und Herr ROTH nicht hier sind; aber lassen Sie mich gleich die Gretchenfrage stellen: Welche Methode verwenden Sie bei Ihren histoautoradiographischen Untersuchungen?

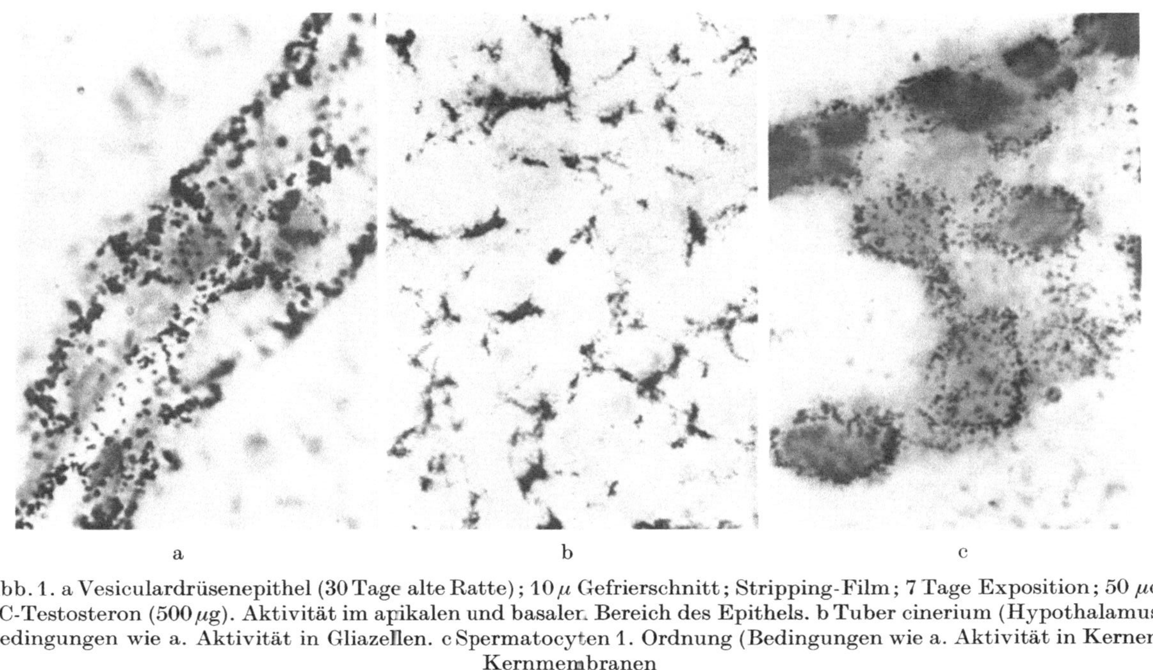

Abb. 1. a Vesiculardrüsenepithel (30 Tage alte Ratte); 10 μ Gefrierschnitt; Stripping-Film; 7 Tage Exposition; 50 μc ^{14}C-Testosteron (500 μg). Aktivität im apikalen und basaler. Bereich des Epithels. b Tuber cinerium (Hypothalamus Bedingungen wie a. Aktivität in Gliazellen. c Spermatocyten 1. Ordnung (Bedingungen wie a. Aktivität in Kernen Kernmembranen

MOSEBACH: Wir haben verschiedene Techniken verwendet; wir haben mit flüssigem Propan abgeschreckt und nach Anfertigung von Gefrierschnitten gefriergetrocknet, wir haben auch mit Formalin in der Gasphase fixiert. Wir haben niemals die Emulsionstechnik angewandt, sondern stets die Stripping-Filmmethode, um Durchtränkung der Schnitte zu vermeiden. Wir haben bei all diesen komplizierten Verfahren gleichartige Bilder gesehen wie bei einfacher kurzer Formalinfixierung des Gewebes und Verarbeitung von Gefrierschnitten. Diese einfache Methode haben wir dann für die längeren Reihenuntersuchungen verwendet.

JUNGBLUT: Ich weiß nicht, ob Sie die Arbeit von STUMPF und ROTH kennen? Danach muß bei derartigen Untersuchungen jeder Kontakt mit Lösungsmitteln vermieden werden. STUMPF und ROTH haben ihr Verfahren u. a. auch auf den Nachweis wasserlöslicher Substanzen, wie z. B. Mesobilirubinogen in den Gallengangskapillaren, ausgedehnt. Ihre Methodik ist so gut, daß die Radioaktivität auf die Kapillaren beschränkt bleibt. Wenn bei der Autoradiographie von nicht kovalent gebundenen Substanzen die Schnitte mit Lösungsmitteln oder auch nur mit Feuchtigkeit in Berührung kommen, können alle möglichen Artefakte auftreten.

Für Oestradiol ist der Zellkern sicherlich der Hauptbindungsort. Nun ist hier zu berücksichtigen, daß die Erfolgsorgane allein das genuine Hormon enthalten. Im Falle des Testosterons befinden sich in den Erfolgsorganen wie Sie sagten, auch Metaboliten. Die Beurteilung von radioautographischen Untersuchungen ist deshalb schwieriger.

WACKER (Frankfurt): Mich hat vor allen Dingen Ihre Bemerkung interessiert, daß diese S-5-Fraktion Ribose und eventuell auch Phosphor enthält. Glauben Sie danach, daß der Repressor ein Nucleoprotein ist?

Wir arbeiten seit längerem über die Enzyminduktion durch Steroide bei Streptomyces hydrogenans und konnten bei diesem Mikroorganismus eine Repressorfraktion isolieren, die gleichfalls Nucleoproteincharakter hat.

JUNGBLUT: Das von der Säule retinierte Material gibt einen positiven Orcintest. Ich bin aber noch nicht sicher, ob daraus auf einen RNS-Gehalt geschlossen werden kann, weil die Basenanalysen fehlen. Die Substanz hatte zwar ein deutliches Absorptionsmaximum bei $260 \, m\mu$, jedoch fand ich später, daß eine Charge des zur Elution verwendeten Desoxycholatpräparates (enzyme grade!) ebenfalls in diesem Bereich absorbierte. Ich möchte deshalb eine Aussage über die mögliche Nucleoproteidnatur der Substanz so lange zurückstellen, bis wir größere Mengen für eine Nucleotidanalyse zur Verfügung haben.

DIRSCHERL (Bonn): Ich wollte ganz kurz eine Arbeit aus unserem Institut erwähnen, die Herr SCHRIEFERS vor 6 oder 7 Jahren gemacht hat. Herr SCHRIEFERS hat Uterusschnitte mit Oestron inkubiert und gefunden, daß das Hormon nur sehr langsam (im Verlauf von Stunden) gebunden wird. Dann ist das Oestron allerdings auch schwer zu entfernen; das gelingt nicht mit Lösungsmitteln, sondern nur mit Agenzien, bei denen man eine hydrolytische

Wirkung vermuten muß (Salzsäure, verdünnte Natronlauge, usw.). Nun ist die Frage: Wie ist das Oestron oder in Ihrem Fall das Oestradiol an das Eiweiß gebunden? Sie haben sich sicher auch darüber Gedanken gemacht.

JUNGBLUT: Darf ich zunächst eine Gegenfrage stellen? Welche Konzentrationen von Oestron wurden verwendet?

DIRSCHERL: Sie waren um Größenordnungen höher, etwa um 10 γ pro ml.

JUNGBLUT: Bei derart hohen Konzentrationen mußten wahrscheinlich Lösungsvermittler verwendet werden oder das Inkubationsmedium war sehr proteinhaltig. Ich glaube nicht, daß ich ohne genauere Kenntnis der Methodik etwas zu den Versuchen sagen kann.

Oestradiol wird unter physiologischen Bedingungen in den Erfolgsorganen nicht kovalent gebunden. Mit organischen Lösungsmitteln läßt sich das Hormon vollständig und unverändert extrahieren. Wichtig für die relativ stabile (aber nicht kovalente) Bindung ist, neben der Steroidstruktur und dem cycloaliphatischen 17-Beta-Hydroxyl, vor allem das phenolische Hydroxyl im A-Ring. 3-methoxylierte Oestrogene wie Mestranol müssen zunächst (außerhalb) in die freien Phenole umgewandelt werden, bevor der Uterus sie bindet [JUNGBLUT, P. W., H. G. NEUMANN, S. SMITH, V. COLUCCI, and E. V. JENSEN: Fed. Proc. **24,** 534 (1965)].

MAURER (Heidelberg): Ich möchte darauf hinweisen, daß wir heute noch keine experimentellen Beweise besitzen, die uns berechtigen, den Oestradiolreceptor einem genetischen Repressor gleichzusetzen. Wir kennen bislang als einzige biologische Funktion der isolierten Proteine nur ihre spezifische Hormonbindungskapazität. In diesem Zusammenhang möchte ich an die Befunde von TALWAR u Mitarb. [Proc. nat. Acad. Sci. (Wash.) **52,** 1059 (1964)] erinnern, wonach Oestradiol-bindende Cytoplasmafaktoren (aus Rattenuteri) die *in-vitro*-RNA-Synthese spezifisch hemmen sollen. Sind derartige Versuche wiederholt worden? Ferner interessiert mich, wie Sie in Ihren Kalbsendometriumversuchen *endogenes* Oestradiol ausschließen. Ich habe nämlich erfahren, daß 95% aller Kälber Amerikas mit Diäthylstilboestrol behandelt werden.

JUNGBLUT: Ich muß Ihnen voll zustimmen. Man kann wirklich noch nicht sagen, daß der „Receptor" ein Repressor ist. Ich sagte in meinem Vortrag nur, daß der 5-S-Faktor *eine* Eigenschaft besitzt, die man auch von einem Repressor fordern müßte.

Die Talwarschen Ergebnisse konnten in unserem Labor weder mit dem Colienzym noch mit der RNS-Polymerase aus B. Lysodeicticus bestätigt werden. Das RNS-Polymerasesystem ist sehr störungsanfällig. Es wird von allen möglichen Organextrakten und auch z. B. von Serumalbumin in unterschiedlichem Maße gehemmt. Weder *in-vivo*-Verabreichung von Oestradiol noch *in-vitro*-Zusatz bewirken eine Aufhebung der Hemmung [DE SOMBRE, E. R., B. FELDACKER, P. W. JUNGBLUT, and E. V. JENSEN: Fed. Proc. **25,** 286 (1966)].

Ihre Frage nach dem endogenen Oestrogengehalt von Kalbsuteri und der Behandlung amerikanischer Kälber mit Diäthylstilboestrol ist sehr berechtigt. Wir verwenden nur Gewebe von sehr jungen Kälbern, die mit einiger Sicherheit noch unbehandelt sind und deren Uteri kaum endogenes Hormon enthalten sollten. Die Brauchbarkeit von Uteri für unsere Zwecke läßt sich leicht testen. TRIS-EDTA-Extrakte von oestrogenhaltigen Organen zeigen eine verminderte Bindung von zugesetztem radioaktivem Oestradiol in der 9-S-Zone.

CHANDRA (Frankfurt): We have done some experiments on testosterone binding to a rat prostatic fraction. This fraction is specific in the sense that an *in-vivo*-treatment of rats with an antiandrogenic steroid (cyproterone-Schering) inhibits the testosterone binding capacity of the fraction. It is not yet purified. However, in our case the bound testosterone gets dissociated on sephadex columns. Have you had any experience in separating your complex by gel filtration?

JUNGBLUT: Wir haben zahlreiche Versuche mit Gelfiltration gemacht. Bei der Passage von frischen TRIS-EDTA- und KCl-Extrakten aus hormonbehandelten Erfolgsorganen über Sephadex G 75 oder G 100 wird Oestradiol fast vollständig mit der makromolekularen Fraktion eluiert. In älteren Extrakten nimmt der Gehalt an nicht makromolekular gebundenem Oestradiol zu. Ich möchte bezweifeln, daß es sich dabei nur um eine allmähliche „Dissoziation" des Hormons von den spezifischen Bindungsfaktoren handelt. Wir haben Grund zu der Annahme, daß auch die (komplexen) Faktoren selbst zum Zerfall neigen.

EBEL: Ich danke Herrn JUNGBLUT nochmals und schließe damit.

Ribosomes and Thyroid Hormones

By J. R. Tata

National Institute for Medical Research,
Mill Hill, London, N. W. 7, England

With 4 Figures

Thyroid hormones are a good example of the shift in emphasis that has taken place in the last five years in work on the action of hormones. Until a few years ago, most hormones were considered to act by modifying the rates of enzymes directly concerned with the ultimate biological effects (Tepperman and Tepperman, 1960), whereas it is now accepted that hormones may act by controlling the synthesis of specific cell constituents. My own feeling is that the same hormone at different doses or in different systems can act in many different ways. This is certainly true of thyroid hormones which have an immediate catabolic or toxic effect by acting at the level of mitochondria but also have a slow action under other conditions in which they control the rate of synthesis of specific constituents. It is this action of thyroid hormones which is manifested in growth and development of the target cell that I shall be dealing with in this paper.

Work on the hormonal control of synthetic activity of the cell received a big impetus from the suggestion made by Karlson (1963) that hormones may regulate the synthesis of RNA at the genetic level. Before that, however, there was already growing evidence from the work of Korner, Williams-Ashman and Mueller that the stimulation by growth hormone and sex steroids of protein synthesis in their target cells could be best explained on the basis of changes in RNA metabolism (Korner, 1965; Mueller, 1965; Williams-Ashman et al., 1964). The initial work on hormones and RNA synthesis focussed attention mainly on the modulation by the hormone of m-RNA synthesis (Tata, 1966a; Sekeris, this issue). However, recently, there is evidence to show that, in nucleated cells of higher animals, protein synthetic activity of the cell could be affected by the transport of RNA from the nucleus to the cytoplasm, the translational activity of the ribosomes (Tomkins, this issue) and the structural disposition of the ribosomes (Henshaw et al., 1963, 1965). It is in the light of these recent suggestions of a regulation of protein synthesis at levels beyond the transcription of the genetic material that I shall try and evaluate some of our recent work on the action of thyroid hormones in two main

Table 1. *Some acute cellular effects of thyroid hormones in the thyroidectomized Rat* (Tata, 1964)

Marked, Early Stimulation	Late or no Change
Synthesis of nuclear and cytoplasmic RNA	G-6-P Lactic and Isocitric dehydrogenases
DNA-dependant RNA polymerase	Increase in mitochondrial ubiquinone
Incorporation of amino acids into protein by mitochondria and microsomes	
Mitochondrial respiration and phosphorylation	NAD pyro-phosphorylase in nuclei
Synthesis of membrane phospholipids	Mitochondrial stability, coupling of phosphorylation and respiration
Sub-mitochondrial cytochrome oxidase and succinoxidase	Creatine phosphokinase
Microsomal G-6-Pase, NADPH-cytochrome C reductase	Amino acid activation Alpha-glucan phosphorylase
Increase in cytochromes a, c and b_5	UDPG-glycogen glucosyl transferase

systems: a) the growth-promoting activity of these hormones in the liver of the thyroidectomized rat and b) the induction of metamorphosis in amphibia.

The main biochemical responses provoked when a young thyroidectomized animal is made to resume its growth by a single dose of thyroid hormone are summarized in Tab. 1. The conditions of the experiments were such that there was a long latent period of several hours or even days before these effects were observed. It can be seen that most of the prompt and marked responses are associated with structural elements of the cell. On the other hand the slow or negative responses, with one exception were not associated with any structural elements. This is not a unique property of thyroid hormones but rather reflects a pattern often observed during development or vigorous growth of a cell when the structural components of the cell are in a state of flux.

Fig. 1 shows how some of these diverse effects of thyroid hormones are related to one another as a function of time after the

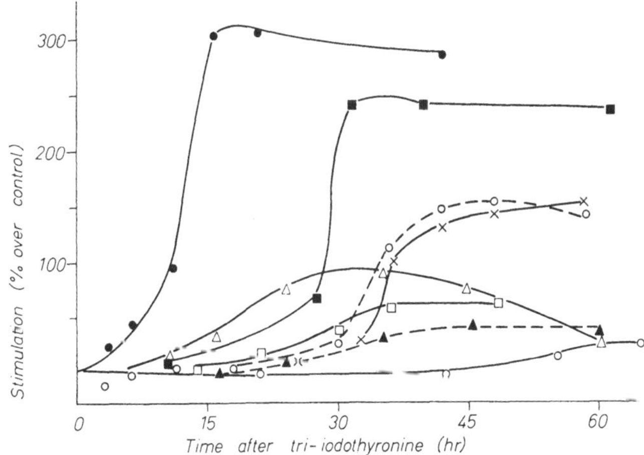

Fig. 1. Summary of data on the sequential stimulations of RNA and protein synthesis in relation to hormonal effect on growth and metabolic activity in the liver of thyroidectomized rat as a function of time after a single injection of 15 to 22 μg of tri-iodothyronine. In most cases two or more parameters were measured in tissues pooled from 6 to 10 rats per group. ● —— ●, Rate of synthesis of rapidly labelled nuclear RNA *in vivo*; Δ —— Δ, Mg^{2+}-activated RNA polymerase; ▲ ······ ▲, Mn^{2+}/(NH$_4$)$_2$SO$_4$-stimulated RNA polymerase; ■ —— ■, accumulation of newly-formed (labelled) ribosomes and polysomes in the cytoplasm; □ —— □, mg total ribosomal RNA/mg DNA; o ······ o, incorporation of ^{14}C-labelled amino acids into protein/mg RNA *in vitro* by ribosomes or microsomes; × —— ×, cytochrome oxidase activity/mg mitochondrial protein; o —— o, increase in weight of the organ. Curves derived from combined data of TATA et al. (1965) and TATA and WIDNELL (1966)

hormone was administered as a single dose to young thyroidectomized rats. Hormone replacement led to abrupt increases in the rates of RNA synthesis by the nucleus and protein synthesis by cytoplasmic particles. The stimulation of nuclear RNA synthesis can be observed some hours before any change in the amino acid incorporation activity of ribosomes in the cytoplasm. Similar sequential stimulations of RNA synthesis followed by that of protein have been recorded for other hormones as well, i.e. growth hormone acting on the liver (WIDNELL and TATA, 1966a) and muscle (FLORINI and BREUER, 1966) testosterone in prostate (WILLIAMS-ASHMAN et al., 1964), oestrogen in the uterus (HAMILTON et al., 1965) and ecdysone in *Calliphora* epidermis (SEKERIS, 1965). The absolute time scale for these sequences however varies considerably according to the hormone and its target tissue.

Ribosomal RNA and protein synthesis

The stimulatory effect of thyroid hormone on nuclear RNA synthesis has been demonstrated both as rapidly-labelled RNA synthesized *in vivo* and as the RNA polymerase activity in isolated nuclei (TATA and WIDNELL, 1966). In many of the earlier studies a hormone-induced increase in pulse-labelling of nuclear RNA or specific activity of RNA polymerase has been interpreted as a stimulation of hormone-specific m-RNA synthesis, perhaps analogous to some well known microbial induction systems (TATA, 1966a). However, careful analysis both of the nuclear RNA labelled *in vivo* and the product of the polymerase has revealed that it is the change in the rates of ribosomal RNA synthesis that were most obvious during the early phase of hormone action. WIDNELL and TATA (1966b) found that incubation of nuclei with $0.4 \text{ M } (NH_4)_2SO_4$, which is commonly used to stimulate nuclear polymerase, leads to a product that is more DNA-like as judged by nearest neighbour analyses and base composition but that the product formed in the absence of high salt concentration is more like ribosomal RNA. It is therefore most significant that it is in the absence of high salt concentrations, i.e. when the product of the polymerase is more like ribosomal RNA, that the early stimulatory effects of thyroid as well as other hormones were best observed. Because of the difference in the nature of products formed and the kinetics of the reaction at low and high ionic strength it is unlikely, as has been suggested for growth hormone (PEGG and KORNER, 1965), that the stimulatory effect of the hormone administered *in vivo* is comparable to that of the addition of salt *in vitro*. Rapidly-labelled RNA synthesized *in vivo* in liver nuclei is predominantly of the ribosomal type and the administration or replacement of thyroid hormone fails to alter markedly the total composition of nuclear RNA. It is noteworthy that the biological activity of thyroid hormones (in common with most growth-promoting hormones) is abolished by low doses of actinomycin D, doses at which ribosomal RNA synthesis is much more susceptible to the inhibitor than m-RNA synthesis (REICH and GOLDBERG, 1964). To what extent an accelerated synthesis of r-RNA or ribosome formation involves hormonal specificity ? To answer this question, WIDNELL and TATA (1966a) designed a series of experiments in which the effect of different growth-promoting hormones, administered separately and in different combination, was studied in the same tissue. For this purpose we studied the effect of growth hormone, thyroid hormone and testosterone on RNA synthesis in rat liver. Taking into account the different latent periods and dose response curves for the three hormones, we found that in every combination of these hormones

Table 2. *Effect of the administration of single injections of tri-iodothyronine and growth hormone to hypophysectomized rats, separately and in combination, on the activity of the two RNA-polymerase reactions of isolated rat liver nuclei*

In both experiments hypophysectomized animals were used. The dose of tri-iodothyronine was 25 μg and of growth hormone 200 μg; the rats weighed about 100 g. The specific activity of RNA polymerase was expressed as $\mu\mu$-moles of [^{14}C] ATP incorporated into RNA/15 min/mg DNA for the Mg^{2+}-activated reaction and as $\mu\mu$-moles, [^{14}C] ATP incorporated/45 min/mg DNA for the $Mn^{2+}/(NH_4)_2SO_4$-activated reaction. Each value is the average of two determinations on nuclei from pooled livers (three rats/group). The RNA product formed in the presence of Mg^{2+}-ions is largely ribosomal RNA and is more DNA-like in the presence of $Mn^{2+}/(NH_4)_2SO_4$. From WIDNELL and TATA (1966a).

Time after single injection (h)		Specific activity of RNA polymerase	
Tri-iodothyronine	Growth hormone	Mg^{2+}-activated	$Mn^{2+}/(NH_4)_2SO_4$-activated
Expt. 1 —	—	646	2040
45	—	861	2710
45	3	1245	2810
—	3	955	2020
Expt. 2 —	—	556	2390
24	—	732	2430
24	12	1030	2410
42	—	795	2920
42	12	1090	3010
—	12	876	2360

the effect on RNA polymerase, assayed under conditions in which the product is chiefly ribosomal RNA, was additive. An example of this is illustrated in Tab. 2. The nuclei of the target cells therefore behaved towards the second hormone in the same way irrespective of whether the first hormone had been previously administered or not. Furthermore, dose-response curves suggested that there was a maximum "quota" of r-RNA or ribosomes that a cell can make specifically in response to a given hormone. A similar type of additive effect was observed when the accumulation of total cellular RNA was determined in the salivary gland of thyroidectomized-castrated rats treated with tri-iodothyronine and testosterone (Tata, 1965). This does not mean that thyroid hormones exclusively regulate the rate of synthesis of r-RNA and to rule out hormonal effect on m-RNA synthesis. What is suggested by these results is that when more or new m-RNA species are synthesized under the influence of a hormone, it is accompanied by a disproportionately greater increase in rate of r-RNA synthesis. This disproportionately higher stimulation of the rate of r-RNA synthesis relative to that of m-RNA may be independent of the initial site of action of hormones and may only reflect a common feature of rapidly growing cells in higher organisms.

Formation, distribution and function of ribosomes and membranes

There are some observations regarding thyroid hormonal action which may be relevant to the processes of ribosome biogenesis and transport. Siegel and Tobias (1966) observed that thyroid hormone added to the culture medium augmented the number and volume of nucleoli of isolated human kidney cells. It is now well established that the nucleolus is the site of maturation and methylation of ribosomal precursor particles (Prescott, 1964). Although the site of synthesis of ribosomal proteins has not yet been established, we found that the stimulation of hepatic growth by thyroid hormone, growth hormone and testosterone provoked a rapid increase in the rate of labelling *in vivo* of both the basic and total nuclear proteins (Tata, 1966b). Although the nature of the basic nuclear proteins was not established it is likely that we were measuring the turnover of a small pool of heavily-labelled ribosomal proteins rather than histones. It is particularly interesting that the temporal pattern of change in the rate of labelling of nuclear proteins *in vivo* induced by tri-iodothyronine (as well as growth hormone) coincided with the stimulation of RNA polymerase under conditions in which the product is chiefly ribosomal RNA (Fig. 2). On the other hand the pattern of changes in the rates of nuclear RNA and protein synthesis *in vivo* did not coincide. The relative proportion of total ribosomes recovered as polyribosomal aggregates is diminished in rat liver after thyroidectomy (TATA and

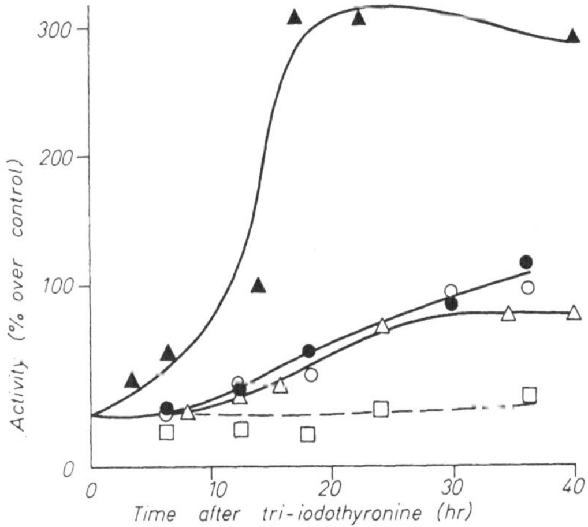

Fig. 2. Effects on the rate of nuclear RNA and protein synthesis of a single injection of 25 μg of tri-iodothyronine to thyroidectomized rats. o —— o, specific radioactivity of total nuclear protein (following a 10 to 15 min pulse of ^{14}C-labelled protein hydrolyzate); ● —— ●, specific activity of basic nuclear protein (pulse of L-lysine-^{14}C); ▲ —— ▲, specific activity of rapidly-labelled nuclear RNA *in vivo*; Δ —— Δ, RNA polymerase assayed at low ionic strength, i.e. when the product is mainly ribosomal RNA. □ ······ □, mg total nuclear proteins/mg DNA. The effects are expressed as a percentage of the increase or decrease in values for preparations from treated animals over those in the untreated controls (time 0). Data from TATA and WIDNELL (1966); TATA (1966b)

WIDNELL, 1966a) as well as hypophysectomy (KORNER, 1965a, b). The effects of hormone-deprivation could be rapidly reversed by the administration of thyroid hormones. Further analysis of polysome formation and *in vitro* activity suggested that thyroid hormones may act at both the levels of transcription and translation of RNA. An example of such a dual effect of tri-iodothyronine on the rate of formation and amino-acid incorporating activity of hepatic ribosomes from thyroidectomized rats is presented in Tabs. 3 and 4. Tab. 3 shows that an enhancement of specific activity of RNA in all the zones of the polyribosome profile accompanied the hormone-induced net increase in cytoplasmic RNA content. The stimulation was observed equally well after a short or long pulse of the radioactive precursor and it was most noticeable in a relatively small fraction of ribosomes that were tightly-bound to

Table 3. *Specific radioactivity of RNA in the 0.4 per cent-sodium deoxycholate-resistant membranes, polyribosomes and monomeric ribosomes from livers of thyroidectomized rats with or without treatment with tri-iodothyronine*

All thyroidectomized rats were given 9.6 μc of [^{14}C] orotic acid 2.2 or 15.2 h before killing; 25 μg of tri-iodothyronine was injected 42.4 h before death to thyroidectomized rats weighing 160 $\pm$ 10 g. Treatment of mitochondria-free supernatant preparation and sucrose-density gradient centrifugation were as described by Tata and Widnell (1966). RNA was estimated from E_{260} after correction for protein content. The ^{14}C determinations were carried out on individual fractions after the addition of 0.2 mg of carrier yeast RNA for precipitation.

Time after [^{14}C] orotic acid (h)	Time after tri-iodo thyronine (h)	Specific activity of RNA (counts/min/E_{260})				
		"Membranes"*	Interface	Polysomes	Monomers	"Top"
2.2	0.0	1250	—	2100	750	1000
	43.0	6700	—	3800	1200	2000
15.5	0.0	5600	7000	11000	7800	2000
	42.2	34500	14300	19800	17300	4500

* The radioactivity recovered in the membranes from control animals was 5 to 8 per cent of the total radioactivity put on the density gradient whereas it was 14 to 17 per cent for the hormone-injected animals.

microsomal membranes. Tab. 4 shows that the overall amino-acid incorporation *in vitro* of ribosomes obtained after hormonal stimulation was not the same in the tri-iodothyronine-treated and thyroidectomized animals. The lower incorporation activity of ribosomes from thyroidectomized rats was accompanied by a slightly increased stimulation of phenylalanine incorporation by poly U (7.7-fold against 5.4-fold in normal). However, the ribosomes from tri-iodothyronine-treated rats exhibited a much higher rate of incorporation per mg. RNA both with endogenous and synthetic messenger.

It can be seen in Fig. 1 that the abrupt rise in amino-acid incorporation by cytoplasmic ribosomes or even mitochondria was accompanied, or immediately preceded, by the appearance of an additional amount of newly-formed ribosomes in hormone-treated animals. It seems therefore that regulation of protein synthesis by thyroid hormones necessitates the formation of new ribosomes, a part or all of the additional ribosomes also being more active in

Table 4. *Stimulation by poly U of the incorporation of [^{14}C] phenylalanine by liver RNP particles from normal, thyroidectomized and tri-iodothyronine-treated thyroidectomized rats*

A dose of 17 μg tri-iodothyronine was injected/100 g body wt. 42 h before killing. RNP particles were prepared by treating the mitochondria-free supernatant with 1 per cent sodium deoxycholate. Each incubation vessel contained RNP particles from 50 mg of liver and cell sap from 40 mg of liver; 125 μg of poly U was added wherever indicated. From TATA and WIDNELL (1966).

Rats	mg RNA/mg protein in RNP particles	Poly U	Radioactivity in protein (counts/min) (mg)	Stimulation by poly U
Thyroidectomized	0.26	—	447	
		+	3420	7.72
Thyroidectomized + tri-iodothyronine	0.36	—	599	
		+	3370	5.61
Normal, untreated	0.46	—	969	
		+	5232	5.43

protein synthesis than those in the hormone-deprived controls. The expression of hormonal activity may perhaps reside in the small population of new ribosomes formed in response to the hormone. It should be emphasized that the modification of amino acid incorporation activity cannot be mimicked by a direct addition of the hormone to already formed ribosomes but occurs at some stage in ribosome formation or localization. The fact that such different hormones as growth hormone, tri-iodothyronine and hydrocortisone induce a burst of ribosome synthesis in rat-liver, yet each hormone regulating the activity or amount of quite different enzymes, itself suggests that some sort of specificity may restrict ribosomes to synthesize only certain proteins. However, it is difficult to provide an experimental demonstration of such a specificity in a fully differentiated tissue like the liver. It is in order to obviate difficulties of this type that I initiated, about 4 years ago, a study of thyroid hormone-induced metamorphosis in anurans (Tata, 1965 b).

Control of amphibian metamorphosis

Amphibian metamorphosis, which is obligatorily dependent on thyroid hormones involves dramatic changes of constitution and structure in almost every type of cell of the larva (Bennett and Frieden, 1962). In tadpole liver the administration of exogenous hormone to pre-metamorphic tadpoles leads precociously to the synthesis *de novo* of many proteins, including urea cycle enzymes, mitochondrial respiratory enzymes and serum albumin. In Fig. 3 are summarized schematically some of our results on the changes in RNA and protein synthesis upon induction of metamorphosis in *Rana catesbeiana*, a species in which the tadpole normally does not undergo spontaneous metamorphosis for 2 to 3 years. It can be seen that, after a lag period of 5 to 6 days following hormone administration, there is an abrupt increase in a number of proteins and enzymes made in the liver. During this lag period there occurs an intense activation of the synthesis of all species of RNA in the nucleus followed by an increasing specific activity of cytoplasmic RNA. A larger fraction of ribosomes from tadpoles induced to metamorphose were present as heavier polysomal aggregates along with a marked increase in the labelling of RNA in the region of ribosomal monomers or ribosomal precursor particles. Such polysome profile analyses also showed that a fraction of newly-formed ribosomes could not be detached from lipid-rich microsomal constituents under conditions in which virtually all polysomes and ribosomes were released from membranous structures in the non-induced control larvae. A pulse of labelled amino-acids given shortly before killing the tadpoles showed that proteins were synthesized on polysomes of all sizes at a more rapid rate in the induced animals. It was however in the small fraction of more tightly-bound ribosomes that the most significant change in the rate of protein synthesis *in vivo* was detected.

The above observation led us to focus attention on the distribution of ribosomes and their attachment to the membranes of the endoplasmic reticulum during hormone-induced growth and development. Examination of the tadpole liver in the electron microscope by Dr. J. A. ARMSTRONG at the National Institute for Medical Research showed that in the pre-metamorphic animal the hepatic ribosomes were distributed rather uniformly throughout the cyto-

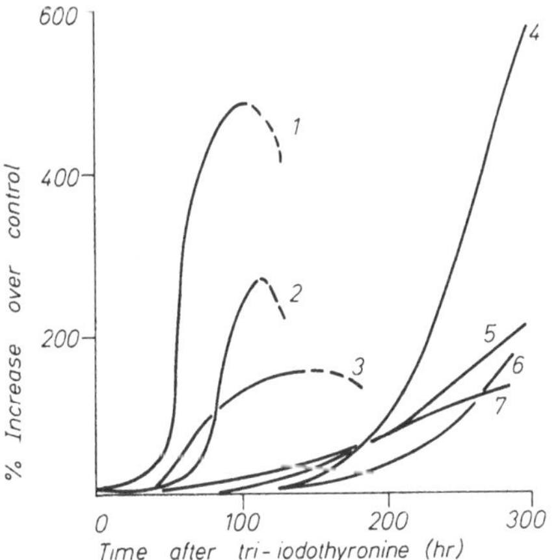

Fig. 3. Schematic representation of sequential stimulation of rates of RNA and phospholipid synthesis in relation to the increases in enzymes or protein synthesized upon the precocious induction of metamorphosis in *Rana catesbeiana* tadpoles with tri-iodothyronine. Curves 1. rate of rapidly-labelled nuclear RNA synthesis; 2. specific activity of RNA in cytoplasmic ribosomes; 3. rate of microsomal phospholipid synthesis; 4. carbamyl phosphate synthetase; 5. cytochrome oxidase per mg mitochondrial protein; 6. appearance of serum albumin in the blood; 7. total liver protein per mg wet weight. The values are expressed as percentage increases over those in the non-induced control tadpoles. The dashed lines in curves 1, 2 and 3 reflect the dilution of specific radioactivity in precursor molecules due to the onset of regression of tissues such as the tail and intestine. Data from TATA (1965b; 1967a, b)

plasm, some of them around simple vesicular membranes. Upon induction of metamorphosis the ribosomes were more often found in pockets of high density and attached principally to complex lamellar membrane structures of the type which form the bulk of rough endoplasmic reticulum in hepatic cells of adult tissues in many mammalian species (TATA, 1967a, b; PALADE, SIEKEVITZ and CARO, 1964).

Attachment and proliferation of ribosomes
and microsomal membranes

Since virtually all protein synthesis *in vivo* may occur on those
ribosomes that are bound to membranes (Henshaw et al., 1963) it
became important to know whether the "new" and "old" ribosomes
were all distributed on existing membranes or bound to newly-
formed structures. We therefore studied the action of thyroid hormones on the rate of synthesis of phospholipids of the smooth and
rough microsomal membranes as well as the rates of formation of

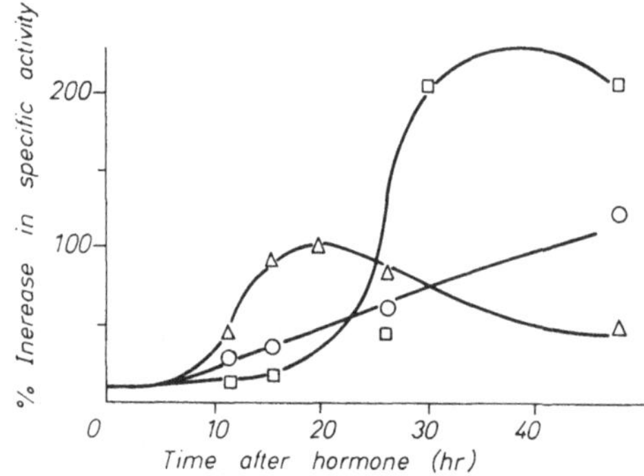

Fig. 4. Co-ordination of the synthesis of microsomal phospholipids with the
appearance of labelled RNA in microsomes and their capacity *in vivo* to
incorporate amino acids into protein following tri-iodothyronine-induced
growth of liver of thyroidectomized rats. The rate of microsomal RNA and
phospholipid synthesis in the liver was determined from the incorporation of
radioactivity into the macromolecules 9 to 11 h after the administration of
60 to 75 μc of ^{32}P and protein synthesis by injecting a mixture of ^{14}C-labelled
amino acids (*Chlorella* protein hydrolysate; 600 μc/mg), 10 min before
killing. The drop in the specific radioactivity of nascent protein at the later
time-intervals is largely due to the substantial increase in membrane and
ribosomal protein content of microsomes. The results are expressed as per
cent increase in specific activity of phospholipid (0), RNA (□) and protein
(Δ) in hormonetreated animals over the values obtained in preparations
from untreated controls. From Tata (1967c)

microsomal RNA and of proteins synthesized *in vivo* on microsomes.
The data summarized in Fig. 4 show that the timing and magnitude
of the effect of the hormone on phospholipid synthesis is co-ordinated with the other biosynthetic activities of the target cells and,
eventually, with the hypertrophy produced. Although the rates of
synthesis of the three types of macromolecules were not coincident

throughout the time span studied, the onset of the increase in the rate of microsomal phospholipids coincided with that of the appearance of additional microsomal RNA. It is at this time of onset that the rate of formation of protein in microsomes was also

Table 5. *Changes induced in the distribution in microsomal subfractions of phospholipids and RNA synthesized during accelerated growth of the livers of thyroidectomized rats induced by a single injection of tri-iodothyronine*

Half of the thyroidectomized rats were treated with 18 μg of T_3 for 32 h before being killed; all animals received 75 μc of ^{32}P 12 h before death. Microsomes were isolated from one aliquot of the mitochondria-free supernatant while smooth and rough membranes were isolated from another. A part of the rough membrane fraction was suspended in 0.18 per cent Na deoxycholate (DOC) (20 mg DOC/3 to 4 mg phospholipid) and the sediment (0.18 per cent DOC pellet) collected for analysis. Each value is the mean of four determinations made on preparations from 3 rat livers. The phospholipid and RNA content of unfractionated microsomes were 6.01 and 4.08 mg/g liver, respectively, and the values in treated animals were 7.45 and 5.10 mg/g liver, respectively. Data from TATA (1967c).

Fractions	Specific activity of Phospholipids (C.P.M/μg P)		Relative distribution of p-lipids (per cent)		Specific activity of RNA (C.P.M./μg P)		Relative distribution of RNA (per cent)	
	control	treated	control	treated	control	treated	control	treated
Microsomes	77.5	185.0	100	100	7.8	15.3	100	100
Smooth membranes	84.0	227.0	40	25	17.5	47.6	6.6	7.9
Rough membranes	79.9	172.0	53	62	6.3	14.9	86.0	74.0
0.18 per cent DOC pellet	95.0	585.0	8	13	3.5	7.9	7.7	18.0

accelerated. Similar results were obtained with growth hormone acting on the liver of hypophysectomized rats and testosterone on the seminal vesicles and liver of castrated rats (TATA, 1967c). The sufficiently different lag periods of action of the three hormones made it possible to demonstrate that an accelerated proliferation

7*

of microsomal membranes was temporally co-ordinated with hormone-induced increases in cytoplasmic RNA and protein synthetic activity *in vivo*.

The increased labelling of membrane phospholipids upon hormonal stimulation of growth took place both in the smooth and rough membranes. In the example in Tab. 5 it can also be seen that a higher proportion of total microsomal phospholipid and RNA was distributed in the rough, than in the smooth membranes, in hormone-stimulated animals. Since the rough membranes are the site of

Table 6. *Synthesis of hepatic microsomal phospholipids, RNA and protein following the precocious induction of metamorphosis in* Rana catesbeiana *tadpoles*

Groups of 12 *Rana catesbeiana* tadpoles (with no hind limbs) were injected with 1 μg of 3,5,3′-tri-iodothyronine at the times indicated before killing. 10 μc of ^{32}P-phosphate were injected 18.0 h before killing the animals and preparing the microsomes as described elsewhere. Each value is the mean of two determinations in duplicate (from Tata, 1967 c).

Time after induction (days)	Incorporation of ^{32}P into		Amount of microsomal (mg/gm Liver)		
	P-lipids (c.p.m./mg P)	RNA (c.p.m./mg P)	P-lipid	Protein	RNA
0	492,000	204,000	2.44	11.0	2.8
1.8	420,000	217,000	2.51	10.1	2.7
3.7	594,000	510,000	2.69	9.6	2.9
6.8	683,000	628,000	3.80	14.2	3.1

protein synthesis, this shift in distribution is compatible with the acceleration in the rate of growth of the liver. It is interesting to note that it was in the fraction resistant to moderate amounts of detergent (0.2 to 0.4 per cent Na deoxycholate) that the most conspicuous effect of hormones on the synthesis of microsomal phospholipids could be observed. We have already seen in Tab. 3 that under similar experimental conditions this fraction of tightly-bound ribosomes exhibited the maximum increase in the specific activity of RNA.

In the amphibian metamorphosis system there was also quite a noticeable accumulation of microsomal phospholipid and protein accompanying the turnover of microsomal RNA upon induction of metamorphosis (Tab. 6). The change in the rate of labelling of phospholipid coincided both with that of RNA as well as with a proliferation of rough endoplasmic reticulum, and a redeployment of ribosomes as observed in the electron microscope. An examination of ribosomes however revealed no difference in the properties or composition of the intact ribosomes or of their RNA or ribosomal structural proteins between preparations from non-metamorphosing and induced tadpoles (TATA, 1967 b).

It seems therefore that the regulation of the rate of protein synthesis by thyroid hormones, as well as other growth and developmental hormones, may involve a simultaneous control of the rates at which cytoplasmic RNA and membranes are generated in the cell. Besides the hormones that we have studied, a pronounced effect of thyrotrophin and oestrogen on phospholipid synthesis in the thyroid and uterus, respectively, also occurs at the time of accelerated RNA synthesis or the onset of biological effects of the hormone (FREINKEL, 1964; MUELLER, 1965). A co-ordinated proliferation of membranes and ribosome attachment geared to the rate of protein synthesis may be common property of higher animal cells.

Conclusions

To conclude, it is quite obvious that the mechanism of action of thyroid hormones has not yet been discovered. What I have described above are late events in the growth-promoting and developmental properties of thyroid hormones. However these late events are closely connected with the expression of the biological effects of the hormones. The effects concern complex changes in the turnover and structural disposition of the ribosomes. They confirm the importance of the transport of RNA from the nucleus into the cytoplasm and attachment of ribosomes to the membranes of the endoplasmic reticulum in the regulation of protein synthesis in cells of higher organisms. The same overall pattern of changes were found in the liver of the thyroidectomized rat or tadpole liver although the type of proteins whose synthesis is accelerated or initiated is quite different in the two systems. Indeed, other developmental hormones such as androgens, oestrogens and pituitary trophic hormones exhibit a similarity in sequences of cellular actions in their target systems. It means that the type of changes induced in the protein synthesizing activity of the cell by the relevant hormone is a reflection of the most sensitive synthetic processes of a growing cell and may not be a unique effect of the hormone.

The specificity of different hormones acting in a similar fashion may depend on the presence in a particular tissue of special sites upon which the hormone initially acts. The identification of the site of action of a hormone is now the central problem in understanding the mechanism of action of hormones. To achieve this it will be necessary to make increasing use of simpler systems *in vitro*, such as tissue and cell cultures in which the biological actions of the hormone *in vivo* are reproduced and at the same time involving the induction or regulation of the synthesis of very specific proteins (TOMKINS, this issue). This can finally lead to the ultimate system of interaction at the molecular level.

References

BENNETT, T. P., and E. FRIEDEN: In Comparative Biochemistry vol. 4, p. 484. Ed. by FLORKIN, M., and H. S. MASON. New York: Academic Press, Inc. 1962.

FREINKEL, N.: In The Thyroid, vol. 1, p. 131. Ed. by PITT-RIVERS, R., and W. R. TROTTER. London: Butterworths 1964.

HAMILTON, T. H., C. C. WIDNELL, and J. R. TATA: Biochim. biophys. Acta (Amst.) **108**, 168 (1965).

HENSHAW, E. C., T. B. BOJARSKI, and H. H. HIATT: J. molec. Biol. **7**, 122 (1963).

—, M. REVEL, and H. H. HIATT: J. molec. Biol. **14**, 241 (1965).

KARLSON, P.: Perspect. Biol. Med. **6**, 203 (1963).

KORNER, A.: Recent Progr. Hormone Res. **21**, 205 (1965a).

— Biochem. J. **92**, 449 (1965b).

MUELLER, G. C.: In Mechanisms of hormone action, p. 228. Ed. by KARLSON, P. Stuttgart: Thieme 1965.

PALADE, G. E., P. SIEKEVITZ, and L. G. CARO: In The exocrine pancreas, p. 234. Ed. by DE REUCK, A. V. S., and M. CAMERON. London: J. & A. Churchill, Ltd. 1962.

PEGG, A. A., and A. KORNER: Nature (Lond.) **205**, 904 (1965).

REICH, E., and I. H. GOLDBERG: Progr. Nucleic Acid Res. and Molec. Biol. **3**, 184 (1964).

SEKERIS, C. E.: In Mechanisms of hormone action, p. 149. Ed. by KARLSON, P. Stuttgart: Thieme 1965.

SIEGEL, E. S., and C. A. TOBIAS: Nature (Lond.) **212**, 1318 (1966).

TATA, J. R.: In Actions of hormones on molecular processes, p. 58. Ed. by LITWACK, G., and D. KRITCHEVSKY. New York: John Wiley and Sons, Inc. 1964.

— In Developmental and metabolic control mechanisms and neoplasis, p. 335. Baltimore: Williams & Wilkins Co. 1965a.

— Nature (Lond.) **207**, 378 (1965 b).
— Progr. Nuc. Acid Res. and Molec. Biol. **5**, 191 (1966 a).
— Nature (Lond.) **212**, 1312 (1966 b).
— Biochem. J. **104**, 1 (1967 a).
— Biochem. J. In press. (1967 b).
— Nature (Lond.) **213**, 566 (1967 c).
—, L. ERNSTER, O. LINDBERG, E. ARRHENIUS, S. PEDERSEN, and R. HEDMAN: Biochem. J. **86**, 408 (1963).
—, and C. C. WIDNELL: Biochem. J. **98**, 604 (1966).
TEPPERMAN, J., and H. M. TEPPERMAN: Pharmacol. Rev. **12**, 301 (1960).
WIDNELL, C. C., and J. R. TATA: Biochem. J. **98**, 621 (1966 a).
— — Biochim. biophys. Acta (Amst.) **123**, 478 (1966 b).
WILLIAMS-ASHMAN, H. G., S. LIAO, R. L. HANCOCK, L. JURKOVITZ, and D. A. SILVERMAN: Recent Progr. Hormone Res. **20**, 247 (1964).

Diskussion

EBEL (Straßburg): Thank you, Dr. TATA, for this interesting lecture. May I open the discussion with a question. You have shown that an important part of hormone action is the synthesis of new ribosomal RNA, new ribosomal particles and new ribosomes. Have you any information if this ribosomal RNA is the same as in the ordinary ribosomes, in other words is there any specialisation among ribosomal RNA's?

TATA (London): We couldn't study this in rat liver because we have a very large population of ribosomes to start with in the control animals, and the newly formed ribosomes are not easily distinguished from these. In metamorphosis there is a dramatic change in the resistence of microsomal membranes to detergents. Therefore, we treated microsomes from metamorphosing tadpoles with a small amount of detergent to enrich the preparation in new ribosomes, and we looked at the base composition, sedimentation constants of RNA and ribosomes, and profiles of ribosomal proteins obtained from subunits. We could't find any differences in their physical properties.

GUDER (München): Wir studierten das Verhalten der HMG-CoA-Reduktase der Rattenleber nach einmaliger Gabe von T_3. Wir beobachteten einen Anstieg der Enzymaktivität nach einer Latenzzeit von 30 Std (Abb. 1). Glauben Sie, Dr. TATA, daß dieser Effekt für ein an die mikrosomalen Membranen gebundenes Enzym ein früher oder später Hormoneffekt ist?

TATA: This is a very interesting result, and an extremely early effect of tri-iodothyrine *in vivo*. However it is not possible to say how the activation of this enzyme would be related to the late events of microsomal membrane proliferation that I have described.

SEIF (Tübingen): Have you done any experiments with thyroxine and tri-iodothyronine indicating qualitatively different effects of these hormones?

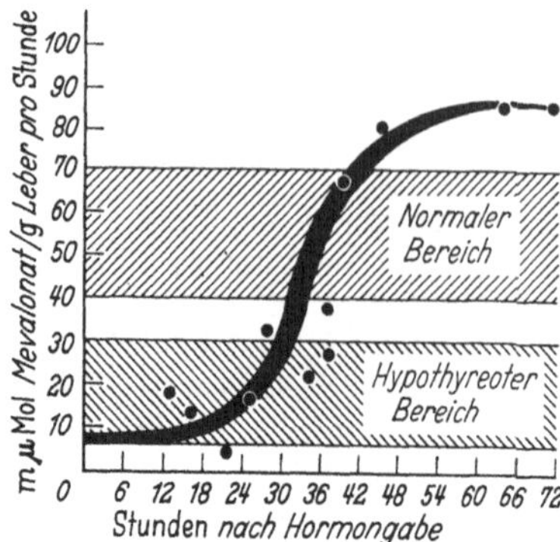

Abb. 1. Verhalten der HMG-CoA-Reductase der Rattenleber nach Injektion
von 50 g Trijodthyronin

TATA: No, there are no qualitative differences. The main difference is a
question of speed. Triiodothyronine is loosely bound to plasma proteins and
therefore reaches the tissues quickly. Whereas the effect of thyroxine,
comes later and lasts longer. The inactive analog, diiodothyronine, is our
usual control to show the specificity for the hormones.

STAUDINGER (Gießen): Am Anfang Ihres Vortrages haben Sie auch auf die
Vermehrung von Cytochrom b_5 in den Membranen unter Triothyroninbe-
handlung hingewiesen. Dann sprachen Sie in anderem Zusammenhang von
der Induktion durch Phenobarbital. Es ist bekannt, daß Cytochrom b_5 nicht
durch Phenobarbital induziert wird. Wie wirken die beiden Induktoren zusammen auf eine Induktion von Cytochrom b_5?

TATA: The effect of triiodothyronine on cytochrome b_5 is a rather small
effect and so is the effect of phenobarbital. We thought we might be misled
by measuring two small effects.

KADENBACH (Marburg): Nachdem uns Herr TATA in einem sehr schönen
Vortrag über die Wirkung wachstumsfördernder Konzentrationen der Schilddrüsenhormone berichtet hat, möchte ich einige Bemerkungen über die Wir-
kungsweise der hohen wachstumshemmenden Konzentrationen dieser Hormone machen. Vor mehr als 10 Jahren hat Herr Prof. MARTIUS an dieser
Stelle über den Wirkungsmechanismus der Schilddrüsenhormone berichtet.
Er kam dabei zu dem Schluß, daß diese Hormone ihre physiologische Funktion, insbesondere die Steigerung des Grundumsatzes, über eine Entkopplung
der oxydativen Phosphorylierung der Mitochondrien bewirken. Später wurde
jedoch, vor allem durch die Untersuchungen von Herrn TATA und seinem
Arbeitskreis gezeigt, daß die Schilddrüsenhormone nach einem anderen
Mechanismus, nämlich über die Steuerung der Enzymsynthese ihre Wirkung
entfalten. Dies galt jedoch nur für die physiologischen, nicht aber für die
hohen, wachstumshemmenden Konzentrationen des Hormons. Für sehr hohe

Konzentrationen wurde nach wie vor der toxische Mechanismus der Entkopplung der oxydativen Phosphorylierung angenommen.

Wir konnten dagegen finden, daß auch isolierte Lebermitochondrien aus thyreotoxischen Ratten, deren Wachstum stark gehemmt war, bei Verwendung der Sauerstoffelektrode an Stelle der allgemein benutzten Warburg-Technik, keine Entkopplung der oxydativen Phosphorylierung oder Verminderung der Atmungskontrolle zeigen [KADENBACH, B.: Biochem. Z. **344**, 49 (1966)]. Das heißt, daß auch in Mitochondrien thyreotoxischer Ratten die Energiesynthese nicht gehemmt ist. Außerdem konnten wir zeigen, daß alle von uns untersuchten Enzyme, die durch physiologische Hormondosen beeinflußt werden, auch mit sehr hohen Dosen linear mit wachsender Hormonkonzentration im Tier ansteigen oder abfallen. Dieser Anstieg setzt sich praktisch fort bis zum Tod des Versuchstieres (der Ratte) und geht der Zunahme des Grundumsatzes parallel. Da die Schilddrüsenhormone sowohl katabole als auch anabole Reaktionen (oder ATP-synthetisierende als auch ATP-verbrauchende) stimulieren — die Erhöhung der Proteinsynthese durch Thyroxin wurde vor allem von Herrn TATA gezeigt — bedeutet dies, daß der Gesamtstoffwechsel des Organismus beschleunigt wird. Die Beschleunigung des „Turnover" fast aller Substrate muß jedoch eine obere Grenze haben. Von Herrn SCHIMASSEK konnte in thyreotoxischen Ratten eine Limitation der Energie nachgewiesen werden. Der ATP/ADP-Quotient liegt in der Leber dieser Ratten bei 1,1, verglichen mit einem Normalwert von 3,3 (MITZKAT, H. J., H. SCHIMASSEK und W. GEROK: Verh. dtsch. Ges. f. inn. Med., 69. Kongreß 1963. Ed. J. F. Bergmann, München). Obwohl in thyreotoxischen Ratten der Sauerstoffverbrauch und damit die ATP-Synthese erhöht ist, kommt es zu einer Hemmung der Proteinsynthese (Abb. 2). Dies wird verständlich durch die Limitation des ATP. Mit wachsendem Stoffwechselumsatz nehmen offenbar die energieverbrauchenden Reaktionen schneller zu als die energiesynthetisierenden. Man kann dieses System mit einem Kraftwagen vergleichen, bei dem die optimale Energieausnutzung bei einer mittleren Geschwindigkeit gegeben ist.

Zusammenfassend kann man sagen, daß die Schilddrüsenhormone auch im toxischen Bereich durch eine physiologische und nicht toxikologische Primärreaktion ihre Wirkung entfalten. Die Thyreotoxikose ist offenbar nur die Folge einer ins Extrem gesteigerten physiologischen Wirkung des Hormons.

TATA: I agree with almost everything you said, but I disagree with your definition of toxicity. The second point is the limitation of the rate of ATP-synthesis. The liver system is a bad one, if one wants to study the effect of thyroxine on calorigenesis and oxygen consumption. Skeletal muscle would be better because its mitochondria responds to very low doses of thyroxine with increased respiration. In muscle it is not possible to produce the same toxic effects even with high doses of thyroxine, because thyroxine is then taken up by the liver, detoxicated and excreted into the bile. Also, the muscle contributes more to the rise in basal metabolic rate than the liver, and it is quite interesting that it is not possible to show an effect of thyroxine *in vivo* or *in vitro* on any simple step of oxidative phosphorylation in muscle mitochondria.

KARLSON (Marburg): Darf ich eine Frage an Herrn KADENBACH richten? Sie haben gesagt, daß die verminderte Proteinsynthese auf einen verminderten ATP-Spiegel zurückgeht. Bekommen Sie eine gute Korrelation zwischen der Proteinsyntheserate und dem ATP-Spiegel? Sie haben den Versuch doch sicher nicht nur einmal gemacht.

KADENBACH (Marburg): Die angegebenen ATP/ADP-Quotienten wurden einer Arbeit von Herrn SCHIMASSEK entnommen (MITZKAT, H. J., H. SCHIMASSEK und W. GEROK: Verh. dtsch. Ges. inn. Med., 69. Kongreß 1963. Ed. J. F. Bergmann, München). Sie wurden an Ratten gemessen, die sich in gleichem Thyroidstatus befanden.

TELLER (Marburg): Ich möchte Herrn KADENBACH noch entgegenhalten, daß der Kliniker, besonders der Pädiater, bei thyreotoxischen Zuständen exzessives Wachstum beobachtet und keinen Wachstumsstillstand.

Furthermore I should like to ask Dr. TATA how cortisol fits into the sequence of RNA synthesis between growth hormone and T_3? Secondly, do tissues from animals of different ages act differently on stimulation by thyroxine?

TATA: I have not done any work with cortisol.

With respect to age, once the tissue reaches a certain chronological age, it becomes refractory to the action of thyroxine. We have always used very young animals and thyroidectomized them about four to five weeks after birth. If you look at the brain, you have to use new-born thyroidectomized animals within two weeks after birth; if you wait longer, the brain will not respond, but the liver will.

TELLER: In this respect it might be interesting to point out that new-born humans have high levels of growth hormone and also high levels of free thyroxine. Maybe this has something to do with the induction of important enzymes.

EBEL: Da keine weiteren Fragen mehr sind, darf ich Herrn Dr. TATA nochmals für seine sehr interessanten Ausführungen danken und schließe die Nachmittagssitzung.

Hormonal Control of Protein Synthesis
at the Translational Level

By G. M. Tomkins and E. B. Thompson

*Laboratory of Molecular Biology NIAMD, National Institute of Health,
Bethesda, USA*

With 9 Figures

In this paper we shall try to do two things: 1. to describe a
system we have been working on for some time, which seems quite
useful in studying the regulation of protein synthesis by hormones
in mammalian cells; and 2. to present some experiments and
tentative ideas on the control mechanisms involved. As you will
see, we have reason to believe that control may be exerted at the
translational level of protein biosynthesis, as well as at the level of
gene transcription.

Possible sites of control of protein synthesis

In Fig. 1 is depicted the current picture of protein synthesis and
its control in bacteria with the operator, the regulator gene and its
product, the repressor, which according to recent work appears to
be a protein. The messenger-RNA and the polysomes are also
indicated.

In this Monod and Jacob model[1], the repressor is believed to act
directly on the DNA of the operon. The evidence for this is not yet
convincing; it could be that the repressor combines with the messenger RNA, thereby inhibiting translation. This is what we would
call regulation at the translational level. In addition, there seems
to be a tight coupling between transcription of the message and its
translation into protein, so that the rate of protein synthesis in-
fluences the rate of production of messenger RNA[2, 3].

Mammalian cells offer some similarities and some contrasts when
compared to this bacterial model (Tab. 1). In mammalian cells,

some messenger-like RNA which is produced is of a large size, corresponding to many genes[4,5,6]. However, the length of the messenger in polysomes is about the length of one gene. There must, therefore, be a step in which the nascent messenger RNA is split into pieces of the proper size for translation. The mechanism by

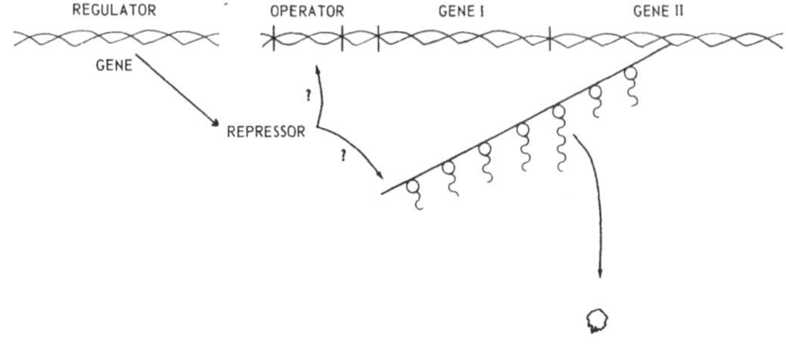

Fig. 1. Regulation of enzyme synthesis in bacteria. The figure depicts the probable mechanism of operon function and its control. Genes I and II are the linked structural genes from which a polygenic messenger is simultaneously transcribed and translated. The repressor, a protein of the unlinked regulatory gene is shown to act either directly on the DNA at the operator or on the messenger RNA, or, conceivably at both levels

Table 1. *Steps in enzyme formation in mammalian cells*

1. Transcription of genes into mRNA
2. Chopping mRNA's to correct size (?)
3. Elimination of some messengers (?)
4. Transport of mRNA to cytoplasm
5. Attachment of mRNA to membrane (?)
6. Initiation of translation
7. Peptide bond formation
8. Folding of Nascent peptide chains
9. Release of completed protein

which this is done is not yet understood. Furthermore, it appears that there is RNA made in the nucleus which never finds its way into the cytoplasm[7,8]. What this means in terms of regulation is completely unknown. It may be that the transport of messenger RNA from the nucleus to the cytoplasm is under control.

On the other hand, the existence of long-lived messenger RNA's in animal cells raises the possibility that messenger RNA's might be stored in the cytoplasm in active form and released for translation only when needed.

Another possible site for regulatory control is at the initiation of protein formation. It is known that formylmethionine is the initiator of protein chains in *E. coli*. In mammalian systems, the situation regarding protein chain initiation is still open; but it is conceivable that control is also exerted at this point.

Protein synthesis may also be regulated by prosthetic groups; for example the synthesis of globin in reticulocytes is stimulated by heme[9]. There is also the possibility that polypeptide chain termination and release are specifically regulated. You see that it is a long way from the information coded in the DNA to a final active enzyme molecule, with a number of possible points of regulation.

Hepatoma cells cultured in vitro as a model system of enzyme induction

For some time we had been studying the induction of tyrosine transaminase in rats *in vivo*. However, *in vivo* experiments are difficult to interpret, so we looked for a simpler system to work with, and fortunately, our attention was directed to the group of rat liver tumors known as "minimal deviation hepatomas". Several of them showed an increase in tyrosine transaminase when the tumor-bearing animals were injected with cortisol. One of these tumors, which has been converted into ascites tumor form, we propagated as a cell culture *in vitro*. These tissue culture cells, called HTC cells, for *hepatoma tissue culture*, still responded with a rise in tyrosine transaminase activity when cortisol was added *in vitro*. We therefore began to work with this system[10].

HTC cells when propagated in tissue culture have a doubling time of approximately 24 h and can easily be cloned.

Now we turn to the kinetics of enzyme induction in these cells. The enzyme we have been working with is tyrosine alpha-ketoglutarate aminotransferase or, in short, tyrosine transaminase. It is a well known pyridoxal phosphate-dependent protein which catalyses the rate-limiting step in tyrosine metabolism in the liver (Fig. 2). It is also involved in glyconeogenesis, because the degradation products of tyrosine can in part find their way to glycogen. Fig. 3 shows an induction experiment in resting cells. (We can induce the enzyme in resting cells as well as in growing cells although the induction seems to proceed faster in growing cells.) In this experiment, we added 10^{-5} M dexamethasone phosphate. The cells were maintained in an "induction medium", that is a medium con-

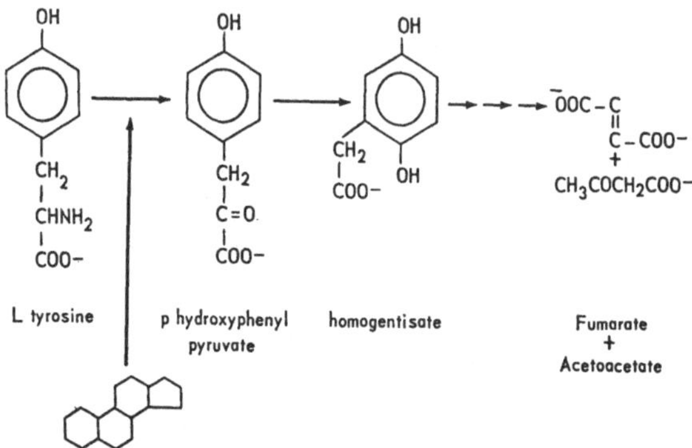

Fig. 2. Pathway of tyrosine degradation. The figure depicts the breakdown of tyrosine in the liver. The first reaction shown is catalyzed by tyrosine amino-transferase, which is induced by adrenal steroid hormones

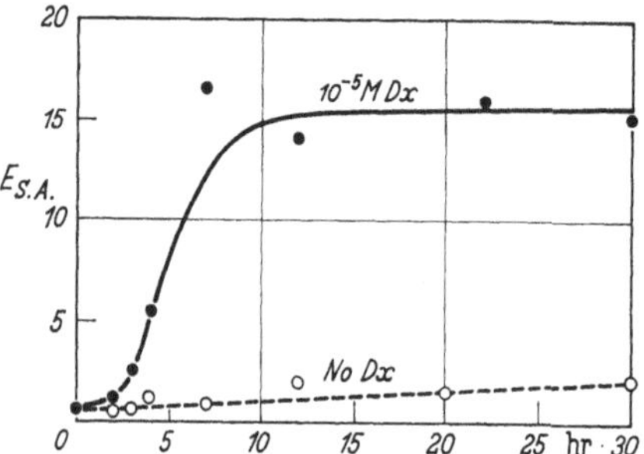

Fig. 3. Induction of tyrosine aminotransferase in a resting suspension of HTC cells. E_{SA} refers to specific activity of the enzyme measured as mμ moles of product formed per minute per mg protein at 25°. "Dx" refers to dexametha-sone phosphate. The figure is reprinted from reference (11) which gives the assay conditions and other details of the experiment

taining amino acids but no serum. In such a medium they survive but will not divide. Within a few hours, we can detect appreciable increases in tyrosine transaminase activity. There is essentially no change without inducer.

As for the time course of induction, the enzyme level begins to rise after $1\frac{1}{2}$ h and reaches a plateau at about 6 to 8 h. This plateau does not mean that the cells cease to produce enzyme; it only means that the rate of breakdown of the enzyme equals its rate of synthesis, so that a steady state is reached. The steady state enzyme level is directly proportional to the rise in the rate of enzyme synthesis, assuming that the rate of enzyme inactivation remains the same in induced and uninduced cells. In the case shown in Fig. 3 the induced enzyme is made fifteen times as rapidly as the basal enzyme.

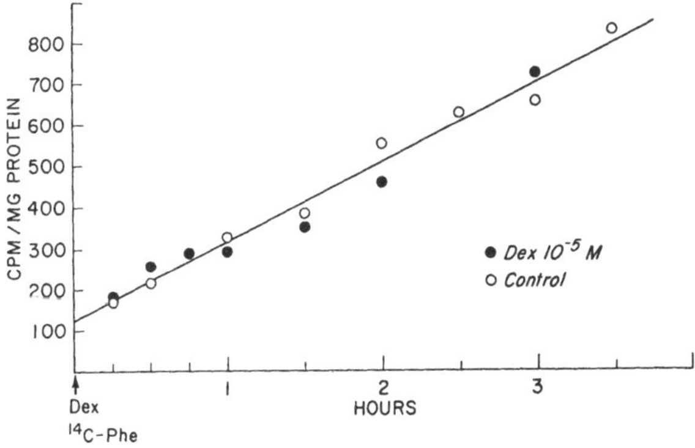

Fig. 4. The incorporation of phenylalanine-^{14}C into protein in HTC cells in the presence and absence of dexamethasone phosphate. The figure is reprinted from (11) which gives further details

The next step was to purify the enzyme. This work was carried out by Drs. HAYASHI and GRANNER[11, 12]. Starting from rat liver, they succeeded in crystallizing the enzyme, with an overall yield of 15%. We have improved our method and can now get pure enzyme in a shorter purification with about 30 to 35% yield. The enzyme is homogeneous in the ultracentrifuge, shows a sedimentation coefficient of 6 S, and we calculated the molecular weight to be 90,000. By immunological techniques, we demonstrated that the enzyme in our tissue culture cells is not distinguishable from the purified enzyme from rat liver[11]. The enzyme is composed of subunits; we don't know yet how many or if they are identical.

The enzyme induction is specific for the glucocorticoid group of steroids. We have mainly used dexamethasone-phosphate which is water soluble, but we can also get good induction with cortisone

 G. M. Tomkins and E. B. Thompson:

and several other glucocorticoids. Estrogens, androgens, and other steroids are inactive.

We now turn to the question how many proteins besides tyrosine transaminase are induced in our tissue culture cells. When we add radioactive amino acids to our tissue culture and precipitate the induced enzyme with antibody, we can demonstrate radioactivity

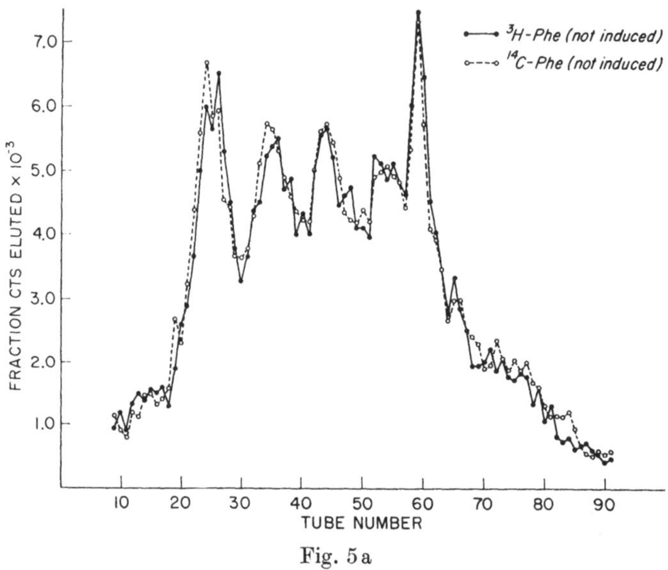

Fig. 5a

Fig. 5a and b. The effect of dexamethasone phosphate on the synthesis of specific proteins in HTC cells. The figures are reprinted from (11) where experimental details are presented. In general, 5a shows the close coincidence of ^{14}C and ^{3}H incorporation when separate flasks of uninduced HTC cells are labelled with radioactive phenylalanine. In 5b, the ^{3}H-labelled cells had been exposed to dexamethasone phosphate and several significant difference peaks appear

in our immunoprecipitate[11]. This shows that indeed new enzyme protein is formed. Another enzyme known to be induced *in vivo* by cortisol is tryptophan pyrrolase. However, in our tissue culture cells we do not find any of this enzyme nor do we find threonine deaminase, another inducible enzyme. Glutamate-pyruvate transaminase, which is also induced in rats by cortisol, occurs in our cells but does not appear to be inducible.

The overall rate of protein synthesis in these cells is not stimulated by corticosteroids. This is born out in Fig. 4 where dexamethasone-phosphate-treated cells have approximately the same incorporation of labelled amino acids into protein as the control. This negative experiment shows us that the induction must be fairly specific; at least it is not an outcome of an overall stimulation

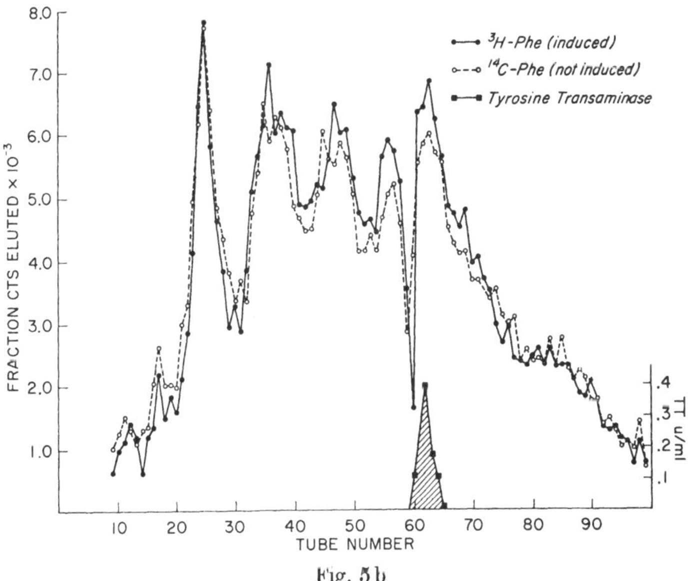

Fig. 5 b

of protein synthesis. We have also done a pertinent double label experiment[11]: one culture was grown for 4 days in the presence of tritiated phenylalanine, another in the presence of C^{14} labelled phenylalanine. One of the cultures was induced with hormone, then the cultures pooled, an extract made and the proteins separated on a DEAE column. As Fig. 5 shows, there is not much difference in the labelling of the two cultures. There are, however, three or four different peaks, one of them containing the tyrosine transaminase.

Synthesis of RNA in induced cells

It is well known from the work of other groups, that cortisol injected into a normal rat stimulates RNA synthesis. It is difficult to get unequivocal proof that messenger RNA is increased in these *in vivo* experiments; ribosomal RNA and transfer RNA are definitely increased. In our tissue culture system the inducer has the following effects on nucleic acid metabolism[13]: 1. DNA and RNA content of the cells are not altered, 2. there is virtually no change in the kinetics of uridine incorporation into RNA under conditions of induction. We have repeated this experiment many times be-

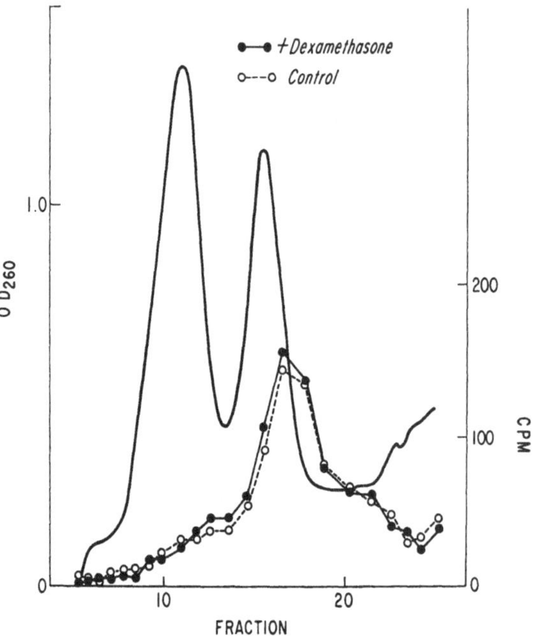

Fig. 6. The incorporation of uridine (^{3}H- and ^{14}C) into RNA isolated from the microsomal fraction of HTC cells. The filled circles represent incorporation in induced, and the open circles in non-induced cells. Reference (11), from which this figure is reprinted, gives further details

cause the result was so different from experiments in the whole animal. However, we could never find any gross difference in the incorporation of any labelled precursor into RNA. Even in double-label experiments, where the control culture was grown without inducer and labelled with tritiated uridine and the inducer culture was grown in the presence of hormone and pulsed with C^{14} uridine, there is a very similar pattern of incorporation of labelled precursors into the RNA profile as shown in the sucrose gradient separation (Fig. 6). There is a small peak at about 14 S which differs; this may be real, since we get it again and again in these experiments. 3. FINKEL et al. have shown that cortisol *in vivo* stimulates the synthesis of RNA carried by the 45 S subribosomal particles[14]. In our tissue culture cells, we do not see much of an increase in the region of the 45 S particle.

Our tentative conclusion from these experiments is that the dramatic increase in RNA synthesis seen after hormone administration in whole animals is not an intrinsic part of the induction mechanism. There seems to be little doubt about the necessity of the synthesis of messenger RNA in the course of enzyme induction; but the amount of messenger is so small that it cannot be detected in these incorporation studies.

The effect of actinomycin D on enzyme induction in HTC cells

As one would expect, actinomycin D at the levels tested in HTC cells does not interfere with DNA synthesis, but does inhibit uridine incorporation into RNA[15]. It also inhibits enzyme induction, as shown in Fig. 7. In this experiment, actinomycin was given at different times after addition of the inducer. When the two are given simultaneously, there is complete inhibition of enzyme induction. When given a little bit later, we find some increase in enzyme synthesis. This experiment suggests that the synthesis of RNA is necessary for enzyme induction; it shows also that the messenger is fairly stable in our system. When the synthesis of new messenger RNA is blocked, the synthesis of enzyme protein continues for about 5 h at the same rate, using the messenger formed during the first few hours.

Cytosine arabinoside inhibits thymidine incorporation into DNA in HTC cells. Uridine incorporation goes on at a normal rate, and so does enzyme induction. This experiment shows what was to be expected, namely that DNA synthesis is not necessary for enzyme induction[15].

We come now to the question of "actinomycin induction". Some time ago, we showed that, in the whole animal, a large dose of actinomycin given somewhat later than the inducer increases the induced levels of tyrosine transaminase and tryptophan pyrrolase[16]. Our conclusion from these experiments was that actinomycin actually stimulates the synthesis of the enzyme rather than inhibiting

its degradation (a possibility always to be kept in mind). That is
what we called the paradoxical effect of actinomycin D. Similar
cases of actinomycin stimulation of protein synthesis have been
reported in other systems.

Fortunately we can see this superinduction of tyrosine transaminase by actinomycin in tissue culture cells as well[10, 11]. This is

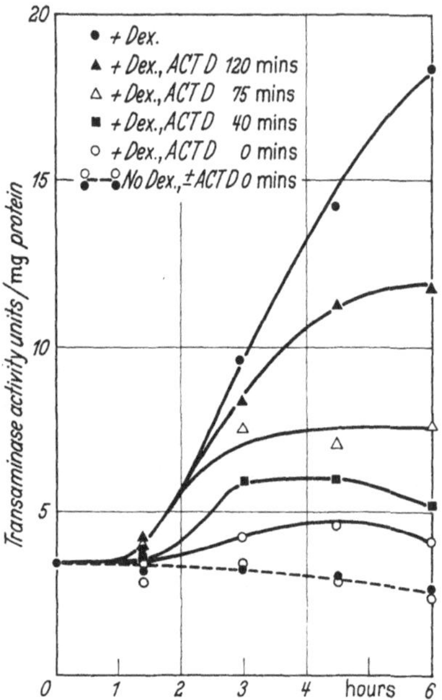

Fig. 7. Reprinted from (11) which gives further details. The inducer (indicated
as Dex) was always added at zero time

shown in Fig. 8. In a similar experiment, we also added labelled
amino acids and compared the specific activity of the total cell
proteins and of the enzyme which was precipitated by our antibody; from these data we concluded that actinomycin increased the
rate of enzyme synthesis about 6 fold.

The amount of actinomycin needed for the superinduction effect
is considerably higher than that needed for inhibition. For example,
we used als little as 0.2 μg/ml of actinomycin for inhibition, and
about 5.0 μg/ml for the superinduction. We can also increase the
basal level of enzyme with actinomycin D[15] alone, which means

that the presence of inducer is not required for this actinomycin effect.

When protein synthesis is blocked with cyclohexamide, the transaminase level decreases at the same rate whether actinomycin D is present or not, i.e., the breakdown of the enzyme is not influenced by actinomycin D.

This experiment, in which large amounts of actinomycin D produced more enzyme, first made us think about control at the translational level[16] because it implies that some factor limits the transla-

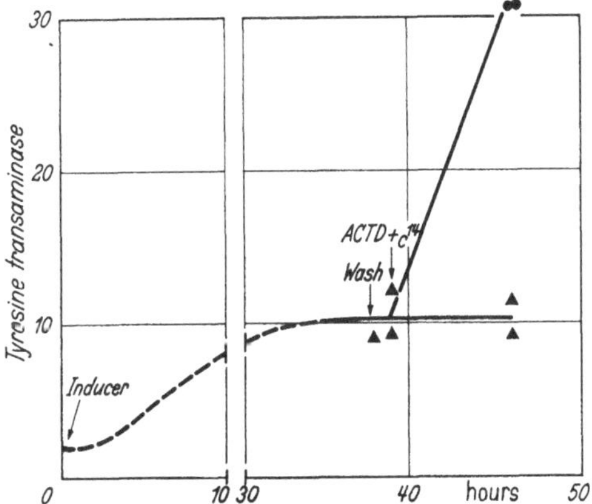

Fig. 8 [from (11)]. Dexamethasone was given at zero time and induction proceeded for 40 hours. The cells were washed and actinomycin D was added to one batch, along with ^{14}C-labelled yeast protein hydrolysate. To the other batch label, but no inhibitor, was added. Six and one half hours later, the transaminase activity was determined and the labelled enzyme was isolated from both groups of cells. Both enzyme activity and isotope incorporation were increased in the actinomycin-treated cells

tion of the messenger in the cytoplasm. The limiting factor is somehow abolished by the presence of actinomycin D.

A second type of experiment which made us think about a control at a later stage than the level of transcription is the following: First enzyme synthesis is induced for some time, and then the inducer is removed (by centrifuging the cells and resuspending them in fresh medium). Enzyme synthesis stops promptly and the enzyme level falls at the normal rate of enzyme inactivation. However, since we have previously established that the RNA for tyrosine transaminase is relatively stable, the inducer must play a role in translating (or stabilizing) the messenger. We also conclude, of course, that the inducer is needed for transcription.

Tab. 2 summarizes the arguments for control of protein synthesis at the translation level.

1. *High concentration of inhibitors of RNA synthesis, stimulate enzyme synthesis.* Our interpretation is that a factor which continually requires the synthesis of RNA is controlling the rate of synthesis of the enzyme at the translational level. We can call this "cytoplasmic repression "or" translational repression". It does not necessarily imply that there is a substance which represses, the

Table 2

Experiment	Interpretation
High concentrations of Actinomycin D, and other inhibitors of RNA synthesis, stimulate tyrosine transaminase formation (Actinomycin Induction)	Factor(s) requiring continued RNA synthesis inhibit the translation of tyrosine transaminase mRNA (translation repressor)
Transaminase mRNA relatively stable (average lifetime $\sim$5 hours) but inducer always required for enzyme synthesis	Inducer required for translation, or protection, of mRNA-as well as for transcription
Inducer is not required for Actinomycin Induction	Inducer antagonizes translation repression in some way

control could be exerted by the physical state of the message, e.g., the attachment to a membrane or something of this kind.

2. *Though the messenger for tyrosine transaminase is stable for at least 5 h, inducer is continuously necessary for enzyme synthesis.* From this experiment we can conclude that the inducer is required either for translation (or for protection) of the messenger as well as for transcription.

3. *The inducer is not required for the actinomycinsuper induction.* This implies that the inducer is somehow antagonizing the translation repressor.

Fig. 9 gives a scheme which illustrates our working hypothesis of the experimental data. On the top you see the classical scheme of how the structural gene makes tyrosine transaminase. On the bottom left there is a gene which produces something — we do not know what —, and this results in translation repression. The translation repressor acts by inhibiting the read-out of the transaminase messenger. A low concentration of actinomycin D will inhibit the formation of the transaminase messenger, but a high concentration is required to inhibit the formation of the translational repressor. When the repressor is degraded, the messenger RNA for tyrosine transaminase is read out more rapidly.

As for the role of the inducer, there can be no doubt that it stimulates transcription although the way in which this is done is

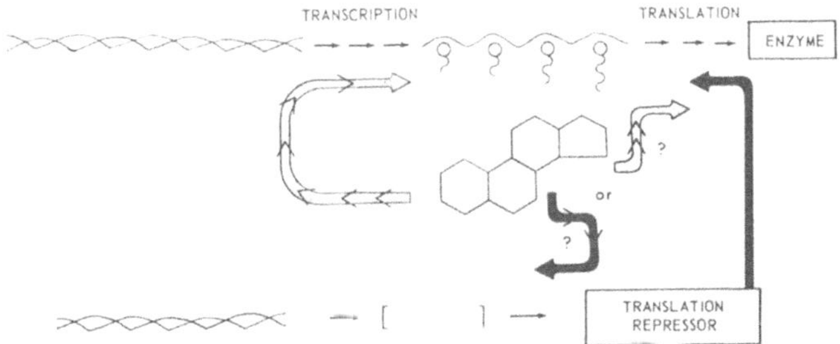

Fig. 9. Schematic representation of the hypothetical role of *the inducer* in both transcription and translation. The upper line depicts the synthesis of tyrosine transaminase (marked "Enzyme") starting with its structural gene, on the left. The lower line depicts the synthesis of translation repressor starting with its gene on the left. The filled arrow leading from the translation repressor represents its inhibition of transaminase synthesis at the level of messenger read-out. The steroid nucleus in the center represents the inducer. The arrow leading from it to the left represents it stimulation of the transcription of the transaminase structure gene. The arrows leading down and to the right represent inhibition of either formation (dark arrow) or action (lighter arrow) of translation repressor

still insufficiently understood. In addition the inducer appears to inhibit either the formation or action of the translation repressor.

In conclusion, it appears that HTC cells are well suited for the study of the control of enzyme synthesis. It is a system which is somewhere in between the isolated nuclei Dr. SEKERIS will talk about, and the system Dr. SUTHERLAND has been working on, and the more complicated systems like the perfused liver which Dr. SCHIMASSEK talked about and the induction of enzymes in the whole animal which is a topic of Dr. SEUBERT.

The schemes presented at the end of the talk are obviously highly speculative and I do not think that one should take them too seriously for the moment. It is more important to do more experiments which will lead to the rejection, modification or acceptance of these models.

References

[1] JACOB, F., and J. MONOD: J. molec. Biol. **3**, 318 (1961).
[2] STENT, G. S.: Science **144**, 816 (1964).
[3] ALPERS, D. H., and G. M. TOMKINS: J. biol. Chem. **241**, 4434 (1966).
[4] BROWN, D. D., and J. D. GURDON: J. molec. Biol. **19**, 399 (1966).
[5] HOUSSAIS, J. F., and G. ATTARDI: Proc. nat. Acad. Sci. (Wash.) **56**, 616 (1966).
[6] SCHERRER, K., L. MARCAUD, D. F. ZAJDELA, I. LONDON, and F. GROS: Proc. nat. Acad. Sci. (Wash.) **56**, 1571 (1966).
[7] HARRIS, H.: In Progress in nucleic acid research, ed. by DAVIDSON, J. N., and W. E. COHN, Vol. 2, p. 19 (1963).
[8] SHEARER, R., and B. MCCARTHY: Biochemistry **6**, 283 (1967).
[9] GRAYZEL, A. I., P. HORCHNER, and I. LONDON: Proc. nat. Acad. Sci. (Wash.) **55**, 650 (1966).
[10] THOMPSON, E. B., G. M. TOMKINS, and J. F. CURRAN: Proc. nat. Acad. Sci. (Wash.) **56**, 296 (1966).
[11] TOMKINS, G. M., E. B. THOMPSON, S. HAYASHI, T. GELEHRTER, D. GRANNER, and B. PETERKOFSKY: Cold Spr. Harb. Symp. quant. Biol. **31**, 349 (1966).
[12] HAYASHI, S.-I., D. K. GRANNER, and G. M. TOMKINS: J. biol. Chem. (In press).
[13] GELEHRTER, T. D., and G. M. TOMKINS: J. molec. Biol. (In press).
[14] FINKEL, H. E., C. HENSHAW, and H. HIATT: Molec. Pharm. **2**, 221 (1966).
[15] PETERKOFSKY, B., and G. TOMKINS: In preparation.
[16] GARREN, L. D., R. R. HOWELL, and G. M. TOMKINS: Proc. nat. Acad. Sci. (Wash.) **52**, 1121 (1964).

Diskussion

DATTA (London): I thank you very much Dr. TOMKINS. I am sure, that many people will have questions to ask, thus I will open the discussion.

KRÖGER (Freiburg): I have two questions: 1. Can you induce the enzyme with the substrate too, and 2.: Do you know something about the stability of your inducer?

TOMKINS: The answer to your first question is no. We can't induce the enzyme with the substrate. To the second question, we have not investigated the metabolism of the inducer in detail. It stays around as active inducer for about a day.

MATTHAEI (Göttingen): Jointly with GERHARD SCHÖCH and PAUL ABADOM, we have developed during the past year a cellfree protein synthetizing system from human placenta. The system contains ribosomes purified by two sedimentations through sucrose, a purified enzyme fraction and transfer RNA from placenta. The ionic environment and energy supply were optimized. This system incorporates 100 amino acid residues per ribosome within the first five minutes of incubation, which may mean 1000 to 2000 residues per *active* ribosome. Protein synthesis is partly inhibited in this system by DNase and actinomycin D. — The initial rate of protein synthesis is doubled by the addition of 10 mg/ml cortisol. At concentrations above 1 mg/ml, estradiol, progesterone, testosterone and nor-testosterone give only inhibitory effects. The action of cortisol has not yet been localized in either the transcription or translation process or both.

BÜCHER (München): With respect to your figure 2, the first points fit very well and the last ones too, buth those in between lie somewhat outside the curve. Thus the question arises: is this an oscillation ?

TOMKINS: This might be. It has only recently occurred to us, that especially in growing cultures, when inducer is present continually, there might be some oscillation phenomena. But this has to be worked out in more detail; we can't say much on this matter at the moment.

SEUBERT (Frankfurt): I have three questions: 1. What is the composition of your incubation medium; 2. what are the antiinducers and 3. are your cells still capable of gluconeogenesis ?

TOMKINS: The composition of our growth medium is swin's S 77 with serum, and the incubation was done mainly with the same medium without serum. This is an ordinary medium with basal salts, amino acids, glucose and so on, very much like Eagle's medium. However, you can leave out a lot of things, for example vitamins. The medium seems not to be too important. — With respect to Gluconeogenesis: We didn't get any accumulation of glycogen. Perhaps we havn't had the best conditions, thus I don't know what the true answer will be. — Our antiinducers were steroids like deoxycorticosterone, progesterone, 17-hydroxy-progesterone etc. However there is a special problem: progesterone is a very good inhibitor of respiration in these cells. It also inhibits protein synthesis and RNA synthesis to a large extent. Only when we go to very low concentrations, do we get normal incorporation of amino acids and uridine and an effect on induction.

VESTER (München): If there is a repressor-like protein, the action of which can only be derepressed by DNA, would you in this case exclude a mechanism by translation repression and take it as a support for transcription repression ? To describe our case: We work on a more selectively acting repressorlike protein, which was isolated from plant material (it inhibits tumor growth in a 10^{-9} dilution). American collegues tried to neutralize the action of this substance. But the only real derepressor was homologous DNA, precursors and RNA had no effect. After what I understand from your results, if our protein were a translation repressor, DNA addition should not neutralize its action. Would you agree with this?

TOMKINS: Well, the fact that there is regulation and repression at the transcriptional level seems very well established in bacterial and in mammalian systems. In our experiments, the question arises if there is regulation and repression at the level of translation in addition to the regulation at the level of transcription. I should mention, that in sea-urchin eggs after fertilization, there is a burst of protein synthesis which is not inhibited by actinomycin. One must conclude therefore, that in this case some messengers which are already present have been unmasked. So there is control at the translation level in higher systems.

NEUBERT (Berlin): Dr. TOMKINS, we have been very much interested in your data on "induction" by actinomycin. Working with isolated mitochondria and measuring incorporation of radioactive amino acids into mitochondrial protein, we found in the presence of rather high concentrations of actinomycin D (1 mg/ml) constantly an increase of "protein synthesis" rather than an inhibition, these findings were discussed at the FEBS-Meeting in Vienna. They may — like yours — represent an "induction" at the translational level. An "induction" by Actinomycin at the transcriptional level seems to be possible too. Studies performed in collaboration with Dr. OBERDISSE showed that if you inject *small* doses of actinomycin (50 to 100 mg/kg) into rats and measure RNA-polymerase in the nuclei isolated from such animals, you find an activity of 120 to 140%, 24 hrs. after the injection of the antibiotic, without a detectable decrease in the activity in the time before (OBERDISSE u. NEUBERT: 30. Tag. pharmakolog. Ges. 1966; 8. Frühjahrstag. Dtsch. pharmak. Ges. 1967).

We are also able to find an increase in RNA-polymerase activity if we add actinomycin and other substances which combine with DNA *in vitro* to isolated nuclei. Those are short time experiments in which incorporation of UTP is studied within 5 min. We find a similar increase when we study DNA-polymerase in isolated nuclei in the presence of *low* concentrations of phleomycin or acriflavin. The increase may be 200% compared with the controls. So one should be aware, that DNA-complexing componds are not only able to inhibit RNA- or DNA-polymerase reactions, but under suitable conditions may also be able to "induce" enzymic reactions. This might have some bearing also on the therapy with such compounds.

TOMKINS: This is very interesting. It seems to me that your experiments with mitochondria are most comparable to ours, since we have doses of actinomycin sufficient to inhibit RNA formation present all the time in our cultures. However, you would have to differentiate between immediate short term effects of actinomycin which at least in our hands (and also from all what I know in the literature) are only inhibitory, and long term effects which might involve induction. Examples of this are known in *in-vivo*-experiments with inhibitors of protein synthesis.

STAIB (Düsseldorf): Which concentration of actinomycin D stimulates the transaminase activity?

TOMKINS: It is between 0.5 mg/ml up to 5 mg/ml. It takes about five times as much to stimulate than to inhibit the induction.

STAIB: In an attempt to examine the discrepacies between FEIGELSON's thesis and LARDY's opinion concerning the role of hepatic enzyme de novo synthesis in cortisol-gluconeogenesis, we were able to depress TKT-induction during the first four hours by actinomycin in a dose of 125 μg/100 g rat. However, in the sixth hour after cortisol-application a depression of TKT-activity by actinomycin to control-level was impossible, notwithstanding the fact, that the actinomycin dose was elevated up to 300 μg/100 g rat and in spite of administering a second dose of the antibiotic at half-time of the six hours-period. The difference between the fourth-hour and sixth-hour-value after Cortisol + Act. D was statistically significant (p < 0,01).

SIEBERT (Hohenheim): Actinomycin D when given in rather high doses to rats causes a serious decrease of ribonuclease activity in liver nuclei. This effect may not be related to an interaction of actinomycin D with DNA. Do you think that such an effect could have something to do with your actinomycin induction?

TOMKINS: We first thought of protection of messenger etc. But purely on logical grounds, this could not be the answer. If you just preserve the messenger which is already there, you will get a continuation of enzyme synthesis at the same level but not an *increase* of enzyme synthesis. That is, your mechanism does not allow for an increase of the amount of messenger present, at least if actinomycin indeed blocks synthesis of new messenger.

TELLER (Marburg): You mentioned in your paper that tyrosine transaminase is not induced by estrogens nor by androgens in your experimental system. Do you have informations about other enzymes such as 5-alpha-reductase being induced by testosterone? I am referring to similar studies you have done together with Dr. Mc QUIRE on the induction of 5-alpha-reductase by thyroxine.

TOMKINS: We have not worked with these hormones.

WITT (Freiburg): GREENGARD reported the induction of tyrosine transaminase in intact rats by pyridoxale phosphate. Did you try to induce the enzyme by pyridoxine in your system? What are your ideas about this so called "coenzyme induction"?

TOMKINS: Well, in our cells there is no induction of transaminase by pyridoxal or pyridoxal phosphate.

EWALD (Frankfurt): Dr. TOMKINS, the kinetics of glycogen deposition and tyrosine transaminase (tyrosin-alpha-oxoglutarate-aminotransferase) in rat liver resulting from treatment with cortisol first show a lag period of almost two hours, reach a maximum after 4 hours and decline again during the sixth and eighth hours. You observed here — as we did in our experiments — an enhancement of the tyrosine transaminase activity after treatment with actinomycin. As an enhancement of enzyme activity may result from other reasons besides an increased *de-novo*-synthesis of enzyme protein I wonder whether you have any idea on the kinetics of this increase after treatment with actinomycin? Have you any indication on whether the enhancement of transaminase activity is due to enzyme induction itself? Finally, can you attribute the enhancement to either of the two templates you distinguished, transcription or translation?

TOMKINS: The main difference in our culture cells is that after induction, the enzyme stays up; it doesn't decline as in the intact animal. Our interpretation is that in the animal, the inducer is gone, it is metabolized; in our culture cells, the inducer is present for a longer time. If we wash the cells and put them in inducer-free medium, the enzyme level also declines in the culture cells.

SCHREIBER (Freiburg): Are your cells capable of the synthesis and secretion of serum proteins?

TOMKINS: No.

SCHREIBER: What is the type of the Morris hepatoma you worked with? Several changes in the inducibility of enzymes in the spectrum of the minimal deviation hepatomas have been described. Some of your data might describe special features of your system and only have small general importance.

TOMKINS: I can't tell you at the moment what the type number of the hepatoma was with which we started out originally; it was one of the minimum deviation hepatomas. What we have done in the meantime, is to try and get some of the functions of the cells back, for example, inducibility of tryptophan pyrrolase. We have tried to make animal passages of our cells and recultivate them and so on. I should say this frankly, because it comes up all the time, that these are abnormal cells in an abnormal situation. We made them deliberately as abnormal as possible to study one function: mechanism of induction of tyrosine transaminase. We expect, however, that this mechanism in our cells will not be fundamentally different from those *in vivo*.

KARLSON (Marburg): I will comment briefly on the question of operons you have mentioned. It is known that ecdysone produces puffs in chironomus and recently Dr. BERENDES at Tübingen has also produced puffs in Drosophila hydei larvae. There is one striking feature: The number of puffs turned on in Drosophila are certainly larger than in chironomus, and they are located on different portions of different chromosomes. So in this case, there are certainly different genes turned on by ecdysone, not only one operon.

TOMKINS: That is very interesting. I thank you, that you mentioned this because I was quite impressed with your findings of gene activations by ecdyson and the time course of puff inductions.

LANG (München): As I understood your talk, only alpha-ketoglutarate-tyrosine-transaminase is inducible by dexamethasone in your hepatoma tissue culture. I wonder if you have been able to induce tryptophane-pyrrolase by dexamethasone in the presence of high doses of actinomycin D, since the corresponding "translation repressor" could be eliminated.

TOMKINS: We have tried it and we have not seen it.

LANG: This would not fit very well in a concept of "translation repressor".

MAURER (Heidelberg): I wonder about the hormone concentration you used in your culture medium. It amounts to 10^{-5} M. However it is known that the blood level of cortisol ranges at 1 to 5.10^{-7} M (WETTSTEIN u. ANNER: Experientia (Basel) **10**, 397 (1954)]. Moreover, dexamethasone has a much higher biological activity than cortisol (FIESER: Steroide, p. 765, 1961). So I wonder if your experiments are really meaningful as far as physiological conditions are concerned.

TOMKINS: We can get about the same level of induction with a 10 times lower concentration of the hormone, say 10^{-6} M. Secondly, in our system there is not much difference between the activity of dexamethasone and cortisol; so I wonder if the difference observed *in vivo* might not be due differences in inactivation and excretion rather than in actual activity. As pointed out in the discussion, our cells are abnormal cells in an abnormal situation. They have lost several capabilities. It might be nicer to work with lower concentration say 10^{-9} M than 10^{-6} M. But I don't think that the molecular mechanisms involved in enzyme induction as we see them in our cells fundamentally differ from those *in vivo*.

DATTA: It is getting late now and I think we have to close the discussion. May I thank all the discussants and also Dr. TOMKINS again.

Wirkung der Hormone auf den Zellkern

Von C. E. Sekeris

Physiologisch-Chemisches Institut der Universität Marburg

Mit 16 Abbildungen

Vor 7 Jahren wurde die Hypothese aufgestellt, daß Hormone durch Beeinflussung des genetischen Materials wirken können[1]. Diese Hypothese war eine Schlußfolgerung aus dem bekannten Experiment von Clever und Karlson[2], die gezeigt haben, daß das Insektenhormon Ecdyson morphologische Veränderungen an den Riesenchromosomen der Speicheldrüsen von Chironomus tentans auslösen kann, Veränderungen, die als Puffs bekannt sind.

Die Auslösung des „Puffing"-Phänomens ist erstens rasch (sie erfolgt innerhalb von 15 min nach Injektion des Ecdyson), zweitens spezifisch (d. h. nur einige bestimmte Puffs werden induziert, und zwar in einer bestimmten Reihenfolge) und drittens durch minimale Mengen Hormon ($10^{-6}\,\mu$g) ausgelöst[3]. Diese Beobachtung fiel zeitlich zusammen mit zwei Ereignissen, die wesentlich dazu beitrugen, die experimentellen Befunde sinnvoll zu deuten. Einerseits erschien 1961 die Theorie von Jacob und Monod[4] über die genetische Regulation der Proteinsynthese, in welcher das Konzept der Messenger-RNA als Träger der in der DNA gespeicherten Informationen zum ersten Mal geprägt wurde; andererseits haben Beermann und seine Schule zeigen können, daß die Puffs als die aktiven Stellen des Genoms angesehen werden können, d. h. Stellen, in welchen eine rege RNA-Synthese stattfindet[5, 6].

So wurde die folgende Hypothese formuliert (Abb. 1)[7]: Das Ecdyson und jedes Hormon mit ähnlichem Wirkungsmechanismus wirkt primär am Genom und führt zur Stimulierung der Messenger-RNA-Synthese, welche anschließend im Cytoplasma als Matrize für die Synthese von spezifischen Proteinen (z. B. Enzymen) dient.

Gemäß Abb. 1 kann man drei Phasen der Hormonwirkung unterscheiden. In der ersten Phase gelangt das Hormon an den primären Wirkungsort, den Zellkern, wird dort vermutlich an einen „Receptor" gebunden (vgl. dazu den Beitrag JUNGBLUT) und setzt den Primärprozeß in Gang. Als Folge davon wird in der zweiten Phase die spezifische Messenger-RNS gebildet und aus dem Kern ausgeschleust. In der dritten Phase wird im Cytopasma das Enzym synthetisiert (Enzyminduktion). Es muß betont werden, daß das Studium der Effekte von Hormonen auf die Regulation der RNA- und Proteinsynthese durch unsere noch lückenhaften Kenntnisse

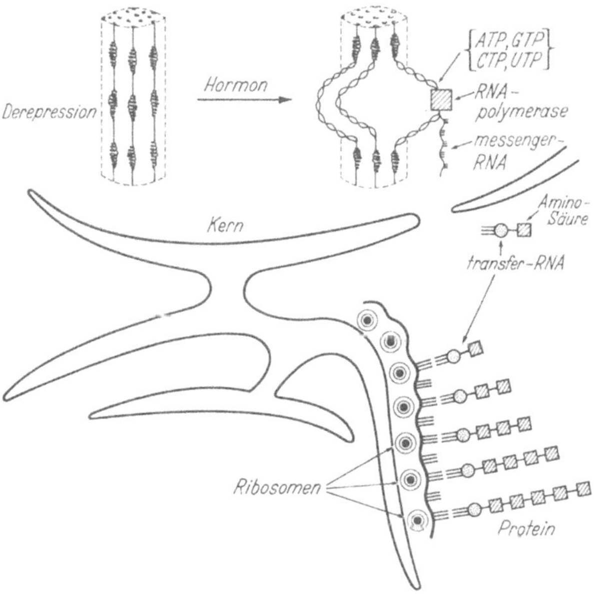

Abb. 1. Schematische Darstellung des Wirkungsmechanismus von Hormonen nach KARLSON

dieser Prozesse erschwert wird. So weiß man z. B., daß nur ein Bruchteil der im Kern synthetisierten RNA im Cytoplasma erscheint, und daß der größte Teil der Kern-RNA im Kern selbst wieder abgebaut wird[8]. Weder die Funktion der im Kern verbleibenden RNA, noch der Mechanismus des Abbaus ist bekannt. Unbekannt ist auch noch der Transportmechanismus der Messenger-RNA vom Kern zum Cytoplasma[9,10] (Transport in Form von Messenger-Partikel oder gebunden an ribosomale Untereinheiten), der Ort der Zusammensetzung der Ribosomen und der

Synthese von ribosomalen Proteinen; und obwohl der Mechanismus der Proteinsynthese im Cytoplasma in großen Linien bekannt ist, bleiben eine Menge Probleme noch offen. Sehr wenig ist über den Mechanismus der Synthese von Kernproteinen bekannt, von Proteinen, welche bestimmt eine wichtige Rolle für die Regulation der RNA-Synthese spielen (vgl. hierzu [11], [12]).

In den letzten Jahren wurde eine große Zahl von experimentellen Daten über die Wirkung der Hormone auf die RNA-Synthese gesammelt, eine Übersicht gibt Tata[13]. Es ist unmöglich, im Rahmen dieses Colloquiums die Fülle der Arbeiten zu referieren. Ich werde mich auf zwei Steroidhormone, das Cortisol und das Ecdyson, beschränken, weil wir auf diesem Gebiet gearbeitet haben. Am Beispiel dieser beiden Hormone will ich die Gültigkeit des Hormon-Gen-Konzepts diskutieren.

1. In-vivo-Versuche mit Cortisol

Eine Hauptwirkung des Cortisols ist die Stimulierung der Synthese von Glykogen in der Leber, die Glykoneogenese. Unter der Wirkung von Cortisol kommt es zur Induktion von Enzymen, die Aminosäuren zu Vorstufen des Glykogens abbauen, und von Enzymen, welche aus diesen Vorstufen Glucose bzw. Glykogen aufbauen.

Die Tab. 1 zeigt einige von Cortisol induzierte Enzyme. Die Steigerung der Aktivität dieser Enzyme stellt eine echte de-novo-Synthese dar. Das wurde im Falle der Tyrosin-Transaminase, Tryptophan-Pyrrolase und Alanin-Transaminase durch direkte Methoden (Einbau von radioaktiv markierten Aminosäuren im Enzymprotein) festgestellt, bei anderen Enzymen indirekt — durch Hemmung der Enzyminduktion mit Hemmstoffen der Protein- und RNA-Synthese. Nach der Hypothese von Karlson sollte der Enzyminduktion eine Stimulierung der Messenger-RNA für diese Enzyme vorausgehen.

Wenn man markiertes ^{3}H-Cortison i.p. Ratten injiziert, findet man schon nach 30 min eine maximale Konzentration in der Leber (Abb. 2). Der größte Teil des Hormons ist im Cytoplasma zu finden, doch dringt auch ein wesentlicher Teil davon in den Kern ein (Abb. 3). Von der im Kern gefundenen Aktivität findet man einen Teil in Methanol-löslicher Form und einen Teil gebunden an Proteine, wie z. B. Proteine des Nucleochromatins aber nicht an RNA und DNA[24]. Im Kern ist der größte Teil der Aktivität in Form der hormonell aktiven Corticosteroide (Cortisol und Cortison) anwesend, dagegen erfolgt im Cytoplasma eine schnelle Umwandlung zu inaktiven Metaboliten (Abb. 4)[25]. Eine der frühesten Wirkungen des Hormons in der Leber ist die Stimulierung der Aktivität der DNA-abhängigen RNA-Polymerase des Kerns[26, 27, 28]. Wir haben die RNA-Polymerase in Form des Weissschen Komplexes isoliert.

Tabelle 1. *Einige mit Cortisol induzierte Enzyme*

Enzyme	Nachweis von de-novo-Synthese		Literatur
	Direkte	Indirekte Hemmung durch Actinomycin, Puromycin, Äthionin usw.	
Tyrosintransaminase	+		KENNEY und FLORA[14]
Tryptophanpyrrolase	+	+	KNOX[15], FEIGELSON und GREENGARD[16] NEMETH und DE LA HABA[17]
Pyruvat-Glutamat-transaminase	+		SEGAL und KIM[18]
Pyruvatcarboxylase		+	HENNING et al.[19]
Phosphoenol-pyruvat-carboxykinase				SHRAGO et al.[20]
Fructose-1,6-diphosphatase		+	KVAM und PARKS[21]
Glucose-6-phosphatase		+	WEBER et al.[22]
Glykogensynthetase		+	HILZ et al.[23]

Die Stimulierung ist maximal nach $1\frac{1}{2}$ Std und nimmt dann allmählich ab (Abb. 5). Als Folge der Stimulierung der RNA-Polymeraseaktivität kommt es zu einer Stimulierung der RNA-Synthese im Zellkern (Abb. 6). ^{32}P wurde gleichzeitig mit Hydrocortison i.p. injiziert und nach verschiedenen Zeiten der Einbau in RNA von Kontroll- und hormonbehandelten Ratten im Kern und im Cytoplasma gemessen. Nach 45 min kommt es zu einer geringen Stimulierung der RNA-Synthese im Kern. Nach $1\frac{1}{2}$ Std ist der Effekt sehr gut sichtbar. Im Cytoplasma kommt es erst etwas später zu einem signifikanten Effekt.

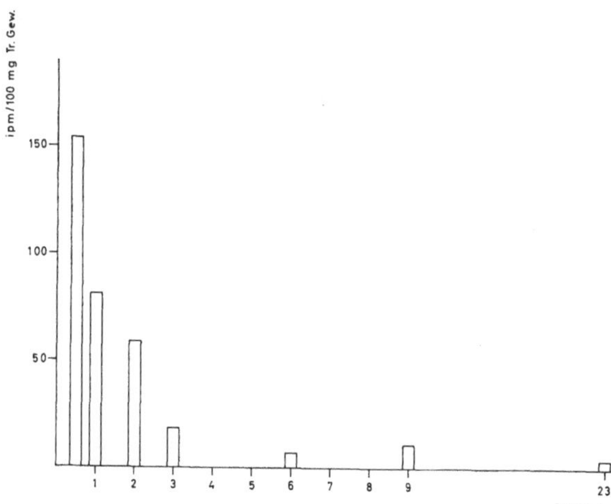

Abb. 2. Aufnahme von ³H-Cortison in der Leber. 10 μC ³H-Cortison (sp. Akt. 259 mC/mM) wurde in Ratten i.p. injiziert und nach ¹/₂, 1, 3, 6 und 9 Std die Radioaktivität in der Leber gemessen

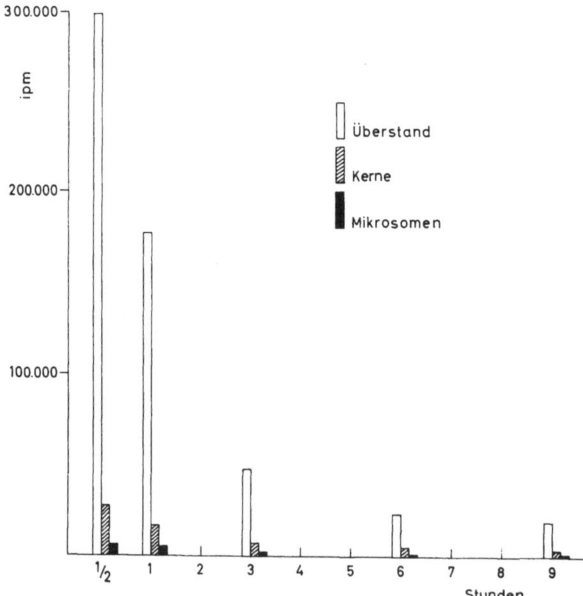

Abb. 3. Verteilung von ³H-Cortison in subcellulären Fraktionen von Ratten-leber

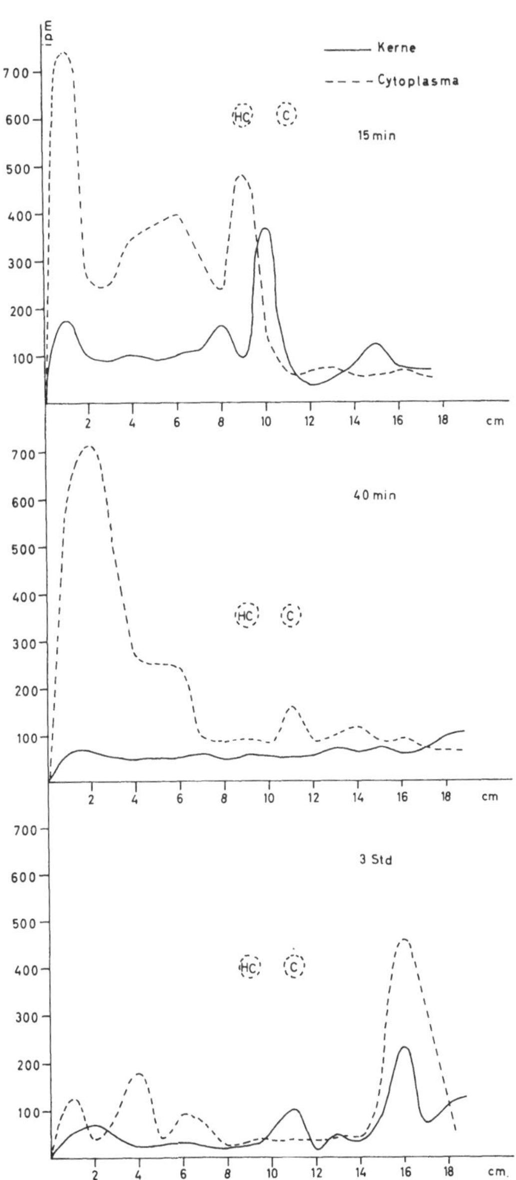

Abb. 4. Dünnschichtchromatographie von Äthanolextrakten aus Leberkernen und Cytoplasma von Ratten, injiziert mit ^{3}H-Cortison. Laufsystem: Dichlormethan 7:Aceton 2:Methanol 1

9*

Die Natur der neusynthetisierten RNA wurde von einigen Arbeitsgruppen mit Hilfe verschiedener Methoden ermittelt. Man kann z. B. die Phenolextraktion der RNA bei verschiedenen Temperaturen durchführen, wobei die Ribosomale und Transfer-RNA bei niedriger Temperatur abgetrennt wird, die Messenger-RNA da-

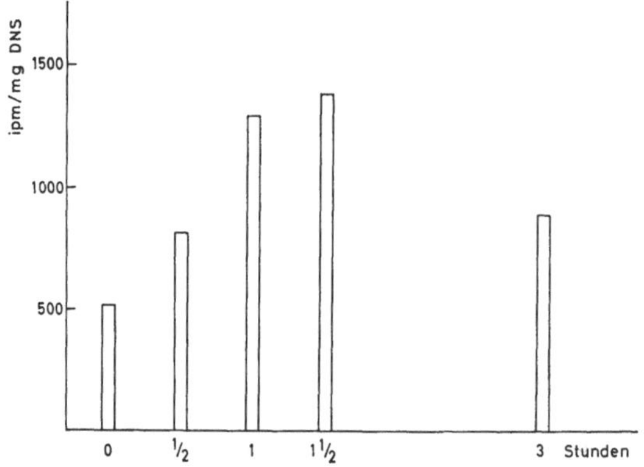

Abb. 5. Stimulierung der RNA-Polymerase der Rattenleber durch Cortisol. Der RNA-Polymerase-DNA-Komplex wurde zu verschiedenen Zeiten nach Cortisolgabe (2,5 mg Cortisol/100 g Ratte) isoliert. Die Polymeraseaktivität wurde durch den Einbau von ^{14}C UTP in RNA gemessen

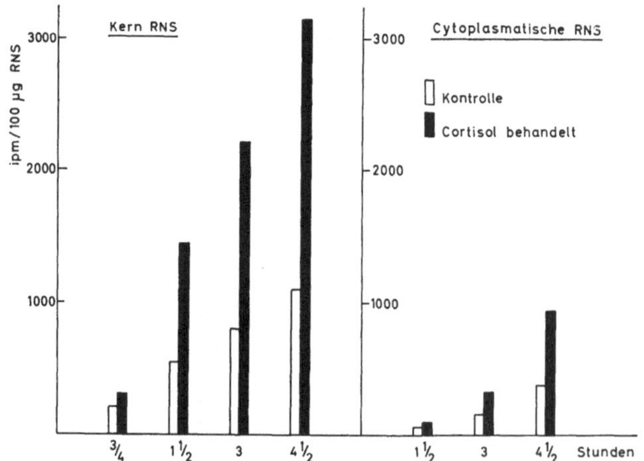

Abb. 6. Stimulierung des Einbaus von ^{32}P in Rattenleberkern- und cytoplasmatischer RNA durch Cortisol (2,5 mg/160 bis 180 g Ratte). Cortisol und ^{32}P wurden gleichzeitig i.p. injiziert und nach 45 min, $1^1/_2$, 3 und $4^1/_2$ Std die RNA aus Kernen und Cytoplasma mit Phenol bei 65° extrahiert

gegen erst bei 65°[36]. Man kann sich auch die unterschiedlichen Molekulargewichte zunutze machen und die RNA mit Hilfe der Dichtegradientenzentrifugation auftrennen[37], oder man kann die Messenger-RNA-Aktivität durch die stimulierende Wirkung in einem *in-vitro*-Protein synthetisierenden System erkennen[30]. Andere Methoden sind die MAK-Chromatographie[38], die Hybridisierungs- technik und die Acrylamidelektrophorese[39]. Verwendet man z. B. die Georgiev-Methode zur Charakterisierung der nach Cortisolgabe

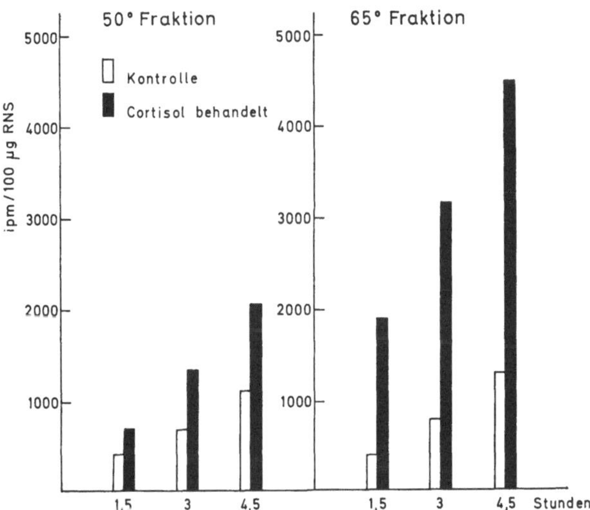

Abb. 7. Stimulierung des Einbaus von ^{32}P in Rattenleberkern-RNA, extra- hiert mit Phenol bei 50° und 65°, durch Cortisol. Experimentelle Bedingungen wie bei Abb. 6

synthetisierenden RNA (Abb. 7), so sieht man, daß es in beiden Kernfraktionen (50°- und 65°-RNA) zu einer Stimulierung der RNA- Synthese kommt[26, 30]. Zu demselben Ergebnis ist man bei Versuchen mit Dichtegradientenzentrifugation[34] bzw. Acrylamid-Gel-Elektro- phorese gekommen[40]. Daß ein Teil der neu produzierten RNA Messen- ger-RNA ist, kann man folgendermaßen beweisen: RNA aus Corti- sol behandelten Ratten wird in einem *in-vitro*-Protein synthetisie- renden System mit RNA aus Kontrolltieren verglichen. Schon nach 30 min stimuliert die RNA aus hormonbehandelten Tieren die Pro- teinsynthese wesentlich stärker als die RNA aus den Kontrolltieren (Tab. 2)[26, 30]. Aus allen diesen Versuchen erkennt man, daß das

Hormon zwar eine Stimulierung der Messenger-RNA-Synthese bewirkt, die ribosomale und Transfer-RNA jedoch gleichfalls vermehrt produziert werden.

Hier kommen wir zu einer der Kernfragen, die für die Richtigkeit der Karlsonschen Theorie von entscheidender Bedeutung ist. Wird nur der Turnover der Messenger-RNA erhöht, oder werden auch

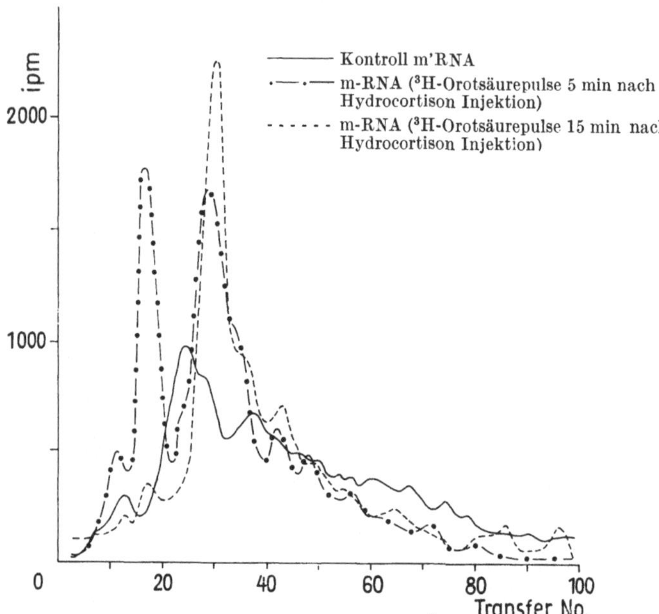

Abb. 8. Gegenstromverteilung von mit Orotsäure markierter m-RNA, isoliert aus Kontroll- und mit Cortisol behandelten Ratten. Dauer des Orotsäurepulses 20 min. Bei den mit Hormon behandelten Tieren wurde das Hormon 5 bzw. 15 min nach Injektion der Orotsäure verabreicht (nach KIDSON und KIRBY, 1965)

neue Messenger-RNA-Moleküle gebildet, d. h. wird neue Information abgerufen? Ist die Wirkung des Hormons auf die RNA-Synthese nur quantitativ oder auch qualitativ? Bis jetzt bleibt diese entscheidende Frage unbeantwortet. Es gibt aber einige Experimente, die für qualitative Effekte sprechen.

a) KIDSON und KIRBY haben Messenger-RNA aus Kontroll- und hormonbehandelten Tieren mit Hilfe der Gegenstromverteilung aufgetrennt[41]. Das Profil der RNA aus hormonbehandelten Tieren zeigte deutliche Abweichungen von dem der Kontrolltiere (Abb. 8).

b) DREWS und BRAWERMAN sind mit der Hybridisierungsmethode zu einem ähnlichen Schluß gekommen (Abb. 9). Sie haben durch kompetitive Hybridisierung von ^{32}P markierter Kern-RNA aus Kontroll- und hormonbehandelten Tieren mit Leber-DNA in Anwesenheit von markierter Kern-RNA gezeigt, daß die nach Hormoninjektion synthetisierte RNA neue Spezies von RNA enthält, die vor der Hormoninduktion nicht da waren.

c) Wir haben eine andere Versuchsanordnung gewählt. Injiziert man einer Ratte Cortisol und verfolgt man anschließend die Aktivität der Tyrosin-, Alanin- und Aspartat-Transaminase in der Leber, so sieht man folgendes: Die Tyrosin-Transaminaseaktivität steigt schnell innerhalb von 2 bis 4 Std an und fällt schnell wieder ab. Dagegen steigt die Aktivität der Alanintransaminase erst nach

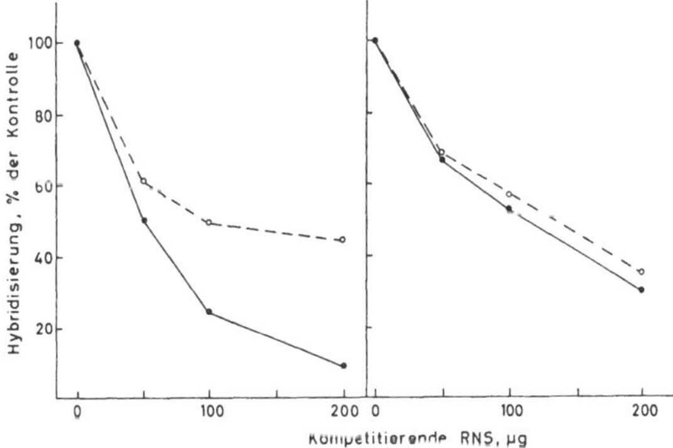

Abb. 9. Hybridisierung von mit ^{32}P markierter RNA aus Kontroll- und mit Cortisol behandelten Tieren mit Leber-DNA in Anwesenheit von a Kern-RNA aus Kontrolltieren und b Kern-RNA aus mit Hormon behandelten Tieren (nach DREWS und BRAWERMAN, 1967)

mehreren Stunden langsam an, und die Aktivität der Aspartattransaminase wird vom Hormon überhaupt nicht beeinflußt. Wenn das Hormon die Synthese der Messenger-RNA für die induzierbaren Enzyme stimuliert, dann sollte die aus 1 oder 2 Std lang Cortisolbehandelten Ratten isolierte Messenger-RNA die Information für die Tyrosintransaminase, aber nicht die Information für die Alanin- und Aspartattransaminase enthalten. Zuerst haben wir geprüft, ob Messenger-RNA aus Cortisol-behandelten Ratten (3 Std nach der Injektion isoliert) die Information für die Tyrosintransaminase trägt. Wir haben die Messenger-RNA einem *in-vitro*-Protein-synthetisierenden System zugesetzt, die Proteinsynthese 40 min ablaufen lassen und die Aktivität der Tyrosintransaminase im System gemessen. Die Tyrosintransaminase wurde mit Hilfe eines radiochemischen Testes gemessen, welcher sehr einfach und gleichzeitig sehr empfindlich ist[43]. Leider zeigt das System auch ohne zugesetzte RNA eine geringe Tyrosintransaminaseaktivität. Immerhin kommt

Tabelle 2. *Stimulierung von Messenger-RNA und der Tyrosintransaminase-Messenger-RNA durch Cortisol*

Adrenalektomierte Ratten (160 g) wurden mit 2,5 mg Cortisol i.p. injiziert. Nach 0, 30 und 90 min wurde die Leberkern-RNA durch Phenolextraktion bei 65° isoliert. 20 γ RNA wurden in dem *in vitro* Protein-synthetisierenden System zugegeben und der Effekt auf den Einbau von ^{14}C-Leucin im Protein sowie auf die Tyrosintransaminaseaktivität des Systems nach 40 min Inkubation gemessen. Das *in-vitro*-System hatte folgende Zusammensetzung: Leberribosomen (100 γ RNA), 105 000 G Überstand, 0,25 μMol ATP, 7,5 10^{-3} μMol GTP, 2,5 μMol MgCl$_2$, 25 μmol Tris, je 0,025 μMol eines Aminosäuregemisches und 0,1 μC ^{14}C-Leucin in einem Endvolumen von 0,5 ml. Die Tyrosintransaminaseaktivität wurde wie in Literatur [43] zit. gemessen.

	^{14}C-Leucin eingebaut in Protein (ipm)	Tyrosintransaminase-aktivität ($\mu\mu$Mol gebildetes p-Hydroxyphenyl-pyruvat/mg Protein)
Ribosomalsystem allein	214	585
Ribosomalsystem + „0" RNA	254	648
Ribosomalsystem + „30 min" RNA	369	783
Ribosomalsystem + „90 min" RNA	404	843

es in den Ansätzen mit m-RNA zu einer erheblichen Steigerung der Tyrosintransaminaseaktivität (Tab. 2)[44]. Nach RNasebehandlung verschwindet die Wirkung der m-RNA; DNase zeigt nur geringen Effekt (Tab. 3). Zusatz von Antibiotica (z. B. Erythromycin, Streptomycin) verhindert die Steigung der Transaminaseaktivität (Tab. 4), daraus schließen wir, daß die gemessene Aktivität auf neu synthetisiertes Enzymprotein zurückzuführen ist. Kürzlich ist es auch Herrn Lang gelungen, mit immunochemischen Methoden eine Markierung des Enzyms mit ^{14}C-Aminosäuren nachzuweisen. Damit hat unsere Interpretation eine wesentliche Stütze gefunden.

Mit der gleichen experimentellen Anordnung haben wir die Wirkung der Messenger-RNA aus Cortisol-behandelten Tieren auf die Aktivität der anderen beiden Transaminasen getestet. Bis jetzt haben wir keinen Effekt von Messenger-RNA, isoliert 30, 60 oder 90 min nach Hormoninjektion, auf die Aktivität der zwei Trans-

Tabelle 3. *Einfluß von Messenger-RNA vorinkubiert mit verschiedenen Enzymen auf die in-vitro-Induktion der Tyrosintransaminase*

	Enzymaktivität (μg gebildetes p-Hydroxyphenyl-pyruvat)
Komplettes System ohne m-RNA	0*
Komplettes System + m-RNA	2,65
Komplettes System + m-RNA behandelt mit RNase	1,35
Komplettes System + m-RNA behandelt mit DNase	2,30
Komplettes System + m-RNA behandelt mit Trypsin	3,10

* Die Enzymaktivität des Systems allein wurde als 0 genommen.

Tabelle 4. *Einfluß verschiedener Antibiotica auf die in-vitro-Induktion der Tyrosintransaminase durch m-RNA*

	Enzymaktivität (μg gebildetes p-Hydroxyphenyl-pyruvat)
Komplettes System ohne m-RNA	15,25
Komplettes System + m-RNA	20,00
Komplettes System + m-RNA + Erythromycin	15,80
Komplettes System + m-RNA + Streptomycin	16,10
Komplettes System + m-RNA + Puromycin	16,75
Komplettes System + m-RNA + Actinomycin D	20,70

(nach LANG, HERRLICH und SEKERIS[44]).

aminasen *in vitro* feststellen können. Diese Versuche sprechen für die Selektivität der Wirkung von Cortisol auf die Messenger-RNA-Synthese[45].

2. In-vivo-Versuche mit Ecdyson

Ecdyson ist das Häutungshormon der Insekten[46, 47]. Es ist ein Steroidhormon folgender Konstitution[48, 49] (Abb. 10): Allein oder in Zusammenarbeit mit einem anderen Insektenhormon, dem Juvenil-

hormon, steuert das Ecdyson die meisten Schritte der postembryonalen Entwicklung der Insekten (Abb. 11). Wir haben vor allem
mit der Schmeißfliege Calliphora erythrocephala gearbeitet. Hier
ist die Hauptwirkung des Ecdysons die Pupariumbildung, äußer-

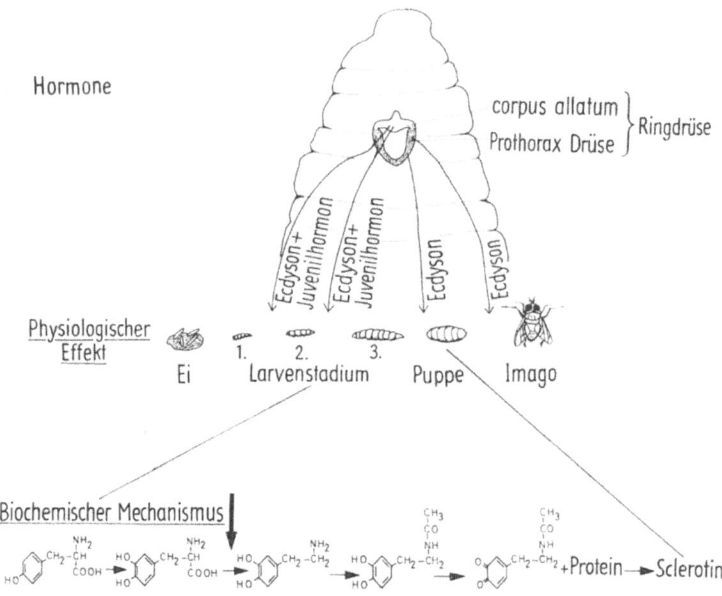

Abb. 10. Formel von Ecdyson

Abb. 11. Schematische Darstellung der Wirkung von Ecdyson bei Calliphora
erythrocephala

lich kenntlich an der Dunkelfärbung und Verhärtung (Sklerotisierung) der letzten Larvencuticula[50]. Dieser Prozeß beruht auf der
Reaktion der Sklerotisierungssubstanz N-Acetyldopamins[51] mit
den Cuticulaproteinen im Sinne einer Chinongerbung[52]. N-Acetyldopamin wird kurz vor der Verpuppung durch Decarboxylierung

von DOPA produziert[43, 53]. Das dazu notwendige Enzym, die DOPA-Decarboxylase, wird vom Ecdyson induziert[54]. Diese Induktion stellt eine de-novo-Synthese vom Enzym dar[55]. Zeitlich früher wird der RNA-Stoffwechsel im Epidermisgewebe, dem Bildungsort der DOPA-Decarboxylase, stark vom Ecdyson stimuliert[56]. Die Natur der neu synthetisierten RNA wurde in Experimenten ähnlich denen bei den Cortisolversuchen, ermittelt. Dabei zeigte sich, daß sowohl die Messenger- als auch die ribosomale und Transfer-RNA stimuliert wird.

Darüber hinaus wurden Hinweise gebracht, daß Teile der neu synthetisierten RNA die Information für die DOPA-Decarboxylase tragen[57]. Die experimentelle Anordnung dieser Versuche ist ähnlich wie die für den Nachweis der Matrizenaktivität der Leber-m-RNS für die Tyrosintransaminase verwendete (Zusammenfassung dieser Versuche s. [58–61]).

Aus den erwähnten *in-vivo*-Versuchen kann man folgendes schließen: Einer der frühesten Effekte vom Cortisol (und Ecdyson) ist die Stimulierung der RNA-Synthese im Kern. Dabei kommt es zur Synthese sowohl von Messenger-RNA als auch von ribosomaler und Transfer-RNA; es gibt Anhaltspunkte dafür, daß es sich bei der Stimulierung von Messenger-RNA nicht nur um quantitative, sondern auch um qualitative Effekte handelt, d. h. daß *die* Messenger-RNA durch das Hormon synthetisiert wird, welche die Information für die an der Gluconeogenese beteiligten Enzyme (bzw. bei *Calliphora* für den die DOPA-Decarboxylase) trägt. Die *in-vitro*-Versuche geben keinen Aufschluß über den primären Wirkungsort des Hormons. Man kann zwar aus den Versuchen mit radioaktiv markierten Hormonen und aus den schnellen Hormoneffekten auf den Kernstoffwechsel einen primären Effekt im Zellkern postulieren, aber man kann nicht ausschließen, daß diese Effekte sekundär sind. Es gibt eine Möglichkeit das zu entscheiden, und zwar die Untersuchung der Hormonwirkung auf isolierte Zellkerne. Findet man z. B. Effekte des Hormons auf die RNA-Synthese isolierter Kerne, dann kann man behaupten, daß die primäre Wirkung des Hormons in bezug auf die RNA-Synthese im Zellkern zu suchen ist.

3. Versuche mit isolierten Zellkernen

Voraussetzung für solche Versuche ist die Herstellung von Kernpräparaten die sauber *und* funktionsfähig sind, z. B. sollen sie noch RNA- und Proteinsynthese zeigen.

     C. E. Sekeris:

Es gibt mehrere Methoden zur Herstellung von Kernen. Zu den meist verwendeten gehören:

1. Die Behrens-Methode, die auf Homogenisierung der Gewebe in organischer Lösung beruht und die saubere, aber funktionsunfähige Präparate gibt[62]; 2. Homogenisierung in isotonischen oder hypertonischen Saccharoselösungen. Diese Methode liefert funktionsfähige Präparate[63–67].

Die von uns verwendete Methode ist in [68–70] beschrieben.

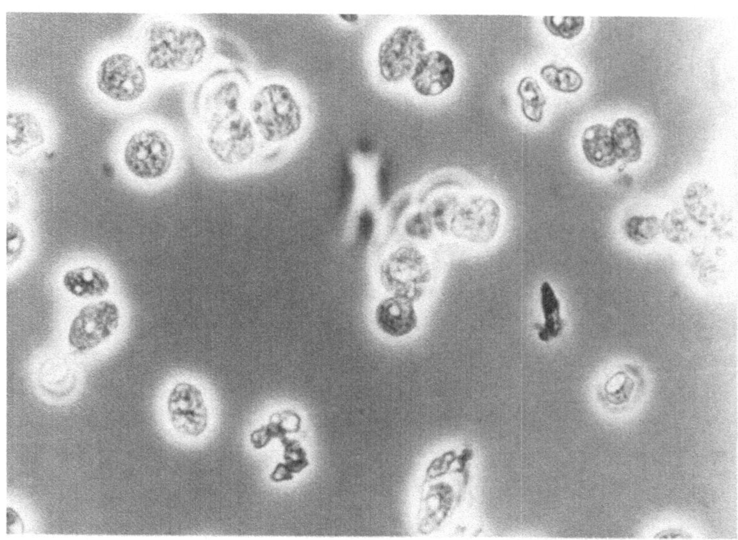

Abb. 12 a. Phasenkontrastmikroskopie von isolierten Rattenleberkernen

Die Leber wird in isotonischer Saccharoselösung homogenisiert. Anschließend werden die Kerne durch Behandlung mit dem Detergens Triton X-100 weitgehend von cytoplasmatischen Bestandteilen befreit. Als Kriterien der Sauberkeit der Kerne haben wir erstens das morphologische Bild (Abb. 12a und 12b), zweitens die Bestimmung der DNA-RNA-Proteinquotienten (Tab. 5) und drittens enzymatische Bestimmungen von Leitenzymen (Tab. 5) herangezogen. Letztere sprechen dafür, daß die Kontamination mit cytoplasmatischen Bestandteilen sehr gering ist. Die nach unserer Methode isolierten Kerne zeigen folgende Funktionen:

1. Sie bauen Aminosäuren in Proteine ein, und zwar hauptsächlich in saure Proteine (Abb. 13)[71]. 2. Sie bauen RNA-Vorstufe in RNA ein (Abb. 13)[72]. 3. Sie können Nucleoside oder Basen in Nucleotide umwandeln (Abb. 13). 4. Sie können Acetylierung und Methylierung von Kernproteinen durch Acetyl-CoA und S-Adenosylmethionin durchführen[73, 74].

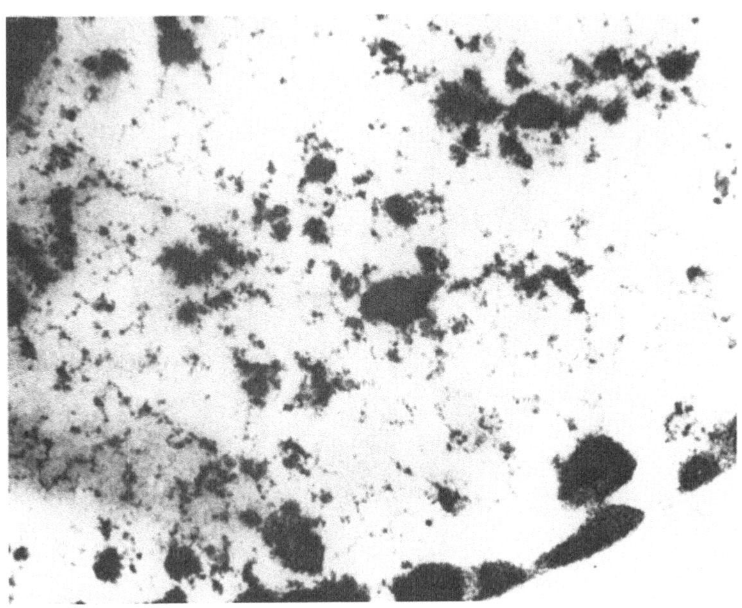

Abb 12 b. Elektronenmikroskopische Aufnahme von isolierten Kernen

Der wesentliche Nachteil des Kernsystems liegt darin, daß die Kerne nur begrenzte Zeit funktionell intakt erhalten bleiben. Entfernt man die Kerne aus ihrer cytoplasmatischen Umgebung, so

Tabelle 5. *DNA, RNA und Proteinquotienten und Gehalt an Leitenzymen von isolierten Rattenleberkernen*

$$\frac{\text{Enzymaktivität/mg lösliches Kernprotein}}{\text{Enzymaktivität/mg Protein des Homogenates}}$$

DNS	RNS	Protein	Glutamatdehydrogenase	Glucose-6-Phosphatase
4,71	1	18,84	0,05	0,28

werden Proteasen und Nucleasen aktiviert, welche allmählich die Kernstruktur und Kernfunktion zerstören.

Inkubiert man die Zellkerne aus Rattenleber mit ^{3}H-Cortisol, so sieht man eine schnelle Aufnahme des Hormons in den Kernen[25] (Abb. 13 und 14). Innerhalb von einigen Minuten ist ein Plateau erreicht. Die gefundene Radioaktivität ist in Form von nicht um-

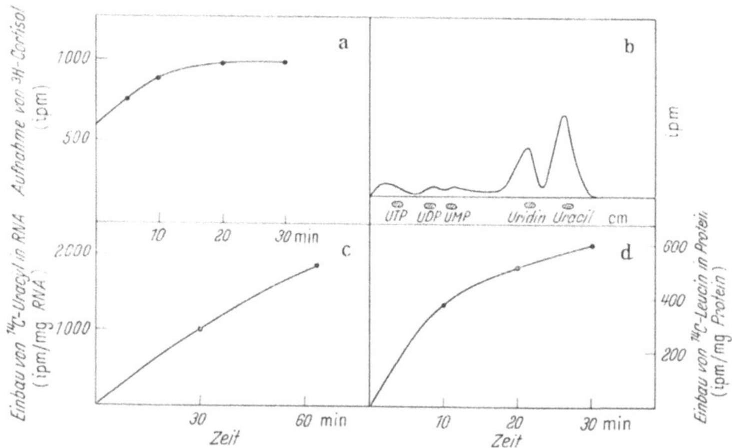

Abb. 13. Einige metabolische Eigenschaften von isolierten Rattenleberkernen. a Aufnahme von ^{3}H-Cortisol, b Umwandlung von ^{14}C-Uracil in Nucleotide, c Einbau von ^{14}C-Uracil in RNA, d Einbau von ^{14}C-Aminosäure in Kernproteinen

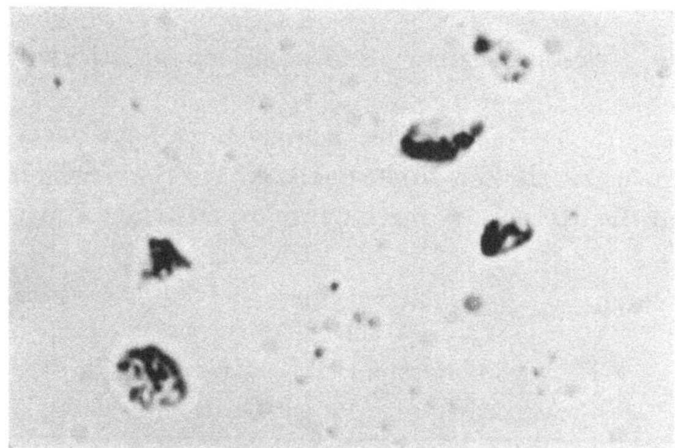

Abb. 14. Autoradiographie von Rattenleberkernen, inkubiert mit ^{3}H-Cortisol (10 min Inkubation bei 37°)

gewandeltem Hormon vorhanden (Abb. 15). Der größte Teil des
Cortisols ist in Methanol-löslicher Form vorhanden, aber ein Teil
ist fest am Chromatin gebunden[75]. Inkubation der Kerne mit Corti-

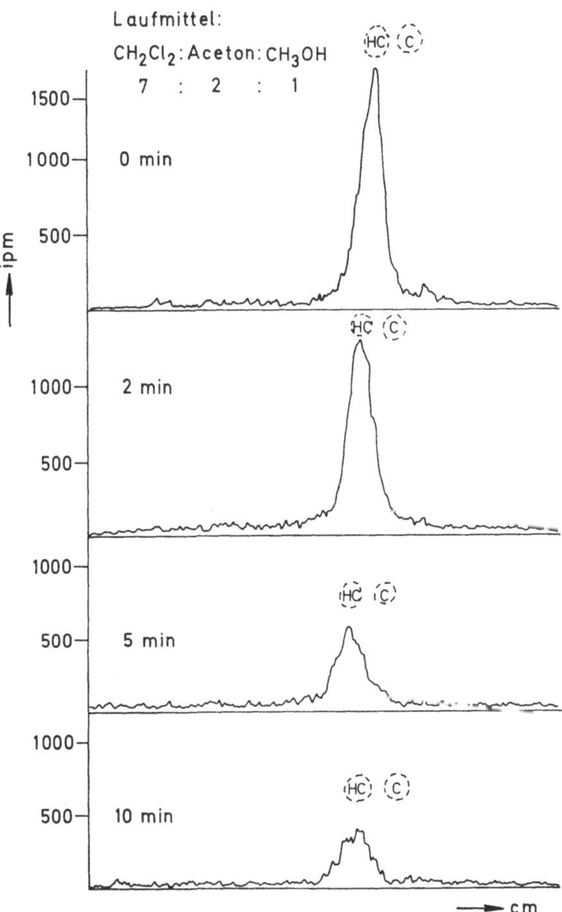

Abb. 15. Dünnschichtchromatographie von Methanolextrakten von Kernen,
inkubiert mit ^{14}C-Cortisol. Laufsystem: Dichlormethan 7:Methanol 2:Ace-
ton 1 (nach TIEMER und SEKERIS)

sol führt zur Stimulierung der RNA-Polymeraseaktivität[76, 77]. Der
Effekt ist fast sofort nach Zugabe des Hormons im Inkubationsmedium sichtbar (Tab. 6). Die Dosisabhängigkeit der Wirkung ist
in der Tab. 7 wiedergegeben.

Daß es sich um eine spezifische Wirkung des Hormons handelt, zeigt die Tab. 8. Von den verschiedenen getesteten Steroiden stimulieren nur Cortisol, Cortison und Testosteron die Polymeraseakti-

Tabelle 6. *Stimulierung der RNS-Polymeraseaktivität von Rattenleberkernen durch Zugabe von 20 Cortisol/ml in vitro*

	Inkubationszeit (min)	Aktivität ($\mu\mu$Mol ^{14}C-UTP eingebaut in RNA/mg DNA/min
Experiment 1	0,5	
Kontrollkerne		89,7
hormonbehandelte Kerne		112
Experiment 2	5	
Kontrollkerne		136
hormonbehandelte Kerne		214
Experiment 3	10	
Kontrollkerne		75
hormonbehandelte Kerne		173
Experiment 4	20	
Kontrollkerne		89,9
hormonbehandelte Kerne		117,4

(nach Lukács und Sekeris[77]).

Tabelle 7. *Dosisabhängigkeit der Wirkung von Cortisol auf die RNS-Polymeraseaktivität von isolierten Rattenleberkernen*

	Cortisol (γ/ml)	Aktivität (μMol ^{14}C-UTP eingebaut in RNA/mg DNA/min)
Kontrollkerne	—	75
hormonbehandelte Kerne	1	124
hormonbehandelte Kerne	10	173
hormonbehandelte Kerne	20	215

(nach Lukács und Sekeris[77]).

vität, dagegen waren verwandte, aber hormonell inaktive Steroide völlig inaktiv. Als Folge der Aktivierung der Polymeraseaktivität kommt es zu erhöhter RNA-Synthese. Das kann man durch den erhöhten Einbau von ^{14}C-Uracil in Kern-RNA zeigen[72] (Tab. 9). Auch die Matrizenaktivität der RNA in einem *in-vitro*-Proteinsynthetisierendem System ist in den Hormonansätzen erhöht (Tab. 10)[69, 78]. Auch hier wirken nur hormonell aktive Steroide stimulierend. Ähnliche Ergebnisse sind nach Einwirkung von Ecdy-

Tabelle 8. *Spezifität der Wirkung verschiedener Steroide auf die RNA-Polymeraseaktivität isolierter Rattenleberkerne*

	Aktivität ($\mu\mu$Mol ^{14}C-UTP eingebaut in RNA/mg DNA/min)
Experiment 1	
Kontrollkerne	111
Kerne + Cortisol	162
Kerne + Testosteron	163
Kerne + Androsteron	116
Experiment 2	
Kontrollkerne	117
Kerne + Cortisol	220
Kerne + Progesteron	146
Kerne + Pregnenolon	71

(nach LUKÁCS und SEKERIS[77]).

Tabelle 9. *Stimulierung des ^{14}C-Uracileinbaus in RNA von isolierten Rattenleberkernen durch 10 γ Cortisol/ml. Inkubationszeit 2 Std bei 37°*

	ipm/mg RNA
Kontrollkerne	5325
hormonbehandelte Kerne	7100

Tabelle 10. *Erhöhung der Matrizenaktivität von Kern-RNA durch Cortisol in vitro*

	^{14}C-Leucineinbau in Protein, ipm/mg ribosomales RNA
Ribosomales System allein	545
Ribosomales System + RNA aus Kontrollkernen	1070
Ribosomales System + RNA aus Kernen, inkubiert mit 15 γ Cortisol/ml	1322

son an isolierten Epidermiskernen zu sehen (ausführliche Beschreibung s. [59]).

Aus den *in-vitro*-Versuchen kann man entnehmen, daß im Bereich des Zellkernes die primäre Wirkung (oder eine der primären

Wirkungen) des Hormons zu suchen ist. Sofort nach seinem Eindringen in den Zellkern muß vom Hormon ein Prozeß ausgelöst werden, der zu einer Steigerung der RNA-Polymeraseaktivität führt.

Der Mechanismus der Steigerung der Polymeraseaktivität, die wahrscheinlich die primäre Wirkung darstellt, *oder* eine Wirkung, die diesen sehr nahe ist, ist noch unbekannt.

Einige der vielen Möglichkeiten sind:

1. De-novo-Synthese der Polymerase. Gegen diese Annahme spricht der schnelle Effekt des Hormons auf die Enzymaktivität. Es wäre unwahrscheinlich, daß innerhalb von einigen Sekunden eine de-novo-Synthese des Enzyms stattfinden könnte. Dazu kommt die Schwierigkeit der Erklärung der Spezifität der Wirkung. Nur wenn man annimmt, daß mehrere Spezies von Polymerasen existieren, dann könnte man die Spezifität der Wirkung auf die Synthese von spezifischen Polymerasen zurückführen. Bis jetzt aber gibt es keine solchen Anhaltspunkte. Die Spezifität der synthetisierten RNA wird durch die Matrize bestimmt.

2. Eine zweite Möglichkeit ist die Aktivierung der Polymerase im Sinne z. B. einer allosterischen Wirkung[79, 80]. Damit könnte man die rasche Steigerung der Polymeraseaktivität erklären, die Spezifität der Wirkung bliebe aber wieder ungeklärt.

3. Eine dritte Möglichkeit wäre die direkte Bindung des Hormons an die Matrizen-DNA. Diese Bindung könnte zu einer Auflockerung der DNA-Doppelhelix und dadurch zu der Möglichkeit der Ablesung der DNA von der Polymerase führen[81]. Die Schwierigkeit dieser Hypothese ist, daß man sich schwer vorstellen kann, wie die vier Basen der DNA spezifische Erkennungsstellen für das Hormon bilden können. Dazu kommt noch der Befund, daß man eine Bindung zwischen radioaktiv-markiertem Hormon und DNA *weder in vivo noch in vitro* finden kann[24, 82].

4. Eine andere Vorstellung ist, daß das Hormon durch einen Derepressionsmechanismus wirkt. Durch eine spezifische Bindung zwischen Hormon und einem Acceptormolekül am Chromatin (in diesem Falle wäre Acceptormolekül gleich Repressor) werden DNA-Abschnitte freigesetzt. Wird die Bindung zwischen Repressor und DNA gelockert, kann diese DNA durch die vorhandene RNA-Polymerase abgelesen werden. Als mögliche Repressoren kommen die Chromatinproteine in Frage.

Daß ein Derepressionsmechanismus eine Rolle spielt, haben
DAHMUS und BONNER nachgewiesen[83] (Abb. 16). Sie haben Chromatin aus Cortisol-behandelten Tieren 4 Std nach der Injektion
isoliert und in einem System mit RNA-Polymerase aus E. coli die
Matrizenaktivität dieses Chromatins verglichen mit der des Chro-

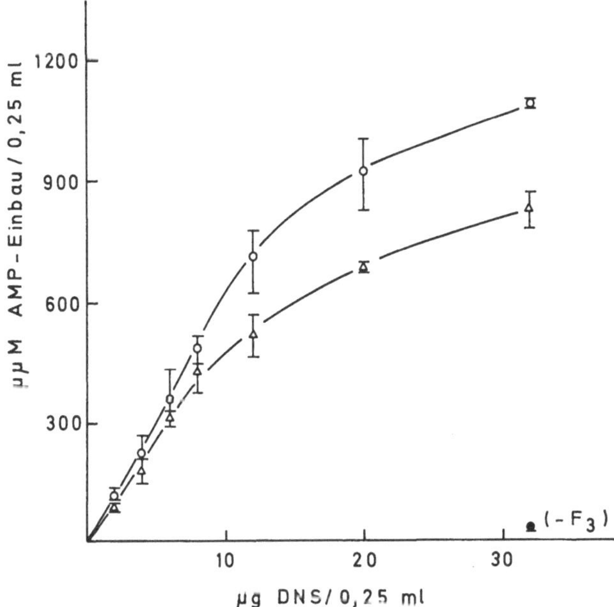

Abb. 16. Erhöhung der Matrizenaktivität von Rattenleberchromatin, isoliert
4 Std nach Injektion von Cortisol. Chromatin aus Kontroll- und mit Hormon
behandelten Tieren wurde mit Escherichia coli-RNA-Polymerase in Anwesenheit von ^{14}C-AMP inkubiert und der Einbau in RNA gemessen.
o—o—o—o Ansatz mit Chromatin aus mit Hormon behandelten Tieren.
Δ—Δ—Δ— Ansatz aus Chromatin aus Kontrolltieren. Der Einbau von
E. coli-Polymerase allein wurde abgezogen (nach DAHMUS und BONNER,
1965)

matins aus Hormon-behandelten Tieren. Das Chromatin aus Hormon-behandelten Tieren führt zu einer höheren Synthese als das
aus Kontrolltieren. Ähnliche Effekte haben wir schon 40 min nach
Hormoninjektion beobachtet[84]. Es gibt analoge Experimente über
den Einfluß von TSH auf die Matrizenaktivität von Schilddrüsen-
chromatin. Dort konnten japanische Forscher eine direkte Wirkung

des Hormons am isolierten Chromatin beobachten[85]. Ähnliche Beobachtungen haben Barker und Warren[86] mit Oestradiol und Endometriumchromatin und wir selbst mit Cortisol und isoliertem Rattenleberchromatin gemacht[87]. Diese Versuche sind noch am Anfang, aber sie stützen die Hypothese einer direkten Wirkung des Hormons an dem genetischen Material.

Zusammenfassend kann man folgendes sagen: Es gibt experimentelle Beweise, daß Hormone im Bereich des Zellkerns ihre primäre Wirkung ausüben. Man hat Anhaltspunkte dafür, daß das Hormon direkt mit dem genetischen Material reagiert. Dadurch kommt es zur Stimulierung der RNA-Polymeraseaktivität und zu erhöhter RNA-Synthese. Es gibt auch *dafür* Anhaltspunkte, daß spezifische Messenger-RNA-Moleküle, die die Informationen für die entsprechenden Proteine tragen, unter Einwirkung des Hormons synthetisiert werden.

Wahrscheinlich wirken Hormone nach verschiedenen Mechanismen — es kann sogar sein, daß *ein* Hormon *mehrere* Wirkungsorte hat. Auf jeden Fall ist die Wirkung der Hormone über die Beeinflussung des genetischen Materials eine Möglichkeit, die inzwischen durch zahlreiche experimentelle Befunde gestützt ist, und die Lücken, die dieses Konzept zeigt, sind hauptsächlich auf die begrenzten experimentellen Möglichkeiten zurückzuführen.

Ein Teil dieser Versuche wurde in Zusammenarbeit mit Dres. N. Lang, P. P. Dukes, W. Schmid, D. Gallwitz, I. Lukács, J. Homoki, K. E. Sekeri und den Herren G. Schütz und K. W. Tiemer durchgeführt. Wir danken Herrn Prof. P. Karlson für die großzügige Förderung unserer Arbeit sowie G. Behrens, U. von Hansen, B. Gallwitz, I. Wieber, D. Langer und I. Hellmann für die wertvolle Hilfe bei der Durchführung der Experimente. Die Deutsche Forschungsgemeinschaft hat die Arbeiten durch Sachbeihilfen gefördert.

Literatur

[1] Karlson, P.: Dtsch. med. Wschr. 86, 668 (1961).
[2] Clever, U., and P. Karlson: Exp. Cell. Res. 20, 623 (1960).
[3] — Chromosoma (Berl.) 13, 385 (1962).
[4] Jacob, F., and J. Monod: J. molec. Biol. 3, 318 (1961).
[5] Beermann, U.: Chromosoma (Berl.) 5, 139 (1952).
[6] Pelling, G.: Nature (Lond.) 184, 655 (1959).
[7] Karlson, P.: Perspect. Biol. Med. 6, 203 (1963).
[8] Harris, H.: Nature (Lond.) 198, 184 (1963).
[9] Georgiev, G. P.: In the cell nucleus, p. 79. Taylor and Francis 1966.
[10] Henshaw, E. C., and H. H. Hiatt: J. molec. Biol. 14, 241 (1965).

[11] SEKERIS, C. E., W. SCHMID, D. GALLWITZ, and I. LUKÁCS: Life Sci. **5**, 969 (1966).

[12] — — — — Ang. Chem. **11**, 601 (1966).

[13] TATA, J. R.: Progr. Nuc. Acid Res. and Molec. Biol. **5**, 191 (1966).

[14] KENNEY, F. T., and R. M. FLORA: J. biol. Chem. **236**, 2699 (1961).

[15] FEIGELSON, P., and O. GREENGARD: J. biol. Chem. **1237**, 3714 (1962).

[16] KNOX, W. E.: Brit. J. exp. Path. **32**, 462 (1951).

[17] NEMETH, A. M., and G. DE LA HABA: J. biol. Chem. **237**, 1190 (1962).

[18] SEGAL, H. L., and X. S. KIM: Proc. nat. Acad. Sci. (Wash.) **50**, 912 (1963).

[19] HENNING, I. V., I. SEIFERT, and W. LEUBERT: Biochim. biophys. Acta (Amst.) **77**, 345 (1963).

[20] SHRAGO, E., H. A. LARDY, R. C. NORDLIE, and D. O. FORSTER: J. biol. Chem. **238**, 3188 (1963).

[21] KVAM, D. C., and R. E. PARKS jr.: Amer. J. Physiol. **198**, 21 (1960).

[22] WEBER, G., G. BANERGEE, and S. B. BRONSTEIN: Biochem. biophys. Res. Commun. **4**, 331 (1961).

[23] HILZ, H., W. TARNOWSKI, and P. AREND: Biochem. biophys. Res. Commun. **10**, 492 (1963).

[24] SEKERIS, C. E., u. N. LANG: Hoppe-Seylers Z. physiol. Chem. **341**, 92 (1965).

[25] —, u. P. P. DUKES: Abstr. II Fed. Europ. Biochem., A. 131, Wien 1965.

[26] LANG, N. u. C. E. SEKERIS: Life Sci. **3**, 391 (1964).

[27] — — Hoppe-Seylers Z. physiol. Chem. **338**, 238 (1965).

[28] BARNABEI, O., B. ROMANO, and G. DI BITONTO: Arch. Biochem. **109**, 266 (1965).

[29] FEIGELSON, M., P. R. GROSS, and P. FEIGELSON: Biochim. biophys. Acta (Amst.) **55**, 495 (1962).

[30] SEKERIS, C. E., and N. LANG: Life Sci. **3**, 169 (1964).

[31] KENNEY, F. T., and F. J. KULL: Proc. nat. Acad. Sci. (Wash.) **50**, 493 (1063).

[32] GARREN, L. D., R. R. HOWELL, and G. M. TOMKINS: J. molec. Biol. **9**, 100 (1964).

[33] WICKS, W. D., D. L. GREENMAN, and F. T. KENNEY: J. biol. Chem. **240**, 4414 (1965).

[34] GREENMAN, D. L., W. D. WICKS, and F. T. KENNEY: J. biol. Chem. **240**, 4420 (1965).

[35] JERVELL, K. F.: Acta endocr. (Kbh.) Suppl. **44**, 88 (1963).

[36] GEORGIEV, G. P., and V. L. MANTIEVA: Biochim. biophys. Acta (Amst.) **61**, 153 (1962).

[37] BRITTON, R. J., and R. B. ROBERTS: Science **131**, 32 (1960).

[38] MANDELL, J. D., and A. D. HERSHEY: Ann. Biochem. **1**, 66 (1960).

[39] TSANEV, R. G., G. G. MARKOV, and G. N. DESSEV: Biochem. J. **100**, 204 (1966).

[40] VENKOV, P. V., E. Z. ANGELOV, L. I. VALEVA, and A. A. HADJIOLOV: Nature (Lond.) **213**, 807 (1967).

[41] KIDSON, C., and K. S. KIRBY: Nature (Lond.) **203**, 599 (1965).

[42] DREWS, J., and G. BRAWERMAN: J. biol. Chem. **242**, 801 (1967).

[43] Sekeris, C. E., and P. Karlson: Biochim. biophys. Acta (Amst.) **62**, 1103 (1962).

[44] Lang, N., P. Herrlich, and C. E. Sekeris: Acta endocr. (Kbh.) (Im Druck).

[45] Sekeris, C. E., W. Schmid und D. Gallwitz: Nicht publizierte Ergebnisse.

[46] Karlson, P.: Vitam. u. Horm. **14**, 227 (1956).

[47] — Abg. Chem. **75**, 257 (1963).

[48] —, H. Hoffmeister, H. Hummel, P. Hocks und G. Spiteller: Chem. Ber. **98**, 2394 (1965).

[49] Huber, R., u. W. Hoppe: Chem. Ber. **98**, 2403 (1965).

[50] Pryor, M. G. M.: Comp. Biochem., IV, Florkin and Mason ed., **1962**, S. 371.

[51] Karlson, P., C. E. Sekeris und K. E. Sekeri: Hoppe-Seylers Z. physiol. Chem. **227**, 86 (1962).

[52] Pryor, M. G. M.: Proc. roy. Soc. **B 128**, 378 (1940).

[53] Karlson, P., and C. E. Sekeris: Nature (Lond.) **195**, 183 (1962).

[54] Sekeris, C. E., and P. Karlson: Biochim. biophys. Acta (Amst.) **63**, 489 (1962).

[55] — — Arch. Biochem. **105**, 483 (1964).

[56] —, N. Lang und P. Karlson: Hoppe-Seylers Z. physiol. Chem. **341**, 36 (1965).

[57] — — Life Sci. **3**, 625 (1964).

[58] — In Mechanisms of Hormone Action, p. 149. P. Karlson ed. Stuttgart: Thieme 1965.

[59] Karlson, P., and C. E. Sekeris: Progr. Hormone Res. **22**, 473 (1966).

[60] — Naturwissenschaften **53**, 445 (1966).

[61] Sekeris, C. E.: In BBA Library Series **10**, 363 (1966).

[62] Siebert, G.: Biochem. Z. **334**, 369 (1961).

[63] Hageborn, G. H., and W. C. Schneider: J. biol. Chem. **197**, 611 (1959).

[64] Rees, K. R., and G. V. Rowland: Biochem. J. **78**, 89 (1961).

[65] Chauveau, J., Y. Monlé, and Ch. Roniller: Exp. Cell. Res. **11**, 317 (1956).

[66] Widnell, C. C., and J. R. Tata: Biochem. J. **92**, 313 (1964).

[67] Blobel, G., and V. R. Potter: Science **154**, 1662 (1966).

[68] Sekeris, C. E., P. P. Dukes und W. Schmid: Hoppe-Seylers Z. physiol. Chem. **341**, 152 (1965).

[69] Dukes, P. P., C. E. Sekeris, and W. Schmid: Biochim. biophys. Acta (Amst.) **126**, 123 (1966).

[70] Gallwitz, D., u. C. E. Sekeris: In Vorbereitung.

[71] Sekeri, K. E., u. C. E. Sekeris: In Vorbereitung.

[72] Dukes, P. P., u. C. E. Sekeris: Abstr. II Fed. Europ. Biochem., 68. Wien 1965.

[73] Gallwitz, D., u. C. E. Sekeris: In Vorbereitung.

[74] Sekeris, C. E., K. E. Sekeri und D. Gallwitz: In Vorbereitung.

[75] Tiemer, K. W., u. C. E. Sekeris: Nicht publizierte Versuche.

[76] Lukács, I., u. C. E. Sekeris: Angew. Chem. **78**, 748 (1966).

[77] — — Biochim. biophys. Acta (Amst.) **134**, 85 (1967).

[78] SCHMID, W., D. GALLWITZ, and C. E. SEKERIS: Biochim. biophys. Acta (Amst.) **134**, 80 (1967).

[79] MONOD, J., J. P. CHANGEUX, and F. JACOB: J. molec. Biol. **6**, 306 (1963).

[80] TOMKINS, G. M., and K. L. YIELDING: Cold Spr. Harb. Symp. quant. Biol. **26**, 331 (1961).

[81] GOLDBERG, M. L., and W. A. ATCHLEY: Proc. nat. Acad. Sci. (Wash.) **55**, 989 (1966).

[82] FEIGELSON, M., and P. FEIGELSON: Advanc. Enz. Regul. **3**, 11 (1965).

[83] DAHMUS, M. E., and J. BONNER: Proc. nat. Ac. Sci. (Wash.) **54**. 1370 (1965).

[84] LUKÁCS, I., J. HOMOKI und C. E. SEKERIS: Nicht publizierte Ergebnisse.

[85] SHIMADA, H., and I. YASUMASU: Gunma Symp. Endocr. **3**, 47 (1966).

[86] BARKER, K. L., and J. C. WARREN: Endocrinology **80**, 536 (1967).

[87] LUKÁCS, I., J. HOMOKI und C. E. SEKERIS: Nicht publizierte Ergebnisse.

Diskussion

SIEBERT (Hohenheim): Ich danke Herrn SEKERIS für seinen Vortrag und darf um Wortmeldungen bitten.

SCHREIBER (Freiburg): Ich möchte drei Fragen stellen und bitte um Entschuldigung, wenn diese zum Teil etwas banaler Natur sind:

1. Zu den Versuchen, in denen der Einbau von ^{32}P in RNS $\pm$ Cortison gemessen wurde: FEIGELSON hat beschrieben, daß Cortison (oder Cortisol) ein gesteigertes Turnover der Purin-Vorläufer bewirkt. Nach Gabe von Cortisol (on) erhält man höhere spezifische Aktivität im Purinprekursorpool bei ^{32}P-Markierung. Der gleiche Effekt wird auch durch NH_4^+ bewirkt. Es könnte so eventuell eine gesteigerte RNS Synthese durch Cortisol vorgetäuscht werden durch die höhere spezifische Aktivität im Prekursorpool. Welche Rolle könnte der Effekt in Ihren Versuchen spielen?

2. Die *in-vitro*-Synthese eines funktionierenden Enzyms in einem zellfreien System stellt einen außerordentlichen Befund dar. Erlauben Sie mir bitte daher die eingehendere Frage nach Ihren Kontrollversuchen. Haben Sie a) eine größere Menge Ihres labels (^{14}C-Tyrosin) allein und b) den Ansatz mit inaktivierten Enzymen chromatografiert?

3. DREWS und BONDY haben für Versuche zur Messung der Inkorporation von markierter Orotsäure in die RNS isolierter Leberzellkerne *in-vitro* darauf hingewiesen, daß

a) es bereits nach sehr kurzer Inkubation ($\sim$ 3 min) zu einem Abbau der hochmolekularen RNS mit entsprechender Änderung des Sedimentationsmusters kommt. Die extrahierte RNS inkubierter Leberzellkerne wandert etwa mit 2 bis 10 S, während die aus nicht inkubierten Kernen extrahierte RNS bedeutend hochmolekularer ist;

b) die Menge der extrahierten RNS schon nach 15 min Inkubation auf die Hälfte absinkt.

Wie ist die physiologische Bedeutung Ihrer Experimente bezüglich der Inkorporation von Orotsäure in isolierte Leberzellkerne *in-vitro* in diesem Zusammenhang zu werten?

SEKERIS (Marburg): Zu Ihrer ersten Frage: Die Veränderungen des Purin-Pools kann man nur als sekundäre Effekte des Cortisols ansehen. Sie erscheinen später als andere Effekte, z. B. die Steigerung der RNS-Polymeraseaktivität in der Leber, die schon 15 bis 30 min nach Cortisolgaben signifikant erhöht ist.

Zu Ihrer zweiten Frage: Wir sind uns völlig der Schwierigkeiten des Nachweises der de-novo-Synthese eines funktionierenden Proteins in einem zellfreien System bewußt. Selbstverständlich haben wir Kontrollversuche mit denaturierten Enzymen und mit dem radioaktiv markierten Tyrosin allein durchgeführt. Man findet dabei ein geringfügiges Auftreten von radioaktiv markierter p-Hydroxy-phenylbrenztraubensäure, welche aus nicht enzymatischer Zersetzung des Tyrosins entsteht. Deswegen laufen immer Kontrollansätze mit, die als 0-Werte dienen.

Zu Ihrer dritten Frage: Es gibt keinen Zweifel darüber, daß es während der Inkubation isolierter Zellkerne zu einem Abbau der Nucleinsäure kommt. Wir haben das auch mit der Dichte-Gradienten-Zentrifugation verfolgt. Interessant ist, daß die RNS überwiegend nur zu Molekülen mit einer Sedimentationsgröße von 12 bis 10 S und nicht bis zu Oligonucleotiden abgebaut wird. Ich habe deswegen in meinem Vortrag betont, daß das Kernsystem, im Moment zumindest, sich nur für kurzdauernde Versuche eignet. Wir sind dabei, die verschiedenen Faktoren, die für die Integrität des Systems eine Rolle spielen, zu untersuchen.

SIEBERT (Hohenheim): Vielleicht darf ich selber noch ein Wort dazu sagen. Die Zellkerne, so wie sie hier isoliert worden sind, sind sicher hochgradig extrahiert. Auf Grund der Isolierungsbedingungen kann man annehmen, daß eine ganze Menge der abbauenden Faktoren, die der intakte Zellkern hat, durch den Isolierungsgang bereits entfernt sind. Ich würde also vermuten, daß autolytische Vorgänge bei dieser Art der Zellkerndarstellung nicht so stark hervortreten.

WALLENFELS (Freiburg): Bei der Auswahl des Modells, wie man sich den Mechanismus der Induktionswirkung vorzustellen hat, ist es besonders wichtig zu wissen, wie die Synthese im nicht induzierten Zustand funktioniert. Sie haben ja ungefähr 50% der Synthese auch ohne daß Sie Cortisol zugeben. Glauben Sie, daß es sich hier um eine durch endogene Cortisolmenge induzierte Synthese handelt, die in Ihren Präparaten vorhanden ist oder hat man es bei dieser Basisniveausynthese mit einer prinzipiell anderen Art der Steuerung der Synthese zu tun?

SEKERIS: Das ist ein sehr wichtiger Punkt. Nicht nur die Corticosteroide, sondern auch andere Hormone beeinflussen den RNS-Stoffwechsel in der Leber. Das, was man einen „unangestellten Zustand" nennt, ist der Ausdruck des Zusammenspieles vieler Faktoren. Die Quantitierung des Effektes,

warum mehr Hormon zu einer größeren Stimulierung führt, ist noch ein
Problem der Spekulation.

WALLENFELS: Es handelt sich bei dieser Diskussion besonders um die
Frage: Wie stellt man die Beziehung her zwischen dem plötzlichen und ganz
eindrucksvollen Auftreten der Puffs und dem Anwachsen der enzymatischen
Aktivität? Wenn beim ruhenden Zustand 50% der Aktivität des induzierten
vorliegt, sollte man ohne Induktion sozusagen „Halbpuffs" haben.

SEKERIS: Ihre zweite Frage gibt uns Gelegenheit, auf den Unterschied der
beiden Systeme, Cortisolleber und Ecdysonepidermis, zurückzukommen. In
der Ratte gibt es immer einen gewissen Corticosteroidspiegel. Wenn man
dagegen den Ecdysonspiegel bei der Entwicklung von Calliphora erythrocephala verfolgt [Abb. bei SHAAYA, KARLSON, J. Ins. Physiol. **11**, 65 (1965)
sieht man, daß bei 6 bis 7 Tagen alten Larven der Ecdysonspiegel praktisch
0 ist, der dann aber rasch kurz vor der Verpuppung ansteigt. Dieser eindrucks-
volle Anstieg der Hormonspiegels führt zu einer Stimulierung der RNS-Synthese in der Epidermis und zu der Induktion der Synthese des Schlüsselenzy-
mes der Sklerotisierung der DOPA-Decarboxylase. Dieses Enzym ist in 6 bis 7
Tagen alten Larven nicht vorhanden. Wir haben Anhaltspunkte dafür, daß das
Ecdyson die Synthese der für die DOPA-Decarboxylase spezifischen Messenger-RNS stimuliert. Hier, im Gegensatz zum Cortisol-Leber-System, kann
man gut Stadien hohen Ecdysonspiegels mit Stadien vergleichen, bei denen
der endogene Hormonspiegel Null ist. Die Schwierigkeit liegt allerdings in
der Präparation der Tiere, da man mehrere Hundert benötigt, um genügend
Ausgangsmaterial zu erhalten.

MOSEBACH (Bonn): Haben Sie entsprechende Versuche gemacht, die die
DNS-Synthese betreffen, eventuell mit regenerierender Rattenleber?

SEKERIS: Nein.

MATTHAEI (Göttingen): Cortisol könnte primär die Synthese einer Initiator-
Transfer-RNS induzieren; dieses Genprodukt würde erst sekundär die Expression aller derjenigen Gene ermöglichen, von denen sich mRNS mit einem
Anfangscodon ableitet, das nur mit Hilfe dieser Initiator-RNS übersetzt
werden kann. Initiator-Transfer-Ribonucleinsäuren könnten in hoch diffe-
renzierten Organismen als Schlüssel zur Induktion der Genaktivitätsmuster
bestimmter Zelldifferenzierungen dienen. Cortisol induziert offenbar ein Gen-
muster und stimuliert die zellfreie RNS-Synthese mit einer Lag-Periode. Wir
kennen auch eine Reihe von Beobachtungen (z. B. TOMKINS), die eine Kop-
pelung der mRNS-Synthese an eine gleichzeitige Übersetzung der neusynthetisierten mRNS durch ein Ribosom vermuten lassen.

SCHEUCH (Göttingen): Sie haben die Cortisolwirkung funktionell an der
Steigerung der Tyrosintransaminaseaktivität gezeigt. Konnten Sie immuno-
logisch nachweisen, daß es sich tatsächlich um eine de-novo-Synthese von
Enzymen handelt? Wenn nicht, so kann die Hormonwirkung auch eine bloße
milieubedingte *Aktivitätssteigerung* der vorgegebenen Enzymmenge bedeuten.

SEKERIS: Ja, vielleicht kann Herr LANG dazu etwas sagen.

LANG (München): Die Induktion von Tyrosintransaminase durch m-RNS *in-vitro* ist nur unter den Bedingungen der Proteinsynthese möglich. Gabe von Hemmern der Proteinsynthese wie z. B. Puromycin oder Erythromycin verhindern diese Enzyminduktion (LANG, HERRLICH und SEKERIS: Acta Endocrinologia, im Druck).

Wir haben außerdem inzwischen spezifische Antikörper gegen Tyrosintransaminase hergestellt, mit deren Hilfe es gelingt, Tyrosintransaminase aus dem *in-vitro*-Inkubationsansatz durch Immunpräcipitation zu isolieren. Diese Präcipitate sind definitiv radioaktiv, was bedeutet, daß wir es mit Neusynthese von Enzymproteinen zu tun haben.

SIEBERT: Vielleicht darf ich selber ein Wort dazu sagen. Es mehren sich jetzt bei uns die Befunde, daß der Zellkern mit Substanzen aus dem extracellulären Raum in Austausch steht, ohne daß dabei das Cytoplasma durchschritten wird. Das kann für Sie interessant sein im Hinblick auf die Frage, wie das Hormon in den Zellkern hinein gelangt, ohne daß es im Cytoplasma metabolisiert wird, und weiter für Ihren Befund, daß einmal im Zellkern befindliches Hormon nicht verändert wird.

SEKERIS: Als ich von diesen Ergebnissen im letzten Jahr gehört habe, hat mich das sehr erfreut. Man könnte allerdings daran denken, daß das Cytoplasma eine regulatorische Funktion ausübt; wenn zuviel Hormon zugeführt wird, dann wird ein Teil im Cytoplasma inaktiviert.

KARLSON (Marburg): Herr JUNGBLUT hat uns gestern über Hormonreceptoren berichtet. Können Sie irgendwelche Parallelen dazu ziehen? Wissen Sie etwas darüber, ob in diesen isolierten Zellkernen eine Substanz vorhanden ist, die das Hormon bevorzugt bindet?

SEKERIS: Ja, man kann Bindungen von Cortisol an Makromoleküle des Zellkerns feststellen. Wenn man Cortisol *in-vivo* injiziert, die Zellkerne anschließend präpariert und das Chromatin daraus gewinnt, dann kann man ein gewisses Verteilungsmuster von Hormon — wir haben markiertes Cortisol verwendet — feststellen. Ein Teil des Hormons ist an basische Proteine gebunden, ein anderer Teil an saure oder neutrale Proteine des Zellkerns. In den Nucleinsäuren findet man praktisch keine Bindung [SEKERIS, C. E., u. N. LANG: Hoppe-Seylers Z. physiol. Chem. **340**, 92 (1965)]. Wenn man jetzt isolierte Kerne mit radioaktivem Hormon inkubiert und dann die gleiche Fraktionierung durchführt, findet man das gleiche Verteilungsmuster. Wenn man isoliertes Chromatin mit Hormon inkubiert, dann sieht man überraschenderweise wiederum dieselbe Verteilung. Was das bedeutet, ist uns noch nicht klar; aus den Verteilungsstudien allein kann man ja über die Spezifität der Bindung nichts aussagen.

KARLSON: Sie haben in einem Ihrer Schemata die RNA-Polymerase gezeigt, die nach ZILLIG aus sechs Untereinheiten besteht. Es ist wohl auch mit

der Möglichkeit zu rechnen, daß einige dieser Untereinheiten allosterische
Proteine sind, die die Repressorfunktion haben. Das würde bedeuten, daß der
Repressor nicht an der DNA angreift, sondern an der DNA-abhängigen RNA-
Polymerase und daß das Repressormolekül eine doppelte Spezifität trägt:
eine allgemeine für RNA-Polymerase und eine spezifische für den jeweiligen
Induktor.

SIEBERT: Herr SEKERIS, Sie haben in Ihrem Vortrag eine Reihe von Daten
gebracht über die Markierung verschiedener RNS-Fraktionen. Welche Ab-
solutbeträge der Markierung liegen vor, vor allem in der Messenger-RNS-
Fraktion? Es ist doch wohl so, daß man für eine „message" mit etwa 10^{-18} g
RNA auskommt. Es würde mich interessieren, wie sich die Mengen Messenger-RNS zu anderen RNS-Arten (ribosomale RNS) verhalten, und welche
absoluten Mengen Sie synthetisieren.

SEKERIS: Das ist schwer abzuschätzen, da wir in unserem System immer
mit einem turnover rechnen müssen und die absoluten Mengen der synthetisierten RNS nicht abschätzen können.

NEUBERT (Berlin): Haben Sie einmal mit adrenalektomierten Tieren gearbeitet? Sie könnten ja dann das endogene Hormon praktisch ausschließen
und vielleicht mit niedrigeren Dosen Cortisol auskommen. Die Cortisoldosen
in Ihren Experimenten waren ja erheblich höher als die physiologischen. Wir
haben in anderem Zusammenhang solche Versuche gemacht und gefunden,
daß bei adrenalektomierten Tieren die RNA-Polymeraseaktivität im Zellkern
erheblich niedriger ist, etwa nur die Hälfte der Kontrolle.

SEKERIS: In einigen dieser *in-vitro*-Versuche haben wir Kerne aus den
Lebern adrenalektomierter Tiere verwendet. Die Empfindlichkeit dieser
Kerne ist tatsächlich erheblich größer, genau wie Sie vermuten. Die Differenz
beträgt mehr als eine Zehnerpotenz [SCHMID, W., D. GALLWITZ, and C. E.
SEKERIS: Biochim. biophys. Acta (Amst.) **134**, 80 (1967)].

GUDER (München): Glauben Sie, daß die zeitliche Beschränkung ihres
isolierten Kernsystems durch Energiemangel begründet ist? Wir haben beobachtet, daß nach Triton-X-100-Behandlung die Kerne nicht mehr in der
Lage sind, ATP zu synthetisieren.

SEKERIS: Ich glaube, daß die Synthese energiereicher Phosphate im Leber-
zellkern praktisch keine Rolle spielt; das ATP wird vom Cytoplasma geliefert. In den Thymuskernen liegen die Verhältnisse allerdings anders.

SIEBERT: Vielleicht darf ich einen Punkt hierzu anfügen: Wir haben in
umfangreichen Studien über den Zellkern der Rattenleber keinen Anhalts-
punkt dafür gefunden, daß das nucleare ATP für irgendwelche metabolischen
Vorgänge im Zellkern limitierend ist.

MAURER (Heidelberg): *In-vivo*-Hormonexperimente bedürfen sorgfältiger Kontrollen, um nicht den Boden der biologischen Wirklichkeit zu verlieren. Eine Kontrolle ist in den eleganten Untersuchungen von C. E. SEKERIS u. Mitarb. aufgezeigt worden: die der *Hormonspezifität*. Eine andere betrifft die *Gewebespezifität*. Es wäre also zu prüfen, wie isolierte Zellkerne anderer Gewebe (als die der Erfolgsorgane) auf Hormongabe reagieren.

SEKERIS: Solche Versuche sind beabsichtigt.

MATTHAEI: Dr. ABERDO beobachtet in dem Placentasystem, daß der partielle Wiederabbau des dort gebildeten radioaktiven Proteins durch das Cortisol selbst offenbar stimuliert wird. Sie haben wenig über langfristige Versuche gesagt. Haben Sie vielleicht doch etwas derartiges beobachtet?

SEKERIS: Den Abbau des Enzyms haben wir nicht untersucht.

MOSEBACH: Wie fallen Ihre Versuche eigentlich aus, wenn Sie das Triton bei der Kernaufarbeitung weglassen, wenn sie also sozusagen membranverschmutzte Kerne haben?

SEKERIS: In diesem Fall ist der Einbau viel geringer und der Hormoneffekt ist nicht so deutlich.

EBEL (Straßburg): Ich möchte noch einmal auf die Synthese dieses Enzyms durch ihre Ribosomen zurückkommen. Es ist sehr schwierig, in einem tierischen System die Stimulierung der Proteinsynthese mit einem natürlichen Messenger zu bekommen. Die Stimulierung ist immer gut mit Poly-U und anderen Polyphosphaten, aber schlecht mit natürlichen Nucleinsäurefraktionen. Ich möchte Sie auch noch fragen, wie hoch die endogene Aktivität Ihrer Ribosomen war. Sie haben die Ribosomen mit Desoxycholat behandelt, so daß sie wahrscheinlich nur noch eine sehr geringe endogene Einbaurate hatten. Wie hoch war dann die Steigerung?

SEKERIS: Zum letzten Punkt zuerst: Mit den von uns verwendeten Desoxycholatkonzentrationen 0,25% wird der natürliche Messenger nicht entfernt. Wir bekommen immer einen Grundeinbau im Kontrollsystem, und der Zusatz von unseren Nucleinsäurefraktionen bewirkt eine Stimulierung um 20 bis 50%.

EBEL: Und in diesem zusätzlichen Protein ist dann das spezifische Enzym?

SEKERIS: Ja, zumindest die Enzymaktivität. Wir haben zunächst Enzymaktivitäten gemessen. Den eindeutigen Nachweis der Enzymsynthese werden hoffentlich die immunologischen Fällungsversuche von Herrn LANG ergeben.

KARLSON: Vielleicht darf ich dazu eine Bemerkung machen. Durch die Einführung des radioaktiven Substrats ist es möglich, noch sehr geringe Umsätze im enzymatischen Test zu erfassen und damit Enzymaktivitäten nachzuweisen, die mit den üblichen Methoden nicht meßbar sind.

STAUDINGER (Gießen): Ich wundere mich, daß Sie und viele andere die mit Rattenleber arbeiten, Cortisol verwenden, das für die Ratte charakteristische Hormon ist nicht Cortisol sondern Corticosteron. Es wäre möglich, daß Corticosteron sehr viel aktiver ist, es auch etwas anders wirkt.

SEKERIS: Wir haben die Frage auch öfter diskutiert. Man sollte wohl wirklich einmal auf Corticosteron umschalten.

SEUBERT (Frankfurt): Herr STAUDINGER, ich kann Ihnen da nicht ganz zustimmen. Dann dürfte man ja auch nicht mit Dexamethason oder anderen synthetischen Corticoiden arbeiten. Es kommt doch hier darauf an, den Hormoneffekt zu studieren.

SEKERIS: Dennoch glaube ich, daß vielleicht von Bedeutung sein könnte, das natürliche Hormon zu nehmen, besonders für solche Experimente, in denen wir die Bindung des Hormons an bestimmte Makromoleküle im Zellkern studieren wollen.

SIEBERT: Wenn keine weiteren Wortmeldungen sind, dann haben wir Herrn SEKERIS und allen Diskussionsteilnehmern noch einmal herzlich zu danken.

Cortisol als Enzyminduktor mit besonderer Berücksichtigung der Gluconeogenese

Von W. Seubert

Institut für vegetative Physiologie,
Chemisch-Physiologisches Institut der Universität Frankfurt am Main

Mit 10 Abbildungen

Als wichtiges Prinzip galt in der Biochemie für viele Jahre die These, daß Auf- und Abbau der Kohlenhydrate, der Fette und des Proteins über die einfache Umkehr einer Reaktionskette verlaufen. Für den Kohlenhydratstoffwechsel galten als experimentelle Stützen dieser These: die von Cori und Cori[1] nachgewiesene Umkehr der Spaltung von Glykogen zu Glucose-1-phosphat durch die Muskelphosphorylase (Gl. 1), die Synthese von Phosphoenolpyruvat (PEP) aus Pyruvat und ATP mit Hilfe der Pyruvatkinase (Gl. 2)[2], oder die Umkehr der Glykolyse bis zur Stufe des Fructose-1,6-diphosphates[3,4] (Gl. 3) mit Hilfe eines künstlich zusammengesetzten Enzymsystems.

$$\text{Glucose-1-P} + \text{Glykogen} \rightleftharpoons \text{Glykogen-Glucose} + \text{P} \qquad (1)$$

$$\text{Pyruvat} + \text{ATP} \xrightleftharpoons{\text{Mg}^{++},\ \text{K}^+} \text{Phosphoenolpyruvat} + \text{ADP} \qquad (2)$$

$$2\,\text{Pyruvat} + 4\,\text{ATP} + 2\,\text{DPNH} + 2\,\text{H}^+ + 2\,\text{H}_2\text{O} \qquad (3)$$
$$\updownarrow$$
$$\text{Fructose-1,6-diphosphat} + 2\,\text{DPN}^+ + 4\,\text{ADP} + 2\,\text{P}$$

In all den erwähnten Fällen war mit Hilfe der beim Kohlenhydratabbau eingeschalteten Enzyme ein partieller Aufbau aus den Abbauprodukten möglich.

Spätere Untersuchungen haben jedoch eindeutig ergeben, daß Synthese und Abbau der Glucose und des Glykogens auf teilweise getrennten Bahnen verlaufen: so wird Glykogen unter physiologischen Bedingungen nicht durch Umkehr der Phosphorylasereaktion, sondern über die UDP-Glucose[5] gebildet (Gl. 3 und 4).

$$UTP + Glucose\text{-}1\text{-}P \rightarrow UDP\text{-}Glucose + P\text{-}P \qquad (3)$$

$$UDP\text{-}Glucose + Glykogen \rightarrow Glykogen\text{-}Glucose + UDP \quad (4)$$

Auch die für den *Aufbau* der Glucose eingeschlagene Reaktionsfolge trennt sich von der Glykolyse auf den Stufen PEP/Pyruvat,
Fructose-6-phosphat/Fructose-1,6-diphosphat und Glucose/Glu-
cose-6-phosphat (Abb. 1): PEP wird durch Kopplung eines Car-

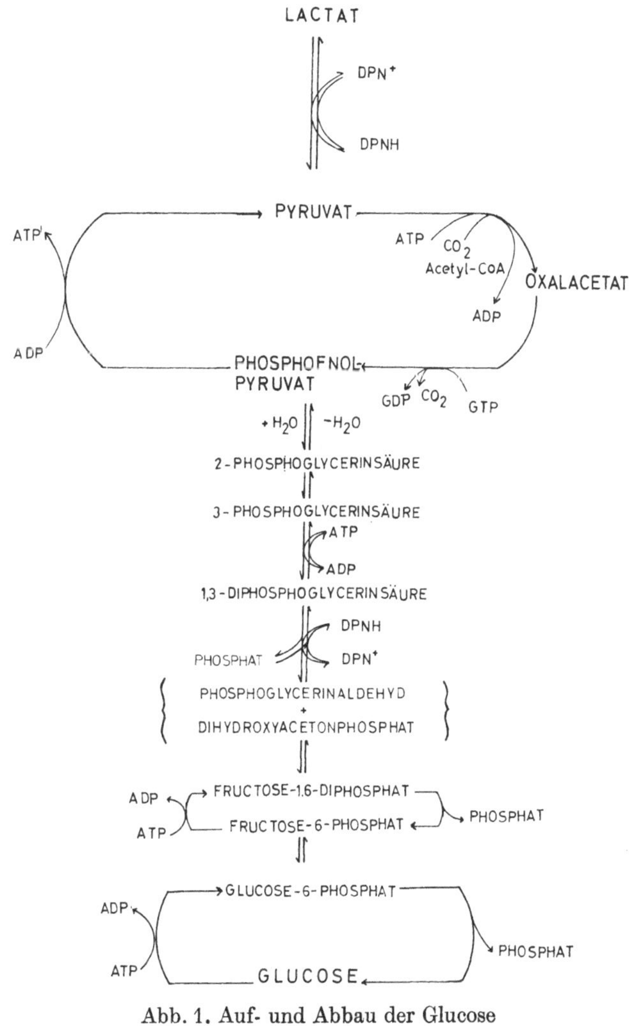

Abb. 1. Auf- und Abbau der Glucose

boxylierungs- und Decarboxylierungsschrittes synthetisiert. Fructose-6-phosphat und Glucose entstehen durch einfache Hydrolyse der Esterphosphatbindung des Fructose-1,6-diphosphates bzw. Glucose-6-phosphates. Der Abbau der Glucose auf diesen Stufen wird durch Kinasen vermittelt.

Die physiologische Bedeutung dieser Trennung von Synthese und Abbau läßt sich aus einem Vergleich der Energiebilanz der PEP-Synthese über die Umkehr der Pyruvatkinasereaktion (Gl. 2) mit der des *in vivo* eingeschlagenen Weges ableiten (Gl. 5 bis 7).

$$\text{Pyruvat} + CO_2 + \text{ATP} \xrightarrow[\text{Acetyl-CoA}]{Mg^{++}} \text{Oxalacetat} + \text{ADP} + P \quad (5)$$

$$\text{Oxalacetat} + \text{GTP} \xrightarrow{Mg^{++},\ Mn^{++}} \text{Phosphoenolpyruvat} +$$
$$+ CO_2 + \text{GDP} \quad (6)$$

$$\text{Bilanz: Pyruvat} + \text{ATP} + \text{GTP}$$
$$\Big\downarrow \begin{array}{l} Mg^{++},\ Mn^{++} \\ \text{Acetyl-CoA},\ CO_2 \end{array}$$
$$\text{Phosphoenolpyruvat} + \text{ADP} + \text{GDP} + P. \quad (7)$$

Dieser Vergleich macht klar, daß der unter physiologischen Bedingungen eingeschlagene Weg energetisch weit mehr begünstigt ist, da mit der Spaltung des GTP ein zusätzlicher Energiebetrag für den Aufbau des PEP zur Verfügung steht. Obwohl sich also die Umkehr der Pyruvatkinasereaktion (Gl. 2) erzwingen läßt, liegt unter physiologischen Bedingungen das Gleichgewicht dieser Reaktion auf Seiten des Pyruvats und begünstigt damit den *Abbau* der Glucose. Erst durch Kopplung zweier Nucleosidtriphosphatspaltender Reaktionen (Gl. 5 und 6) wird das Gleichgewicht zugunsten des PEP verschoben und damit der *synthetische* Prozeß begünstigt. Dieses Prinzip hat Gültigkeit nicht nur für den Kohlenhydratstoffwechsel, sondern auch für den Stoffwechsel der Fettsäuren, der Triglyceride, der Phosphatide, der Sphingophosphatide, der Terpene und des Proteins.

In vielen Fällen bestimmen derartige „one way"-Schritte als langsamste Teilschritte einer Reaktionsfolge die Geschwindigkeit eines komplexen Stoffwechselprozesses. Sie bieten somit einen idealen Angriffspunkt, Stoffwechselvorgänge zu steuern. Durch Drosselung oder Beschleunigung eines derartigen Schrittes kann z. B. die Synthese einer Verbindung verlangsamt bzw. beschleunigt werden, ohne daß dabei der Abbau beeinträchtigt wird. Dieser Prozeß kann sogar im entgegengesetzten Sinne gesteuert werden. So ist z. B. eine beschleunigte Lipogenese, Fettsäuresynthese, Gluconeogenese, Glykogenese stets mit einer Drosselung des Abbaues gekoppelt.

Hinsichtlich der Mechanismen, die den Umsatz von „one way"-Schritten steuern, hat man — abgesehen von der Steuerung von Transportvorgängen — folgende Möglichkeiten entdeckt[6]:

1. Die Induktion oder Repression der Synthese der bei diesen Schritten eingeschalteten Enzyme.

2. Die Aktivierung oder Hemmung dieser Enzyme.

3. Die Steuerung des Enzymabbaues.

Sämtliche Mechanismen können durch Hormone bzw. Metabolite in Gang gesetzt werden.

Die Steuerung des Kohlenhydratstoffwechsels durch das Cortisol, eines Vertreters der Glucocorticoide, ist ein Beispiel für die erwähnten ersten beiden Möglichkeiten der Stoffwechselsteuerung. Auch dieses Hormon stimuliert die Gluconeogenese durch *Induktion* geschwindigkeitsbestimmender Enzyme der Gluconeogenese und durch Aktivierung bzw. Hemmung von Enzymen des Kohlenhydratstoffwechsels.

Der tägliche Bedarf an Glucose für die Aufrechterhaltung der Lebensprozesse im Gehirn, den Nervenzellen und den Erythrocyten beträgt nach KREBS[7] beim Menschen etwa 130 g. Während das Herz, der ruhende Muskel und andere Organe ihren Energiebedarf bevorzugt aus der Verbrennung von Fettsäuren decken[8], sind die Nervenzellen und die Erythrocyten unbedingt von der Glucosezufuhr abhängig. Die Glucose ist weiterhin notwendig zur Deckung des Energiebedarfs des Muskels unter anaeroben Bedingungen, als Vorläufer von Aminozuckern, Uronsäuren und der Ribosebausteine der Nucleinsäuren. Im Ruhezustand wird dieser Bedarf an Glucose aus dem mit der Nahrung zugeführten Kohlenhydrat ausreichend gedeckt. Bei kohlenhydratarmer Kost, nach längerem Hunger oder körperlicher Betätigung, wenn die Glykogenreserven erschöpft sind, muß die Glucose durch die Gluconeogenese aus Protein oder Lactat bereitgestellt werden. Das für die Gluconeogenese verantwortliche Enzymsystem in Leber und Niere kann sich den jeweiligen Anforderungen anpassen. Die glucogene Kapazität dieses Enzymsystems steht unter der Kontrolle der Glucocorticoide und anderer Faktoren.

Dies geht aus dem vermehrten Einbau von markiertem CO_2, Alanin oder Pyruvat in Glucose und Glykogen hervor, wie er schon wenige Stunden nach Gaben von Cortisol oder nach längerem Hunger bei der Ratte beobachtet werden konnte[9, 32, 33]. Als eine der Ursachen dieser erhöhten Kapazität von Niere und Leber kam eine Beschleunigung der Umwandlung von Pyruvat bzw. Oxalacetat in PEP durch eine Hormon- bzw. hungerbedingte Aktivierung oder Induktion der PEP-carboxykinase (Gl. 6) und Pyruvatcarboxylase (Gl. 5) in Betracht. Eine erste experimentelle Stütze für diese Annahme lieferten zunächst Lardys[10, 11] und unsere[12, 13] Arbeitsgruppen (Tab. 1). So finden sich die Aktivitäten der löslichen PEP-carboxykinase und Pyruvatcarboxylase 6 Std nach einer einmaligen Dosis von Cortisol und nach Hunger etwa auf das Doppelte erhöht. Die Aktivitäten der mitochondrialen Pyruvatcarboxylase steigen erst nach längerem Hunger[13, 14] oder mehrtägiger Hormonbehandlung[15] an. Die beiden Enzyme der Niere verhalten sich ähnlich[13].

Die Aktivitäten der *löslichen* Pyruvatcarboxylase lassen sich optimal nur bei Verwendung salzhaltiger, isotoner Lösungen extrahieren[13]. Sie entsprechen etwa 40 bis 50% der gesamten in der Leber nachgewiesenen Pyruvatcarboxylase. Glutaminsäuredehydrogenase, ein ausschließlich in den Mitochondrien lokalisiertes Enzym, wird durch diese Behandlung nur zu einem geringen Prozentsatz ($< 3\%$) aufgeschlossen. Dieses unterschiedliche Verhalten von Pyruvatcarboxylase und Glutaminsäuredehydrogenase beim Aufschluß der Leber konnte zwei Ursachen haben:

1. unterschiedliche Löslichkeit der beiden *mitochondrialen* Enzyme,
2. eine mögliche Lokalisation der Pyruvatcarboxylase außer in den Mitochondrien auch in einem extramitochondrialen Kompartment.

Ein unterschiedliches Verhalten der mitochondrialen Pyruvatcarboxylase und Glutaminsäuredehydrogenase konnte ausgeschlossen werden. Extrahiert man isolierte Mitochondrien mit salzhaltigen, isotonen Lösungen, so ist der prozentuale Anteil der in Lösung gegangenen Enzymaktivitäten für beide Enzyme in der gleichen Größenordnung (Tab. 2). Die unterschiedliche Verteilung

Tabelle 1. *Wirkung von Cortisol und Hunger auf PEP-carboxykinase und Pyruvatcarboxylase der Rattenleber*

Autoren	Vorbehandlung der Versuchstiere	PEP-carboxykinase löslich mμMole/min/mg Protein	Pyruvatcarboxylase löslich μMole/min/g Leber	mitochondrial μMole/min/g Leber
LARDY et al. (1964)[10]	Adrenalektomiert	48	—	—
FOSTER et al. (1966)[11]	Adrenalektomiert + Cortisol (6 Std)	78	—	—
	Adrenalektomiert, Hunger (12 Std)	97,4	—	—
	Normal, Hunger (24 Std)	96,2	—	—
	μMole/min/g Leber			
HENNING et al. (1963)[12]	Adrenalektomiert	2,6	1,3	1,9
HENNING et al. (1966)[13]	Adrenalektomiert + Cortisol (6 Std)	5,7	2,3	2,0
	Normal, Hunger (72 Std)	6,3	3,1	6,9
FREEDMAN und KOHN (1964)[14]	Normal	—	—	1,41
	Normal, Hunger (24 Std)	—	—	2,44
WAGLE (1966)[15]	Normal	—	—	1,28
	Normal, Cortisol (5 Tage)	—	—	3,08

Tabelle 2. *Extraktion von Pyruvatcarboxylase und Glutaminsäuredehydrogenase aus Mitochondrien**

| Homogenisation min | Pyruvatcarboxylase | | Glutaminsäuredehydrogenase | | Verhältnis % Pyruvatcarboxylase/ % Glutaminsäuredehydrogenase |
	E/g Leber Feuchtgewicht	% des Gesamtgehaltes	E/g Leber Feuchtgewicht	% des Gesamtgehaltes	
1	0,97	25,3	0,55	15,7	1,6
2	1,42	36,8	1,05	30,0	1,2
4	1,94	50,5	1,80	51,4	0,98
6	2,22	57,8	2,10	60,0	0,96

* Sediment 1000 bis 12000 g. Aufschlußmedium (1:10) nach Henning et al.[13]. Die Gesamtaktivitäten der Enzyme wurden nach Aufschluß mit 0,1% Desoxycholsäure (DOC) ermittelt. Pyruvatcarboxylase: 3,85 E/g Leber; Glutaminsäuredehydrogenase: 3,5 E/g Leber. Die für die Aktivitätsbestimmung in den Versuchsansatz eingeschleppten DOC-Mengen zeigten keine Hemmwirkung.

Tabelle 3. *Intracelluläre Verteilung von Glutamatdehydrogenase und Pyruvatcarboxylase*
Nach Ohly und Seubert[31]

Fraktion*	DNA-Gehalt μg/mg N	Pyruvatcarboxylase E/mg Protein	Glutamatdehydrogenase E/mg Protein	Verhältnis Pyruvatcarboxylase/ Glutamatdehydrogenase
900 bis 12000 g (Leichte Mitochondrien)	14,8	0,072	0,100	0,72
12000 g (Zellkerne, leichte und schwere Mitochondrien)	162,0	0,06	0,038	1,58
900 g (Zellkerne und schwere Mitochondrien)	265,0	0,053	0,032	1,66

* Die einzelnen Fraktionen wurden nach Aufschluß der Leber mit 0,25 M Sucrose (1:4) durch Zentrifugieren bei den angegebenen g-Zahlen erhalten. Zur Extraktion der Enzyme wurden die Präparationen gefriergetrocknet und das Trockenpulver mit m/10 Tris-Puffer (pH 7,2 und m/1000 Glutathion) extrahiert.

von Pyruvatcarboxylase und Glutaminsäuredehydrogenase auf verschiedene Zellfraktionen spricht dagegen für ein zusätzliches extramitochondriales Vorkommen der Pyruvatcarboxylase. Wie aus Tab. 3 ersichtlich, verschiebt sich das Aktivitätsverhältnis Pyruvatcarboxylase/Glutaminsäuredehydrogenase mit zunehmendem DNA-Gehalt der Fraktionen zugunsten der Pyruvatcarboxylase. Inwieweit dieses Verhalten auf das Vorkommen der Pyruvatcarboxylase im Zellkern zurückzuführen ist, wird augenblicklich untersucht. Aber schon auf Grund der bisher vorliegenden Befunde können wir uns keinesfalls dem Postulat verschiedener Arbeitsgruppen anschließen[24, 18, 19, 23], wonach die Carboxylierung des Pyruvats ausschließlich auf die Mitochondrien beschränkt sein soll.

Als *einer* der Primäreffekte des Cortisols sollte der Anstieg der Pyruvatcarboxylase und PEP-carboxykinase in Leber und Niere auch zur erhöhten glucogenen Kapazität der beiden Organe beitragen. Um den Einfluß der *in vivo* in der Peripherie mobilisierten Fettsäuren auf den Kohlenhydratstoffwechsel auszuschließen (vgl. S. 170), ermittelten wir die glucogene Kapazität nach verschiedener Vorbehandlung der Tiere an Nierenrindenschnitten, da diese nach KREBS[24] gegenüber der Leber erhebliche Vorteile für das Studium der Gluconeogenese aufweisen. Succinat und Pyruvat kamen als Substrate zur Verwendung, da beide Verbindungen über die durch PEP-carboxykinase und Pyruvatcarboxylase katalysierten Schritte in die Glucose übergehen und außerdem die Zellmembran der Nierenrindenzellen leicht passieren. Das Ergebnis dieser Messungen ist in Abb. 2 zusammengefaßt: die glucogene Kapazität ist bei *optimalem* Angebot an Substrat 6 Std nach einer einmaligen Dosis von Cortisol und nach Hunger eindeutig erhöht. Die beobachtete Stimulierung der Gluconeogenese liegt etwa im Bereich der für die PEP-carboxykinase und die Pyruvatcarboxylase (170 bis 210% der Kontrollen) in Leber und Niere gefundenen Aktivitätsanstiege.

Da die beiden anderen Schlüsselenzyme der Gluconeogenese, Fructose-1,6-diphosphatase und Glucose-6-phosphatase, erst nach längerer Hormoneinwirkung ansprechen (vgl. hierzu [25]), haben wir aus den diskutierten Befunden auf die Umwandlung von *Oxalacetat in PEP* als geschwindigkeitsbestimmenden Schritt der Gluconeogenese bei einem Überangebot an Substrat geschlossen. Diese Annahme wird durch Untersuchungen von EXTON und PARK[26] an der perfundierten Leber gestützt (Tab. 4). Diese Autoren bestimmten den Gehalt der Leber an verschiedenen Zwischenprodukten der Gluconeogenese nach Durchströmung mit Lactat. Aus der *Konstanz* der Werte für Glucose-6-phosphat, Fructose-1,6-diphosphat und Triosephosphat nach Perfusion bei *steigenden* Konzentrationen an Lactat schlossen sie auf einen Schritt *vor* dem Triosephosphat als limitierenden Schritt der Gluconeogenese bei optimalem Substratangebot. Durch spätere Messungen von PEP, Malat und Pyruvat[28]

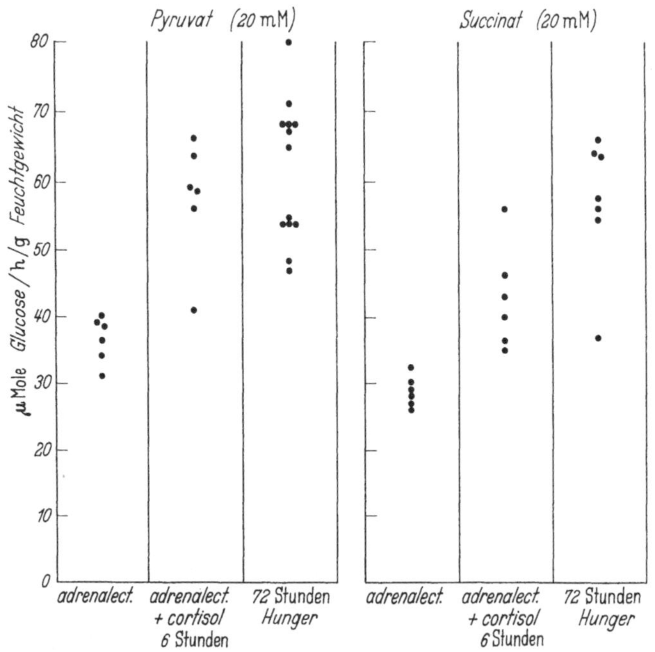

Abb. 2. Wirkung von Cortisol und Hunger auf die glucogene Kapazität der Nierenrinde (nach Henning et al.[13])

unter gleichen Bedingungen konnten sie diesen Schritt zwischen PEP und Pyruvat lokalisieren: Malat und Pyruvat fanden sich erhöht, PEP unverändert.

Die Synthese des PEP wird jedoch nicht nur durch die Aktivitäten der PEP-carboxykinase gesteuert, sondern auch durch die Pyruvatcarboxylase. Nur so ist die Drosselung der Gluconeogenese aus *Fumarat* (Tab. 5[34]) durch Hemmstoffe der Pyruvatcarboxylase verständlich. Nach allgemein gültigen Vorstellungen sollte wegen der hohen Aktivitäten an Fumarase und Malatdehydrogenase in

der Leber die Umwandlung von Fumarat in PEP unabhängig von
der Pyruvatcarboxylase sein. Ein Einfluß der Pyruvatcarboxylase
auf diesen Prozeß erschien uns nur über eine Steuerung der Oxalacetatspiegel denkbar, vorausgesetzt, daß die in der Zelle vorliegenden Konzentrationen des Oxalacetats zur Sättigung der PEP-
carboxykinase nicht ausreichen. Wir bestimmten daher den Ein-

Tabelle 4. *Glucoseproduktion aus Lactat und Spiegel glykolytischer Zwischen-
produkte in der perfundierten Rattenleber*
Nach EXTON und PARK[26]

Substrat	Glucose μMole/g Leber/h	Substratspiegel mμMole/g Trockengewicht		
		Glucose-6-phosphat	Fructose-1,6-diphosphat	Dihydroxyaceton-phosphat
—	14	130	25	63
Lactat 10 mM	60	257	56	110
Lactat 20 mM	62	262	51	105

Tabelle 5. *Wirkung von Phenylpyruvat auf Oxalacetat und Gluconeogenese aus
Fumarat in Nierenrindenschnitten*
Nach SEUBERT, HUTH und OHLY[31, 34]

Fumarat M	μMole Glucose/h/g Feuchtgewicht		mμMole Oxalacetat/100 mg Schnitte	
	ohne Hemmstoff	Phenylpyruvat 10^{-3} M	ohne Hemmstoff	Phenylpyruvat 10^{-3} M
10^{-3}	10,2	5,9	—	—
$2,5 \times 10^{-3}$	20,1	16,5	—	—
$5,0 \times 10^{-3}$	32,7	22,0	—	—
10^{-3}	39,9	25,2	0,57*	0,61*

* Mittelwerte aus sieben bzw. vier Versuchen

fluß von Phenylpyruvat, eines Hemmstoffes der Pyruvatcarboxylase, auf den Oxalacetatgehalt von Nierenrindenschnitten in Gegenwart von Fumarat als glucogenem Substrat. Eine Senkung der
Oxalacetatspiegel in Gegenwart des Hemmstoffes ließ sich nicht
demonstrieren (Tab. 5). Der Drosselung der Gluconeogenese aus
Fumarat durch Hemmstoffe der Pyruvatcarboxylase muß demnach
ein anderer, noch unbekannter Mechanismus zugrundeliegen.

Ursprünglich hatte man die Wirkung der Glucocorticoide auf den Kohlenhydratstoffwechsel ausschließlich auf die Enzyminduktion in der Leber zurückgeführt. Daß diese Annahme nicht gerechtfertigt ist, geht aus Untersuchungen der Arbeitsgruppe um Lardy[36] hervor. Lardy u. Mitarb. untersuchten den Einfluß des Actinomycins, eines Hemmstoffes der Proteinsynthese, auf die durch Cortisol ausgelöste Induktion der PEP-carboxykinase und die vermehrte Glykogensynthese in der Leber. Bei *vollständiger* Unterdrückung des Enzymanstiegs durch eine Dosis von 175 μg fanden

Tabelle 6. *Wirkung von Cortisol und Actinomyzin auf Blutglucose, Glykogen und PEP-carboxykinase in der Leber adrenalektomierter, gehungerter (24 Std) Ratten*

Nach Ray et al.[36]

Vorbehandlung	8 Std nach Behandlung		
	Blutglucose mg-%	Leberglykogen mg-%	PEP-carboxykinase mμMole/min/mg Protein
NaCl-Lösung	52	etwa 10	100
Actinomycin (174 μg/Tier)	53	etwa 10	81
Cortisol (10 mg/Tier)	76	2100	109
Cortisol + Actinomycin	68	830	81

sie jedoch nur eine *partielle* Unterdrückung (60%) der Glykogensynthese (Tab. 6). Dieser Befund ließ außer auf die Steuerung der Enzymsynthese auch auf die Steuerung der Aktivitäten schon vorhandener Enzyme durch das Hormon schließen. Zwei Mechanismen kommen hierfür in Betracht: eine Hemmung beim Glucoseabbau eingeschalteter Enzyme bzw. eine Aktivierung glucogener Enzyme. Nach bisher vorliegenden Befunden erscheint eine Aktivierung glucogener Enzyme *in vivo* nur auf der Stufe Pyruvat/Oxalacetat eingeschaltet. Die Pyruvatcarboxylase wird nicht nur induziert, sondern auch durch einen Metaboliten, das Acetyl-CoA, aktiviert[37, 38, 39]. Nach Messungen von Henning und Seubert[39] liegt der Aktivierungsbereich des Enzyms der Rattenleber zwischen 3×10^{-5} bis etwa 10^{-4} M Acetyl-CoA (Abb. 3). Die tierische Pyruvatcarboxylase zeigt im Gegensatz zum bakteriellen Enzym eine absolute Abhängigkeit von Acetyl-CoA (Abb. 3).

Die von SCHONER et al.[9] mit Hilfe einer Isotopenmethode[41] ermittelten Acetyl-CoA-Werte in der Leber liegen im Bereich des Anstiegs der Acetyl-CoA-Aktivierung der Pyruvatcarboxylase. Cortisol und Hunger erhöhen Acetyl-CoA in der Leber [9, 60, 63, 64]. Der Effekt des Hormons ist schon 3 Std nach Gabe des Hormons schwach sichtbar und erreicht nach 6 Std bis zu 150% der Kon-

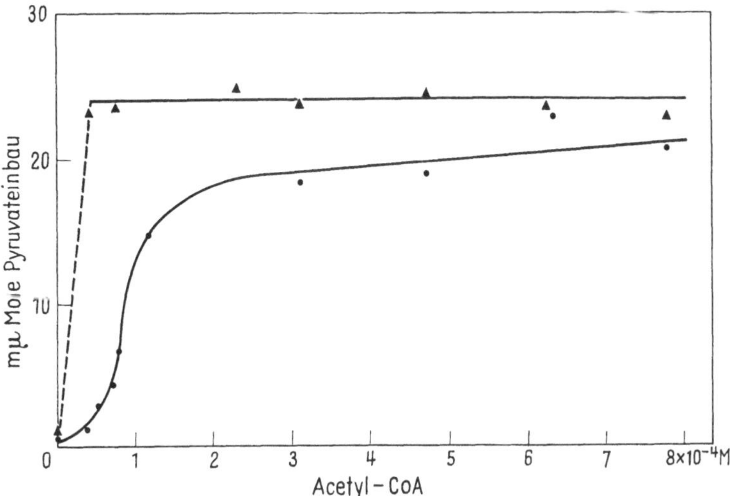

Abb. 3. Abhängigkeit des Pyruvateinbaus von der Acetyl-CoA-Konzentration im Carboxylasetest nach HENNING und SEUBERT[39]. Untere Kurve: tierische Carboxylase. Obere Kurve: Pyruvatcarboxylase aus *Pseudomonas citronellolis*

trolltiere. Die Fixierung von radioaktivem CO_2 in Glykogen entspricht etwa der vermehrten Glykogenablagerung in der Leber (Abb. 4). Der Anstieg des Glykogens unter Cortisol ist demnach nicht nur ein Maß für eine beschleunigte Umwandlung von Glucose in Glykogen, sondern auch ein Maß für eine vermehrte Gluconeogenese.

Inwieweit Änderungen des Acetyl-CoA-Gehaltes der Leber in den beobachteten Bereichen (40 bis 70 mμMole/g Feuchtgewicht) die Gluconeogenese beeinflussen, geht aus Untersuchungen von KREBS et al.[42] hervor. Nach Tab. 7 wird die Neusynthese der Glucose aus

Lactat in Nierenrindenschnitten des Kaninchens durch Acetessigsäure um das Fünffache beschleunigt. Ähnliche Befunde erhielt
Krebs beim Studium der Wirkung von Acetat, Propionat, Crotonat
und anderer Fettsäuren auf die Gluconeogenese aus Lactat in

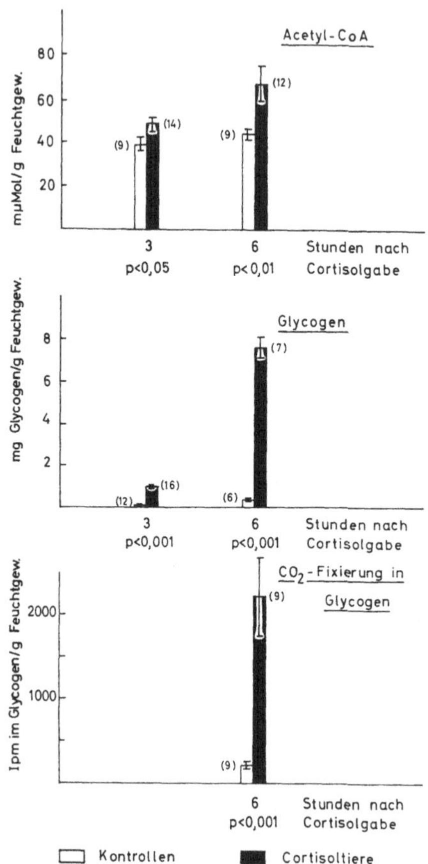

Abb. 4. Wirkung von Cortisol auf Acetyl-CoA, Glykogen und die CO_2-
Fixierung in das Glykogen in der Rattenleber (nach Schoner et al.[9]). Versuchsbedingungen: adrenalektomierte Tiere, 18 Std Hunger, 5 mg Cortisol/
100 g Körpergewicht mit Schlundsonde. Kontrollen 0,5 ml 0,9% Kochsalz-
lösung. Entnahme der Lebern 3 bzw. 6 Std nach Versuchsbeginn in Pheno-
barbitalnarkose (150 mg/100 g Körpergewicht) mit Gefrierstopptechnik. Zur
Ermittlung der CO_2-Fixierung wurden 30 min vor Entnahme der Lebern
6,5 μC KHC^{14}O$_3$ (spez. Akt. 0,53 μC/μMol) intraperitoneal injiziert. Die an-
gebenen Werte sind Mittelwerte $\pm$ σ_M

Nierenrindenschnitten. Die stimulierende Wirkung dieser Fettsäuren konnte auch bei Sättigung des glucogenen Enzymsystems mit Substrat beobachtet werden. Ein „sparing"-Effekt der Fettsäuren als Ursache der vermehrten Gluconeogenese scheidet demnach unter diesen Bedingungen aus. KREBS hat daher auch eine Aktivierung der Pyruvatcarboxylase durch vermehrt gebildetes Acetyl-CoA als Ursache der Stimulierung der Glucosebildung diskutiert. Er konnte diese Annahme experimentell stützen: gleichlaufend mit der vermehrt gebildeten Glucose fand sich auch Acetyl-CoA in der Zelle erhöht (Tab. 7).

Die glucogene Kapazität der Leber wird aber nicht nur durch eine hormonbedingte Aktivierung und Induktion glucogener En-

Tabelle 7. *Einfluß von Acetessigsäure auf Gluconeogenese und Acetyl-CoA-Spiegel in Nierenschnitten des Kaninchens*
Nach KREBS et al.[42]

Substrat	μMole/g Feuchtgewicht (nach 90 min, 40°)	
5 mM	Glucose	Acetyl-CoA
ohne Substrat	0,4	0,016
Lactat	1,9	0,016
Acetacetat	0,7	0,036
Lactat + Acetacetat	10,1	0,023

zyme, sondern auch durch Drosselung des Glucoseabbaues erhöht[94]. Zwei Mechanismen kommen hierfür in Betracht:

1. die Repression der Synthese glykoloytischer Schlüsselenzyme (Glucokinase, P-Fructokinase, Pyruvatkinase),
2. die Hemmung dieser Enzyme durch Metabolite.

Im Hungerzustand und bei der diabetischen Stoffwechsellage sind vermutlich beide Mechanismen für die Drosselung der Glykolyse in der Leber verantwortlich[43, 44]. Die Wirkung des Cortisols äußert sich lediglich in der *Hemmung* von Enzymaktivitäten durch Metabolite. Die möglichen Effektoren und deren Angriffspunkte sind in Tab. 8 zusammengefaßt: nach Messungen von TARNOWSKI[49, 50] bleiben ATP, ADP und AMP unter dem Einfluß von Cortisol unverändert. Eine Steuerung der Glykolyse in der Leber an der P-Fructokinase durch Variation der ATP-, ADP- und AMP-Spiegel unter dem Einfluß von Cortisol ist daher unwahrscheinlich. Inwieweit Änderungen des Gehaltes an cyclischem AMP die Aktivitäten der P-Fructokinase beeinträchtigen, kann vorläufig nicht entschieden werden, da entsprechende Messungen bis jetzt noch nicht vorliegen. Eine Hemmung glykolytischer Enzyme durch unveresterte Fettsäuren im Blut ist ebenfalls auszuschließen, da die zur Hemmung erforderlichen Konzentrationen[46] weit oberhalb der physiologischen Konzentrationen liegen[51, 52]. Der von Tarnowski et al.[50] gefundene und von uns bestätigte[9] Anstieg des Citrats unter dem Einfluß des Cortisols in der Leber wäre dagegen für eine Hemmung der P-Fructokinase ausreichend[50, 53–55]. Das Gleiche gilt auch für die Hemmung der Pyruvatdehydrogenase und Pyruvatkinase

Tabelle 8. *Kontrolle von Schlüsselenzymen des Glucoseabbaues in der Leber*

Enzym	Inhibitor	Aktivator
Glucokinase	Fettsäuren[46]	—
P-Fructokinase	ATP[45]	ADP, 5′-AMP
	Citrat[45]	3′, 5′,-AMP, anorganisches P
	Fettsäuren[46]	Fructose-6-P[45]
Pyruvatkinase	ATP[34]	
	Fettsäuren[46]	Fructose-1,6-diphosphat[40]
	DPNH[82]	
	Acetyl-CoA[9]	
Pyruvatdehydrogenase	Acetyl-CoA[47]	—
	DPNH[48]	

durch Acetyl-CoA, das — wie schon berichtet — ebenfalls unter dem Einfluß von Cortisol in der Leber ansteigt (K_I der Pyruvatdehydrogenase für Acetyl-CoA = $1{,}25 \times 10^{-5}$ M)[56]. Inwieweit erhöhte DPNH-Spiegel nach Gaben von Glucocorticoiden[30] die Aktivitäten der Pyruvatdehydrogenase beeinträchtigen, kann erst nach Ermittlung der für die Hemmung des tierischen Enzyms erforderlichen DPNH-Konzentrationen entschieden werden.

Ein Anstieg des Citrats wurde unter verschiedenen, durch vermehrte Gluconeogenese charakterisierten Stoffwechselsituationen beobachtet (Fluoracetatvergiftung[57, 76, 77], Diabetes[54, 58], Hunger[54, 58, 60], Cortisolbehandlung[9, 49, 50], Fettsäuren *in vivo* und *in vitro*[54, 58, 61, 62], Glucagon[60, 65]). In all den untersuchten Fällen war er begleitet von einem Anstieg der Hexosephosphate. Dieser Anstieg wurde mit einer Drosselung der Glykolyse auf der Stufe Fructose-6-phosphat/Fructose-1,6-diphosphat durch Citrathemmung der P-

Fructokinase erklärt. Wie aus Abb. 5 ersichtlich, findet sich mit einer Verzögerung von etwa 2 Std der Gehalt der Leber an Glucose-6-phosphat ebenfalls unter dem Einfluß des Cortisols erhöht. Gleichzeitig steigt auch Citrat in der Leber an. Auf den ersten Blick scheint demnach eine Beziehung zwischen Glucose-6-phosphat- und

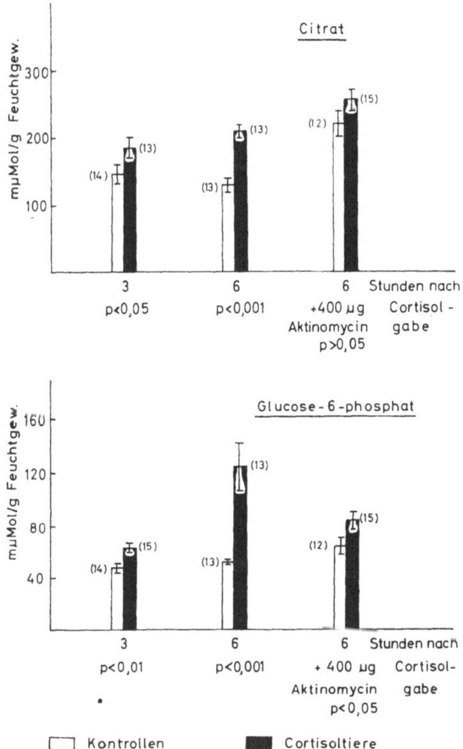

Abb. 5. Wirkung von Cortisol und Actinomycin D auf den Citrat- und Glucose-6-phosphatgehalt der Rattenleber (nach Schoner et al.[9]). Versuchsbedingungen wie bei Abb. 4. Zur Untersuchung der Actinomycinwirkung wurde der Hemmstoff 30 min vor Cortisolgabe intraperitoneal injiziert. Die angegebenen Werte sind Mittelwerte $\pm \sigma_M$

Citratanstieg zu bestehen. Beim späteren Studium der Wirkung von Actinomycin auf den Gehalt der Leber an verschiedenen Metaboliten fanden wir jedoch bei unverändertem Glucose-6-phosphat weit höhere Citratspiegel (Abb. 5). Die von verschiedenen Autoren diskutierte Rolle des Citrats als Regulator der Glykolyse erscheint uns daher — zumindest unter der Wirkung des Cortisols — als fraglich. Zur gleichen Schlußfolgerung kamen auch Tarnowski[95] und Wieland[96] auf Grund anderer Untersuchungen.

Die in der Leber nachgewiesenen Aktivitäten der Pyruvatkinase (Gl. 2) und PEP-carboxykinase (Gl. 6) liegen bei 55 E/g Feuchtgewicht[100] bzw. 3,9 E/g Feuchtgewicht[101]. Beide Enzyme finden sich bei der Ratte im Cytoplasma der Zelle. Die Aktivitäten der Pyruvatkinase wären also ausreichend, um auch unter Cortisol ver-

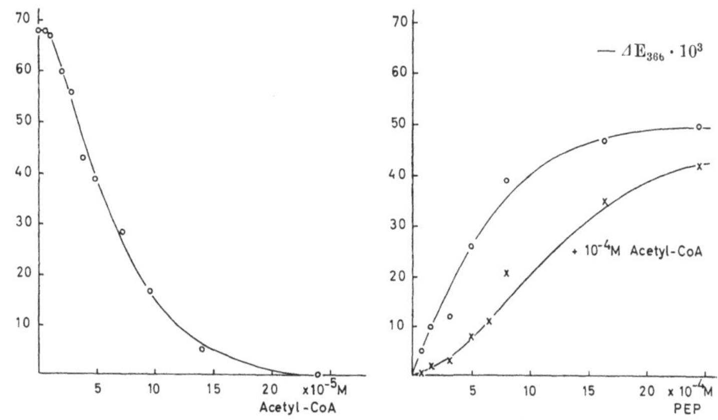

Abb. 6. Hemmung der Pyruvatkinase der Rattenleber durch Acetyl-CoA (nach Schoner et al.[9]). Versuchsbedingungen: Acetyl-CoA-Abhängigkeit: 0,04 ml Cytosol, 66 mM Tris pH 7,2, 6,6 mM MgSO₄, 6,6 mM KCl, 0,33 mM DPNH, 1,12 mM ADP, 0,49 mM PEP, 6 Einheiten Lactatdehydrogenase. Phosphoenolpyruvatabhängigkeit: 0,02 ml Cytosol, 66 mM Tris pH 7,2, 6,6 mM MgSO₄, 6,6 mM KCl, 0,33 mM DPNH, 1,68 mM ADP, 6 Einheiten Lactatdehydrogenase. Leberaufschluß 1:10 mit 0,25 M Sucrose

mehrt gebildetes PEP wieder zu Pyruvat abzubauen. Die kombinierte Wirkung der Pyruvatcarboxylase (Gl. 5), PEP-carboxykinase (Gl. 6) und Pyruvatkinase (Gl. 2) würde damit letztlich in der Spaltung von GTP zu GDP und anorganischem Phosphat resultieren. Unter dem Einfluß des Cortisols wird die Synthese der Pyruvatkinase nicht reprimiert[44]. Es mußte also ein anderer Mechanismus unter diesen Bedingungen für eine Inaktivierung der Pyruvatkinase Sorge tragen. Wir haben daher eine mögliche Hemmung der Pyruvatkinase durch diejenigen Substrate untersucht, die unter dem Einfluß des Cortisols in der Leber ansteigen. Von den bisher untersuchten Metaboliten erwies sich das Acetyl-CoA als Hemmer des Enzyms der Leber. In Abb. 6 ist die Geschwindigkeit der Spaltung von PEP in Abhängigkeit von der PEP-Konzentration mit und ohne Acetyl-CoA graphisch dargestellt. Wie ersichtlich, beeinflußt Acetyl-CoA die Bindung des Substrates durch das Enzym. Die Hemmung der Pyruvatkinase durch Acetyl-CoA zeigt sich im Bereich der Konzentrationen (Abb. 6), die zur Aktivierung der Pyruvatcarboxylase erforderlich sind (Abb. 3). Der von TARNOWSKI et al.[50] nachgewiesene Anstieg des PEP ist daher vermutlich nicht nur das Resultat einer beschleunigten PEP-Synthese, sondern auch eines gedrosselten Abbaues dieser Verbindung.

Tabelle 9. *Wirkung von Caprylat auf den Stoffwechsel von $C^{14}O_2$ und Pyruvat in Mitochondrien*
Nach WALTER, PAETKAU and LARDY[21]

	μMole/min/g Leber	
	ohne Caprylat	mit Caprylat (2 mM)
Pyruvat verbraucht	3,62	2,13
$C^{14}O_2$-Einbau in Malat, Citrat und Fumarat	1,40	1,73
Verhältnis Pyruvat verbraucht/ $C^{14}O_2$-Einbau	2,59	1,23

Auch für eine Drosselung des Glucoscabbaues auf der Stufe der Pyruvatdecarboxylierung unter dem Einfluß des Cortisols liegen inzwischen experimentelle Beweise vor: der Anstieg des Acetyl-CoA nach Cortisolgaben ist sicherlich die Folge der lipolytischen Wirkung des Cortisols am Fettgewebe[60, 63, 64, 71]. WALTER et al.[21] fanden bei Inkubation von Mitochondrien mit Pyruvat und radioaktiver Kohlensäure in Gegenwart von Caprylsäure die CO_2-Fixierung in Malat, Citrat und Fumarat erhöht, gleichzeitig aber die Pyruvatverwertung erniedrigt (Tab. 9). LARDY schloß aus diesem Befund auf einen Pyruvat-„sparing"-Effekt der Fettsäuren, der die Pyruvatcarboxylierung und damit die Synthese der C_4-Dicarbonsäuren begünstigt. Es ist möglich, daß dieser Wirkung der Fettsäuren auf die Oxydation der Brenztraubensäure ebenfalls eine Hemmung der Pyruvatdehydrogenase durch erhöhtes Acetyl-CoA zugrunde liegt.

Die vermehrte Bereitstellung von Reduktionsäquivalenten aus der Beta-Oxydation der mobilisierten Fettsäuren begünstigt außerdem durch Verschiebung des Redoxpotentials des DPN[+]-Systems nach negativeren Werten die Synthese der Glucose auf der Stufe der Triosephosphatdehydrogenase[70, 81]. Eine ähnliche Funktion übernimmt bei der Gluconeogenese aus den Aminosäuren die mit der Harnstoffsynthese gekoppelte Regenerierung von Aspartat aus

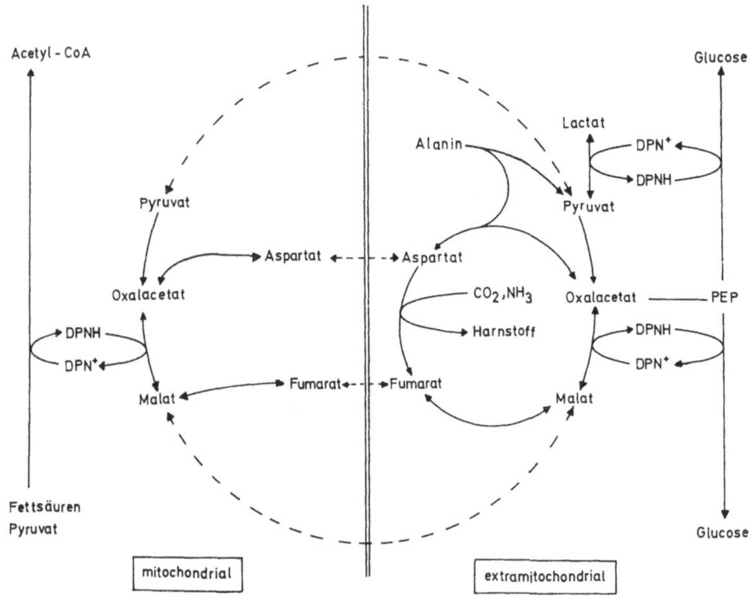

Abb. 7. Extra- und intramitrochondriale Regenerierung des für die Gluconeogenese erforderlichen Wasserstoffes

Fumarat[21]. Intramitochondrial in Freiheit gesetzter Wasserstoff wird mit dem Malat in den extramitochondrialen Raum transportiert. Die für die Regenerierung und den Transport des Wasserstoffs von verschiedenen Autoren postulierten Mechanismen sind in Abb. 7 zusammengefaßt[21, 70, 73, 75]. Das Schema unterscheidet sich prinzipiell in einem Punkt, nämlich einer möglichen Carboxylierung von Pyruvat auch im extramitochondrialen Raum. Diese Annahme erscheint nicht nur auf Grund unserer Untersuchungen über die Lokalisation der Pyruvatcarboxylase gerechtfertigt. Auch ältere Isotopenversuche lassen sich mit einer ausschließlich mitochondri-

alen Carboxylierung des Pyruvats nicht vereinbaren: Asparaginsäure und Malat, die als Transportformen der Oxalessigsäure beim
Passieren der mitochondrialen Membran diskutiert werden, setzen
sich vor Umwandlung in die Glucose vollkommen ins Gleichgewicht
mit der symmetrisch aufgebauten Fumarsäure, d. h. die Radioaktivität einer 3-C^{14}-Asparaginsäure bzw. -Äpfelsäure findet sich
gleichmäßig auf die 1,2- und 5,6-Position der Glucose verteilt[34, 97, 98].
Die Äquilibrierung des entsprechenden C-Atoms der Milchsäure ist
dagegen unvollständig. Auch aus diesem unterschiedlichen Verhalten geht hervor, daß die Oxalacetatbildung aus Lactat im Cytoplasma nicht unbedingt intramitochondrial über Aspartat und
Malat und deren Transport in den cytoplasmatischen Raum verlaufen muß, sondern auch extramitochondrial möglich ist (Abb. 7).
Dies erscheint sinnvoll, wenn man berücksichtigt, daß der für die
Gluconeogenese aus Lactat und den Aminosäuren Alanin, Serin,
Cystein und Glycin erforderliche Wasserstoff an der Lactatdehydrogenase vollständig bzw. über die mit der Harnstoffsynthese gekoppelte Regenerierung der Asparaginsäure aus Fumarat zum
größten Teil extramitochondrial regeneriert wird.

Wie schon erwähnt, mobilisieren die Glucocorticoide die Fettsäuren in der Peripherie. Die negative bzw. positive Kontrolle
der Glykolyse und Gluconeogenese durch die unveresterten Fettsäuren steht demnach in enger Beziehung zur lipolytischen Wirkung des Cortisols am Fettgewebe. Nach Untersuchungen von
FAIN[72] wird die vom Wachstumshormon und den Glucocorticoiden abhängige Freisetzung von Fettsäuren an isolierten
Fettzellen im Gegensatz zur ACTH-Wirkung durch Actinomycin
vollständig unterdrückt (Tab. 10). Dieser Befund läßt wiederum
auf eine Enzyminduktion im Fettgewebe als primäre Ursache der
Fettsäuremobilisierung unter der Wirkung des Cortisols schließen.
Er stand im Widerspruch zu unserem Postulat, nach dem die nach
Gaben von Actinomycin beobachtete Stimulierung der Gluconeo-
genese durch Cortisol als Stimulierung durch mobilisierte Fettsäuren gedeutet wurde. Um diesen Widerspruch zu klären, untersuchten wir den Einfluß steigender Actinomycindosen auf den
Citrat- und Acetyl-CoA-Anstieg in der Leber nach Cortisolgaben.
Nach Abb. 8 reichen die von LARDYS Arbeitsgruppe verabreichten
Mengen an Actinomycin jedoch nicht aus, um den Anstieg der verschiedenen Metaboliten in der Leber vollständig zu unterdrücken.

Erst bei einer Verdoppelung bleibt der Anstieg des Acetyl-CoA und des Citrats aus; der Anstieg des Glucosephosphats ist nur mehr schwach signifikant. Das gebildete Glykogen fällt auf etwa 10% des unter Cortisol beobachteten Wertes ab. Ähnlich verhält sich die Fixierung radioaktiver Kohlensäure in das Glykogen. Wir möchten uns daher dem Postulat von Fain anschließen, demzufolge auch die Fettsäurefreisetzung aus dem Fettgewebe unter Cortisol von einer Enzyminduktion abhängt. Die — verglichen mit der Leber — zur Unterdrückung dieser Wirkung des Cortisols erforderlichen höheren

Tabelle 10. *Wirkung von Actinomycin auf die Fettsäurefreisetzung durch Dexamethason, Wachstumshormon und ACTH an isolierten Fettzellen* Nach Fain, Kovacev, Scow[72]

Hormone	Actinomycin M	Freisetzung von Fettsäuren nach 4 Std μMole/mMol Triglycerid
Keines	0	332
Dexamethason (0,016 μg/ml)	0	397
Wachstumshormon (1 μg/ml)	0	487
Dexamethason (0,016 μg/ml) + Wachstumshormon (1 mg/ml)	0	731
Dexamethason + Wachstumshormon	$1,4 \times 10^{-6}$	81
ACTH (0,001 μg/ml)	0	1128
ACTH	$1,4 \times 10^{-6}$	1127

Actinomycindosen sind wahrscheinlich auf die geringere Durchblutung des Fettgewebes zurückzuführen.

Zusammenfassung: Für die erhöhte glucogene Kapazität der Leber und Niere innerhalb von 6 Std nach Applikation von Cortisol sind die folgenden Wirkungen des Hormons verantwortlich:

1. Induktion der PEP-carboxykinase und Pyruvatcarboxylase,

2. Aktivierung der Pyruvatcarboxylase durch erhöhte Acetyl-CoA-Spiegel,

3. Drosselung des Glucoseabbaues auf den Stufen Fructose-6-phosphat/Fructose-1,6-diphosphat, PEP/Pyruvat und der Pyruvatdecarboxylierung. Für die Drosselung des Glucoseabbaues auf den beiden letzten Stufen sind vermutlich ebenfalls die erhöhten Acetyl-

CoA-Spiegel verantwortlich. Der dem Glucose-6-phosphatanstieg zugrunde liegende Steuermechanismus bedarf noch der Aufklärung. Die unter 2. und 3. aufgeführten Effekte sind die Folge der lipolytischen Wirkung des Cortisols und daher Sekundäreffekte des Hormons.

Voraussetzung für eine vermehrte Synthese der Glucose ist jedoch nicht nur eine erhöhte glucogene Kapazität der für die Gluco-

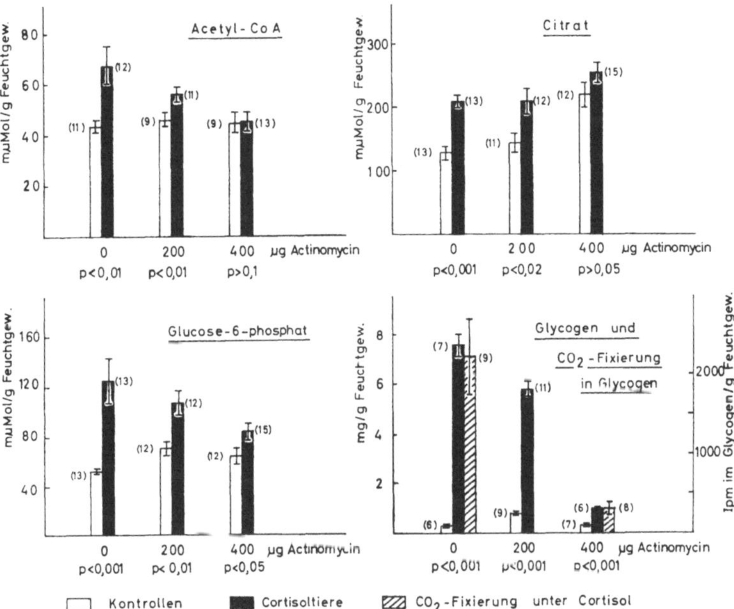

Abb. 8. Wirkung von Actinomycin *in vivo* auf den Cortisol-abhängigen Anstieg von Acetyl-CoA, Citrat, Glucose-6-phosphat, Glykogen und die CO_2-Fixierung in das Glykogen 6 Std nach Cortisolgabe in der Rattenleber (nach SCHONER et al.[9]). Versuchsbedingungen s. Abb. 4. Actinomycin wurde 30 min vor Cortisolgabe intraperitoneal injiziert. Die angegebenen Werte sind Mittelwerte $\pm\ \sigma_M$

neogenese verantwortlichen Organe, sondern auch ein ausreichendes Angebot an Substrat. Diesen Bedarf decken die glucogenen Aminosäuren, die unter dem Einfluß der Glucocorticoide innerhalb weniger Stunden vermehrt in der Peripherie in Freiheit gesetzt werden. Dies geht aus Messungen individueller Aminosäuren in Leber, Thymus, Muskel und Plasma hervor, wie sie von BETHEL

12*

et al.[83] 4 Std nach Applikation von Cortisol in der Ratte vorgenommen wurden. Von den untersuchten Aminosäuren fanden sich alle in Muskel und Thymus erhöht (Tab. 11). In der Leber und im Plasma stiegen bevorzugt Glutaminsäure, Asparaginsäure und Alanin an. Die übrigen Aminosäuren zeigten ein unterschiedliches Verhalten. Die hohen Aktivitäten an Glutamat-Pyruvat- und Glutamat-Oxalacetat-Transaminase in der Leber gewährleisten eine rasche Bildung von Alanin und Asparaginsäure aus Glutamat und den entsprechenden Alpha-Ketosäuren. Verschiedene Autoren[84, 85, 87] bringen daher den bevorzugten Anstieg von Glutaminsäure, Alanin und Asparaginsäure sowie die Aminosäuremobilisierung in der Peripherie in enge Beziehung zur Induktion der Tyrosin-Ketoglutarat-Transaminase (Gl. 8) und Tryptophanpyrrolase (Gl. 9).

$$\text{Tyrosin} + \text{Alpha-Ketoglutarat} \rightleftharpoons \text{Hydroxyphenylpyruvat} + \\ + \text{Glutamat} \tag{8}$$

$$\text{Tryptophan} + O_2 \rightarrow \text{Formylkynurenin.} \tag{9}$$

Diese beim Aminosäureabbau eingeschalteten Enzyme steigen innerhalb von 5 bis 6 Std nach Cortisolgaben auf das Drei- bis Zehnfache an. Ähnliche Ergebnisse erhielt Schole beim Studium der Aktivitäten der L-Aminosäureoxydase unter dem Einfluß des Cortisols[86]. Einige der erwähnten Autoren[84, 85, 87] nehmen an, daß es infolge eines vermehrten Abbaues von Tyrosin, Tryptophan und Phenylalanin unter Cortisol zu Änderungen der relativen Konzentrationen freier Aminosäuren zugunsten der Glutaminsäure kommt. Dieses Mißverhältnis soll die Proteinsynthese in der Peripherie beeinträchtigen und zur vermehrten Freisetzung der Aminosäuren führen. Folgende Befunde stützen diese Annahme (Tab. 12):

1. Die Oxydation von C^{14}-Tryptophan findet sich nach Erhöhung der Tryptophanpyrrolase *in vivo* durch Verabreichung von unmarkiertem Tryptophan auf das Siebenfache beschleunigt[87, 88]. Der oxydative Abbau dieser Aminosäuren wird demnach durch die Aktivitäten der Tryptophanpyrrolase limitiert. Da der Gehalt der Zelle an Tryptophan verhältnismäßig gering ist, kann eine Senkung der Tryptophanspiegel zu einer verminderten Proteinsynthese führen.

2. Der Anstieg des Aminosäure-N in der Leber durch Cortisolgaben wird durch Actinomycin reprimiert[89]. Dieser Befund steht

Tabelle 11. *Wirkung von Cortison auf Aminosäurespiegel in verschiedenen Geweben der Ratte*
Nach Bethel et al.[83]

Aminosäuren	μMole/100 g Gewebe oder Plasma							
	Leber		Plasma		Muskel		Thymus	
	Kontrolle	Cortison (4 Std)	Kontrolle	Cortison (4 Std)	Kontrolle	Cortison (4 Std)	Kontrolle	Cortison (4 Std)
Glu, Asp, Ala	371	483	38	46	265	328	721	824
Gly, His, Isol, Leu, Lys, Phe, Pro, Ser, Thre, Tyr, Val	—	—	—	—	1017	1125	434	474

Tabelle 12. *Wirkung von Tryptophan, Actinomycin und Glucocorticoiden auf den Aminosäurestoffwechsel und die Lympho-cytolyse in vivo*

Autoren	Behandlung	Untersuchte Wirkung	%
Moran und Sourkes[88]	Keine	Oxydation von C^{14}-Tryptophan (0,2 mg) nach 1 Std	6 (C^{14} im CO_2 ausgeatmet)
	Tryptophan (1 mM)	Oxydation von C^{14}-Tryptophan (0,2 mg) nach 1 Std	35 (C^{14} im CO_2 ausgeatmet)
Weber et al.[89]	Keine	Aminosäure-N der Leber	100
	Triamcinolon, 6 Std (10 mg/100 g Körpergewicht)	Aminosäure-N der Leber	164
	Triamcinolon + Actinomycin (100 μg/100 g Körpergewicht)	Aminosäure-N der Leber	112
Hofert und White[90]	Laparatomie + Salzlösung	Lymphocyten	125 des Ausgangswertes (nach 3 Std)
	Laparatomie + Cortisol (5 mg/100 g Körpergewicht)	Lymphocyten	37 des Ausgangswertes (nach 3 Std)
	Totale Evisceration + Cortisol	Lymphocyten	85 des Ausgangswertes (nach 3 Std)
	Totale Evisceration + Cortisol	Lymphocyten	111 des Ausgangswertes (nach 6 Std)

ebenfalls mit einer Enzyminduktion als Ursache der Aminosäure-freisetzung in Einklang.

3. Die nach Cortisolgaben beobachteten Gewichtsverluste von Thymus und Lymphknoten sowie die Lymphocytolyse steuern nach Feigelson und Feigelson[84] ebenfalls zu den erhöhten Aminosäurespiegeln in der Peripherie bei. Diese Wirkung des Cortisols ist am eviscerierten Tier fast vollständig aufgehoben (Tab. 12). Hofert und White[90] schließen daher auf einen in der Leber gebildeten Faktor als Ursache der vermehrten Lymphocytolyse unter Cortisol. Diese Annahme wurde von Feigelson und Feigelson[84] gestützt: Viele der Wirkungen des Cortisols (verminderte Protein-

Tabelle 13. *Immunchemische Titration der Tyrosintransaminase und Tryptophanpyrrolase in der Rattenleber*

Autoren	Enzym	Verhältnis: Hormon beh. Tier/Kontrolle	
		Enzymaktivitäten	Antigentiter
Feigelson und Greengard[93]	Tryptophanpyrrolase	4,6	5,0
Kenney[74]	Tyrosintransaminase	10,4	10,4

synthese im lymphatischen Gewebe) lassen sich durch Applikation von Glutaminsäure nachahmen.

Als Kriterium einer vermehrten *de-novo*-Synthese eines Enzyms wird in vielen Fällen die Repression des Aktivitätsanstiegs durch einen Hemmstoff der Proteinsynthese wie z. B. Actinomycin oder Puromycin, bewertet. Im Falle der Cortisolwirkung liegen aber auch direkte Beweise für eine Rolle dieses Hormons als Enzyminduktor vor. Mit Hilfe serologischer Methoden ließ sich am Beispiel der Tyrosin-Transaminase und der Tryptophanpyrrolase eindeutig demonstrieren, daß der Aktivitätsanstieg nach Cortisolgaben einem Anstieg der Enzymmenge entspricht: titrierte man die Leberextrakte von Normal- und Hormon-behandelten Tieren mit Antiseren, die durch Immunisierung von Kaninchen mit dem jeweiligen gereinigten Enzym hergestellt worden waren, so fand sich eine gute Übereinstimmung zwischen Antigentiter und Aktivitätsgehalt (Tab. 13). Mit Hilfe der Isotopentechnik konnte außerdem gesichert werden, daß der Anstieg der Tryptophanpyrrolase und Tyrosintransaminase nach Cortisolgaben das Ergebnis einer vermehrten Synthese und nicht eines gedrosselten Abbaues der beiden Enzyme ist[79, 85, 99].

Mit der *in vivo* beobachteten Wirkung des Cortisols auf die Aktivitäten verschiedener Enzyme in Leber und Niere war aber noch nicht geklärt, ob das Hormon primär die *de-novo*-Synthese von Enzymen steuert, oder ob der beobachtete Anstieg der verschiedenen Enzyme das Resultat einer veränderten Stoffwechsellage ist. Verschiedene Arbeitsgruppen haben daher versucht, an der isoliert perfundierten Leber oder an Leberschnitten Hormoneffekte zu demonstrieren. Derartige Bemühungen waren erfolgreich. So ließ sich an der perfundierten Leber und an Leberschnitten in Gegenwart von Glucocorticoiden ein vermehrter Einbau von radioaktivem CO_2[107], Alanin[66, 102–107, 120] und Pyruvat[107, 120] in die Glucose demonstrieren. Erhöhte Enzymaktivitäten nach Hormonbehandlung *in vitro* ließen sich jedoch nur an der perfundierten Leber[108, 109], nicht dagegen an Leberschnitten nachweisen[110]. Auch eigene Versuche, nach Inkubation von Leberschnitten mit Cortisol oder Triamcinolon einen Anstieg der Pyruvatcarboxylase zu demonstrieren, schlugen fehl. Kinetische Messungen ergaben sogar einen raschen Abfall der Pyruvatcarboxylaseaktivitäten, bedingt durch die zunehmende Schädigung des Gewebes mit zunehmender Inkubationsdauer[111]. Unsere Bemühungen, *in vitro* einen Hormoneffekt zu demonstrieren, konzentrierten sich daher auf den Nachweis eines derartigen Effektes an Nierenrindenschnitten[112]. Dieses Gewebe ist widerstandsfähiger als die Leber und übersteht eine mehrstündige Inkubation bei hoher Schüttelfrequenz (150/min) weit besser. In Abb. 9 ist das Ergebnis eines Versuches mit Nierenrindenschnitten graphisch dargestellt[113]: nach 2stündiger Vorinkubation der Schnitte und anschließender weiterer Inkubation (2 Std) mit Pyruvat (20 mM) und radioaktivem Bicarbonat (10 mM) fand sich in Gegenwart von Cortisol die Fixierung von radioaktivem CO_2 signifikant erhöht. Die Wirkung des Hormons ist im Bereich der in der Leber nachgewiesenen Konzentrationen sichtbar[25]. Cortisol kann durch Triamcinolon und Dexamethason ersetzt werden. Das *in vivo* unwirksame Tetrahydrocortisol und 11-Desoxycorticosteron erwies sich auch in unserem System als unwirksam[113]. Die vermehrte CO_2-Fixierung wurde in Gegenwart von Puromycin, eines Hemmstoffes der Proteinsynthese, vollständig unterdrückt. Dieser

Befund stand mit einer Enzyminduktion als Ursache der vermehrten CO_2-Fixierung im Einklang. Diese Annahme ließ sich auch durch Messungen der Pyruvatcarboxylaseaktivitäten bestätigen: die Aktivitäten fanden sich nach 2stündiger Inkubation von Nierenrindenschnitten in Gegenwart von Cortisol erhöht. Puromycin

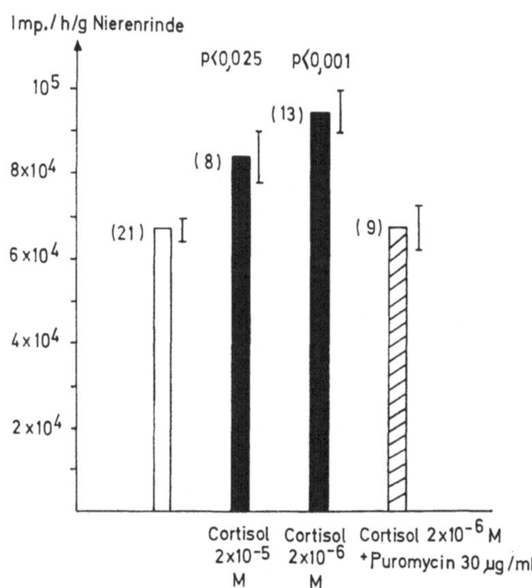

Abb. 9. Wirkung von Cortisol und Puromycin auf die Fixierung von radioaktivem CO_2 durch Nierenrindenschnitte (nach L'AGE et al.[113]). Zusammensetzung der Versuchsansätze: 8 ml Phosphatpuffer nach KREBS[119], 200 mg Nierenrindenschnitte adrenalektomierter, gehungerter (12 bis 15 Std) Ratten, 0,6 mg Pepton/ml, übrige Komponenten wie angegeben. Die Schnitte wurden zunächst 2 Std bei 37° unter Schütteln (150/min) inkubiert, anschließend weitere 2 Std mit 20 mM Pyruvat, 5 mM Glucose und 10 mM $KHC^{14}O_3$ (0,1 µC/µMol). Gasphase: 100% O_2. Nach Inkubation wurden die Schnitte abzentrifugiert und nach Austreiben des CO_2 die Radioaktivität bestimmt. Die Zahl der Versuche ist in Klammern angegeben. Die angegebenen Werte sind Mittelwerte $\pm$ σM

hob die Wirkung des Hormons auf (Tab. 14). Insulin konnte den Enzymanstieg in Gegenwart des Cortisols nicht unterdrücken. Eine Repressorfunktion dieses Hormons, wie sie neuerdings diskutiert wird[114], ist daher unwahrscheinlich.

Die *in vitro* beobachteten Effekte des Cortisols auf die CO_2-Fixierung liegen nach 4 Std bei 135 bis 140% der Kontrollen, der Anstieg der Pyruvatcarboxylase nach 2 Std bei 120 bis 125% der Kontrollen (bei Ermittlung der Enzymaktivitäten mußten die Versuche wegen zunehmender Schädigung der Schnitte früher abgebrochen werden). Hinsichtlich des Ausmaßes der Cortisoleffekte sind also Schnitte dem *in-vivo*-System unterlegen. Trotzdem erwiesen sich die Nierenrindenschnitte als wertvolles Studienobjekt, da bei Ermittlung der optimalen Zusammensetzung des Inkuba-

Tabelle 14. *Wirkung von Cortisol, Insulin und Puromycin auf die Aktivitäten der Pyruvatcarboxylase in Nierenrindenschnitten adrenalektomierter Ratten* Nach M. L'AGE, H. V. HENNING, B. OHLY und W. SEUBERT[113]

Versuchsgruppe	In vitro (2 Std)			
	n	E/g Feuchtgewicht*	P	% der Kontrolle
Kontrolle	26	2,98 $\pm$ 0,09		100
Puromycin (30 μg/ml)	7	3,04 $\pm$ 0,19	> 0,1	102
Insulin (250 mE/ml)	13	3,03 $\pm$ 0,09	> 0,1	102
Cortisol (2 $\times$ 10^{-5} M)	19	3,49 $\pm$ 0,15	< 0,005	117
Cortisol + Insulin	7	3,68 $\pm$ 0,08	< 0,001	123,5
Cortisol + Puromycin	8	2,92 $\pm$ 0,18	> 0,100	98

Zusammensetzung der Versuchsansätze: 8 ml Krebs-Henseleit-Puffer; 400 mg Nierenrindenschnitte adrenalektomierter, gehungerter (12 bis 15 Std) Ratten*; 5 $\times$ 10^{-3}M Pyruvat; 0,6 mg Pepton/ml; übrige Komponenten wie angegeben. Die Schnitte wurden 2 Std unter Schütteln (150/min) inkubiert. Gasphase: 95% O_2, 5% CO_2. Zur Bestimmung der Pyruvatcarboxylase wurde bei 15000 g zentrifugiert und der lösliche Anteil der Pyruvatcarboxylase, wie früher beschrieben, extrahiert[13].

---

* Mittelwerte $\pm$ σM.

tionsmediums sich neue Hinweise hinsichtlich des Wirkungsmechanismus des Cortisols ergaben. Um optimale und vor allen Dingen reproduzierbare Ergebnisse bei den *in-vitro*-Versuchen zu erhalten, erwies sich der Zusatz eines Aminosäuregemisches als unbedingt erforderlich (Abb. 10). Das Aminosäuregemisch vermag Pyruvat nicht zu ersetzen, denn ohne den Zusatz dieses Substrates ist die Pyruvatcarboxylase nicht mit Substrat gesättigt und der Cortisoleffekt nicht sichtbar. Demnach sind die Aminosäuren beim Induktionsprozeß direkt beteiligt. Zu einem ähnlichen Ergebnis kamen

Peraino et al.[115] beim Studium der Induktion von Tyrosintransaminase *in vivo*. Nach mehrtägiger Verabreichung einer protein-
freien Diät an Ratten konnten diese Autoren einen Cortison-
abhängigen Anstieg der Tyrosintransaminase nur bei gleichzeitiger
Verfütterung eines Caseinhydrolysates beobachten.

Es war von Interesse, zu prüfen, ob das Aminosäuregemisch
durch einzelne Aminosäuren ersetzt werden kann. Nach vorläufigen

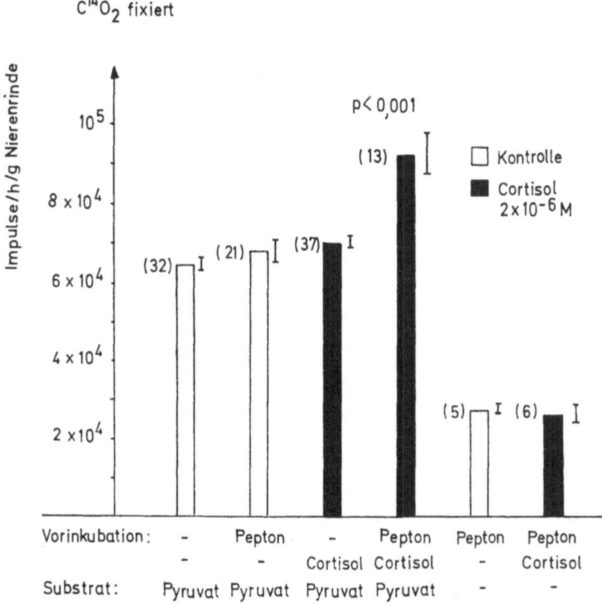

Abb. 10. Einfluß von Aminosäuren auf die Cortisolwirkung *in vitro* (nach
L'age et al.[113]). Zusammensetzung der Versuchsansätze und Inkubations-
bedingungen vgl. Abb. 9

Ergebnissen mit Gemischen von drei bis fünf Aminosäuren scheint
das der Fall zu sein. Die Aufschlüsselung nach einzelnen Aminosäuren ist jedoch noch nicht abgeschlossen. Auch Befunde von
Kröger et al.[116, 117] sprechen für eine Beteiligung einzelner Aminosäuren beim Induktionsprozeß: bei Verabreichung geringer Hormondosen an Ratten (2,5 mg/kg) konnten diese Autoren einen Anstieg der Tyrosintransaminase nur bei gleichzeitiger Verabreichung
von Tyrosin beobachten (Tab. 15). Der von Substrat *und* Hormon
abhängige Anstieg der Enzymaktivitäten wurde durch Actinomycin

Tabelle 15. *Kombinierte Wirkung von Cortisol und Substraten auf Enzymaktivitäten in der Leber adrenalektomierter Ratten*

Autor	Enzym	Substrat bzw. Hormon	E/g Leber
KRÖGER et al.[116]	Tyrosintransaminase	Kein	0,25
KRÖGER und GEUER[117]	Tyrosintransaminase	Cortison	0,22
	Tyrosintransaminase	L-Tyrosin	0,32
	Tyrosintransaminase	Cortison + Tyrosin	0,74
	Tyrosintransaminase	Cortison + Tyrosin + Actinomycin	0,36
MARVER et al.[119]	δ-Aminolävulinsäuresynthetase	Kein	18,8
	δ-Aminolävulinsäuresynthetase	Allylisopropylacetamid	41,7
	δ-Aminolävulinsäuresynthetase	Cortisol	32,2
	δ-Aminolävulinsäuresynthetase	Allylisopropylacetamid + Cortisol	167,6

reprimiert. Ursache des Enzymanstiegs ist demnach eine vermehrte Synthese des Emzyms. Zu einem ähnlichen Ergebnis kamen Marver et al.[118] beim Studium der Induktion der Delta-Aminolävulinsäuresynthetase durch Allylisopropylacetamid. Auch in diesem Falle konnte erst die kombinierte Verabreichung des Substrates mit dem Hormon den Enzymanstieg auslösen (Tab. 15). Die Rolle des Cortisols bei diesem Induktionsprozeß ist noch völlig unklar. Ohne Zweifel bieten sich aber hier gänzlich neue Ansatzpunkte, an zellfreien Systemen mögliche Wirkungen des Cortisols zu studieren.

Die in der vorliegenden Arbeit zitierten Untersuchungen aus dem eigenen Arbeitskreis wurden in großzügiger Weise von der Deutschen Forschungsgemeinschaft unterstützt.

Literatur

1 Cori, G. T., and E. F. Cori: J. biol. Chem. **134**, 733 (1940).

2 Lardy, H. A., and J. Ziegler: J. biol. Chem. **159**, 343 (1945).

3 Krimsky, I.: J. biol. Chem. **234**, 232 (1959).

4 Mendicino, J., and M. F. Utter: J. biol. Chem. **237**, 1716 (1962).

5 Leloir, L. F., and C. E. Cardini: J. Amer. chem. Soc. **79**, 6340 (1957).

6 Karlson, P.: Dtsch. med. Wschr. **86**, 668 (1961).

7 Krebs, H. A.: Proc. roy. Soc. B, **159**, 545 (1963).

8 Fritz, I. B.: Physiol. Rev. **41**, 52 (1961).

9 Schoner, W., W. Prinz, U. Haag und W. Seubert: In Vorbereitung.

10 Lardy, H. A., D. O. Foster, E. Shrago, and P. D. Ray: Advanc. Enzymol. **2**, 35 (1964).

11 Foster, D. O., P. D. Ray, and H. A. Lardy: Biochemistry **5**, 555 (1966).

12 Henning, H. V., I. Seiffert, and W. Seubert: Biochim. biophys. Acta. (Amst.) **77**, 345 (1963).

13 —, B. Stumpf, B. Ohly, und W. Seubert: Biochem. Z. **344**, 274 (1966).

14 Freedman, A. D., and L. Kohn: Science **145**, 58 (1964).

15 Wagle, S. R.: Diabetes **15**, 19 (1966).

16 Siebert, G., and G. B. Humphrey: Advanc. Enzymol. **27**, 232 (1965).

17 L'age, M., H. Brod, and W. Seubert: Unveröff. Versuche.

19 Keech, D. B., and M. F. Utter: J. biol. Chem. **238**, 2609 (1962).

18 Lardy, H. A., V. Paetkau, and P. Walter: Proc. nat. Acad. Sci. (Wash.) **53**, 1410 (1965).

20 Shrago, E., and H. A. Lardy: J. biol. Chem. **241**, 663 (1966).

21 Walter, P., V. Paetkau, and H. A. Lardy: J. biol. Chem. **241**, 2523 (1966).

22 Haynes, R.: J. biol. Chem. **240**, 4103 (1965).

23 Struck, E., J. Ashmore, and O. Wieland: Advanc. Enzymol. **4**, 219 (1966).

24 Krebs, H. A., D. A. H. Benett, P. de Gasquet, T. Gascoyne, and T. Yoshida: Biochem. J. **86**, 22 (1963).

25 HÜBENER, H. J.: Dtsch. med. Wschr. **87**, 438 (1962).

26 EXTON, J. H., and C. R. PARK: J. biol. Chem. **240**, 955 (1965).

27 DI PIETRO, D. L.: J. biol. Chem. **239**, 4051 (1964).

28 EXTON, J. H., and C. R. PARK: Fed. Proc. **24**, 537 (1965).

29 WIELAND, O., u. G. LÖFFLER: Biochem. Z. **339**, 204 (1963).

30 YOUNG, D.: Arch. Biochem. **114**, 309 (1966).

31 OHLY, B., u. W. SEUBERT: Unveröff. Versuche.

32 FRIEDMAN, B., E. H. GOODMAN, Jr., and S. WEINHOUSE: J. biol. Chem. **240**, 3729 (1965).

33 ASHMORE, I., A. B. HASTINGS, F. B. NESBETT, and A. E. RENOLD: J. biol. Chem. **218**, 77 (1956).

34 SEUBERT, W., u. W. HUTH: Biochem. Z. **343**, 176 (1965).

35 HERNBROOK, K. R., H. B. BURCH, and O. H. LOWRY: Biochem. biophys. Res. Commun. **18**, 206 (1965).

36 RAY, P. D., D. O. FOSTER, and H. A. LARDY: J. biol. Chem. **239**, 3396 (1964).

37 UTTER, M. F., and D. B. KEECH: J. biol. Chem. **235**, 17 (1960).

38 KEECH, D. B., and M. F. UTTER: J. biol. Chem. **238**, 2609 (1963).

39 HENNING, H. V., u. W. SEUBERT: Biochem. Z. **340**, 160 (1964).

40 TAYLOR, C. B., and E. BAILEY: Biochem. J. **102**, 32c (1967).

41 PRINZ, W., W. SCHONER, U. HAAG und W. SEUBERT: Biochem. Z. **346**, 206 (1966).

42 KREBS, H. A., R. N. SPEAKE, and R. HEMS: Biochem. J. **94**, 712 (1965).

43 WEBER, G., R. L. SINGHAL, N. B. STAMM, M. A. LEA, and E. A. FISHER: Advanc. Enzymol. **4**, 59 (1966).

44 WEBER, G., M. A. LEA, G. A. FISHER, and N. B. STAMM: Enzym biol. clin. **7**, 11 (1966).

45 STADTMAN, E. R.: Advanc. Enzymol. **28**, 41 (1966).

46 WEBER, G., H. W. CONVEY, M. A. LEA, and N. B. STAMM: Science **154**, 1357 (1966).

47 GARLAND, P. B., and P. J. RANDLE: Biochem. J. **91**, 6c (1964).

48 HANSEN, R. G., and U. HENNING: Biochim. biophys. Acta (Amst.) **122**, 355 (1966).

49 TARNOWSKI, W., M. KITTLER, and H. HILZ: Abstracts of the 2nd Meeting of the Federation of Biochem. Soc., p. 87, Wien 1965.

50 TARNOWSKI, W.: Habilitationsarbeit, Universität Hamburg, 1966.

51 WALES, C. N., and G. C. KENNEDY: Biochem. J. **90**, 620 (1964).

52 WILLIAMSON, J. R., D. H. WRIGHT, W. J. MALAISSE, and J. ASHMORE: Biochem. biophys. Res. Commun. **24**, 765 (1966).

53 UNDERWOOD, A. H., and E. A. NEWSHOLME: Biochem. J. **95**, 868 (1965).

54 PARMEGGIANI, A., and R. H. BOWMAN: Biochem. biophys. Res. Commun. **12**, 268 (1963).

55 POGSON, C. I., and P. J. RANDLE: Biochem. J. **100**, 683 (1966).

56 GARLAND, P. B., and P. J. RANDLE: Biochem. J. **91**, 6c (1964).

57 BOWMAN, R. N.: Biochem. J. **93**, 13c (1964).

58 GARLAND, P. B., and P. J. RANDLE: Biochem. J. **93**, 678 (1964).

59 NEWSHOLME, E. A., and P. J. RANDLE: Biochem. J. **93**, 641 (1964).

[60] Williamson, J. R., B. Herczeg, H. Coles, and R. Danish: Biochem. biophys. Res. Commun. **24**, 437 (1966).

[61] Newsholme, E. A., and A. H. Underwood: Biochem. J. **99**, 24c (1966).

[62] Williamson, J. R.: J. biol. Chem. **240**, 2308 (1965).

[63] Bortz, W. M., u. F. Lynen: Biochem. Z. **339**, 77 (1963).

[64] Wieland, O. H., and L. Weiss: Biochem. biophys. Res. Commun. **10**, 333 (1963).

[65] Williamson, J. R.: Biochem. J. **101**, 11c (1966).

[66] Eisennstein, A. B.: Advanc. Enzymol. **3**, 121 (1965).

[67] Haft, D. E.: Biochim. biophys. Acta (Amst.) **104**, 581 (1965).

[68] Koeppe, R. E., G. A. Mourkides, and R. J. Hill: J. biol. Chem. **234**, 2219 (1959).

[69] Freedman, A. D., D. R. Rimsay, and S. Graff: J. biol. Chem. **235**, 1854 (1960).

[70] Struck, E., J. Ashmore, and O. Wieland: Advanc. Enzymol. **4**, 219 (1966).

[71] Fain, J. N., R. O. Scow, and S. S. Chernick: J. biol. Chem. **238**, 54 (1963).

[72] —, V. P. Kovacev, and R. O. Scow: J. biol. Chem. **240**, 3522 (1965).

[73] Krebs, H. A., T. Gascoyne, and B. M. Notton: Biochem. J. **102**, 275 (1967).

[74] Kenney, F. T.: J. biol. Chem. **237**, 1611 (1962).

[75] Haynes, R. C.: J. biol. Chem. **240**, 4103 (1965).

[76] Williamson, J. R., E. A. Jones, and G. F. Azzone: Biochem. biophys. Res. Commun. **17**, 696 (1964).

[77] Peters, R. A.: Advanc. Enzymol. **18**, 113 (1957).

[78] Rosell-Perez, M., C. Villar-Palasi, and J. Larner: Biochemistry **1**, 763 (1962).

[79] Kenney, F. T.: J. biol. Chem. **237**, 3495 (1962).

[80] Struck, E., J. Ashmore und O. Wieland: Biochem. Z. **343**, 107 (1965).

[81] Williamson, J. R., R. A. Kreisberg, and P. W. Felts: Proc. nat. Acad. Sci. (Wash.) **56**, 247 (1966).

[82] Weber, G., R. L. Singhal, N. B. Stamm, and S. K. Srivastava: Fed. Proc. **24**, 745 (1965).

[83] Bethel, J. J., M. Feigelson, and P. Feigelson: Biochim. biophys. Acta (Amst.) **104**, 92 (1965).

[84] Feigelson, M., and P. Feigelson: J. biol. Chem. **241**, 5819 (1966).

[85] Knox, W. E., and O. Greengard: Advanc. Enzymol. **3**, 247 (1965).

[86] Schole, J.: Autoreferate, Tagung der Ges. für Biologische Chemie, Marburg, Okt. 1966.

[87] Sandhoff, I., and T. L. Sourkes: Canad. J. Biochem. **40**, 739 (1963), zitiert von Knox, W. E., and O. Greengard in Advanc. Enzymol. **3**, 247 (1965).

[88] Moran, J. F., and T. L. Sourkes: J. biol. Chem. **238**, 3006 (1963).

[89] Weber, G., S. K. Srivastava, and R. L. Singhal: J. biol. Chem. **240**, 750 (1965).

[90] Hofert, J. F., and A. Whit: Endocrinology **77**, 574 (1965).

[91] Shimizu, C. S. N., and S. A. Kaplan: Endocrinology **74**, 709 (1964).

[92] PENA, A., B. DVORKIN, and A. WHITE: J. biol. Chem. **241**, 2144 (1966).

[93] FEIGELSON, P., and O. GREENGARD: J. biol. Chem. **237**, 3714 (1962).

[94] LEWIS, R. A., D. KUHLMAN, D. C. DELBUE, G. F. KOEPF, and G. W. THORN: Endocrinology **27**, 971 (1940).

[95] TARNOWSKI, W.: Pers. Mittlg.

[96] WIELAND, O.: Pers. Mittlg.

[97] HOBERMANN, H. D., and A. F. D. ADAMO jr.: J. biol. Chem. **235**, 934 (1959).

[98] BLOOM, B., and D. O. FOSTER: J. biol. Chem. **237**, 2744 (1962).

[99] SCHIMKE, R. T., E. W. SWEENEY, and C. M. BERLIN: J. biol. Chem. **240**, 322 (1965).

[100] VON FELLENBERG, R., H. EPPENBERGER, R. RICHTERICH und A. AEBI: Biochem. Z. **336**, 334 (1962).

[101] LARDY, H. A.: Harvey Lectures, Series 60, p. 261. New York: Academic Press 1966.

[102] HAYNES, R. C.: Endocrinology **71**, 399 (1962).

[103] OKUNO, G.: Med. J. Osaka Univ. dent. Soc. **10**, 483 (1960), zitiert von UETE, T., and J. ASHMORE in J. biol. Chem. **238**, 2906 (1963).

[104] EISENSTEIN, A. B., E. BERG, D. GOLDENBERG, and B. JENSEN: Endocrinology **74**, 123 (1964).

[105] —, M. RONA, and B. BRUMMER: Endocrinology **76**, 191 (1965).

[106] —, S. SPENCER, S. FLATNESS, and A. BRODSKY: Endocrinology **79**, 182 (1966).

[107] UETE, T., and J. ASHMORE: J. biol. Chem. **238**, 2906 (1963).

[108] GOLDSTEIN, L., E. J. STELLA, and W. E. KNOX: J. biol. Chem. **237**, 1723 (1962).

[109] BARNABEI, O., and F. SERENI: Biochim. biophys. Acta (Amst.) **91**, 239 (1964).

[110] CIVEN, M., and W. E. KNOX: J. biol. Chem. **234**, 1787 (1959).

[111] HENNING, H. V., and W. SEUBERT: Unveröff. Versuche.

[112] —, W. HUTH, and W. SEUBERT: Biochem. biophys. Res. Commun. **17**, 496 (1964).

[113] L'AGE, M., H. V. HENNING, B. OHLY, and W. SEUBERT: In Vorbereitung.

[114] WEBER, G., L. SINGHAL, and S. K. SRIVASTAVA: Advanc. Enzymol. **3**, 43 (1965).

[115] PARAINO, C., C. LAMAR, and H. C. PITOT: J. biol. Chem. **241**, 2944 (1966).

[116] KRÖGER, H., J. PHILLIP und A. WICKE: Biochem. Z. **344**, 227 (1966).

[117] —, and B. GREUER: Nature **210**, 200 (1966).

[118] MARVER, H. S., A. COLLINS, and D. D. TSCHUDY: Biochem. J. **99**, 31c (1966).

[119] KREBS, H. A., and P. DE GASQUET: Biochem. J. **90**, 149 (1964).

[120] HAYNES, R. C.: Advanc. Enzymol. **3**, 111 (1965).

Diskussion

STAUDINGER (Gießen): Ich danke Herrn SEUBERT vielmals für seine ausgezeichnete Darstellung seiner Versuche und eröffne damit die Diskussion.

CLAUSER (Paris): Glauben Sie, Dr. SEUBERT, daß bei der Phosphorylierung und Reduzierung der 3-Phosphoglycerinsäure, die ja energetisch noch ungünstiger liegt als die Phosphorylierung der Brenztraubensäure, die Einwirkung eines separaten Enzymsystems schon vollkommen ausgeschlossen ist?

SEUBERT (Frankfurt): Es ist richtig, daß die Phosphorylierung und Reduktion der 3-Phosphoglycerinsäure energetisch ungünstig liegt. Man darf aber nicht diesen Schritt für sich betrachten, sondern muß das gesamte Gleichgewicht vom Phosphoenolpyruvat bis zum Fructose-1,6-diphosphat in Rechnung ziehen. UTTER hat sich damit sehr eingehend befaßt und gezeigt, daß das Gleichgewicht weitgehend auf Seiten des Fructose-1,6-diphosphates liegt. Mit anderen Worten: Die ungünstige Phosphorylierung und Reduktion des Phosphoglycerats ist gekoppelt mit anderen Reaktionen, die das Gleichgewicht in Richtung Synthese drängen. Die Gleichgewichtskonstante, die UTTER ausgerechnet hat, liegt etwa bei 300.

CLAUSER: Bei den Pflanzen kennen wir ja ein NADPH-abhängiges Enzym, das diese Reduktion bewirkt. Halten Sie es für völlig ausgeschlossen, daß auch im Säugetier solche oder ähnliche Enzyme existieren, die einen getrennten Weg gehen?

SEUBERT: Nein, für ausgeschlossen würde ich das nicht halten. Ausgeschlossen ist wohl nichts in der Biochemie.

SUMM (Frankfurt-Höchst): Von ASHMORE (J. ASHMORE, in R. W. MC-GILVERY and B. M. POGELL, Fructose-1,6-diphosphatase and its role in Gluconeogenesis, 1964, Washington DC, Libr. of Congress Catalogue 64—13412) wurde gezeigt, daß die Inkorporation von $^{14}CO_2$ in das Leberglykogen der Ratte nach Cortisolgabe bereits nach 30 bzw. 60 min bis 10fach erhöht ist; d. h. die Induktion der Schlüsselenzyme für die Umschaltung auf Gluconeogenese müßte bereits erfolgt sein. Meine Frage an Sie ist: nach welcher Zeit sehen Sie die Erhöhung der Pyruvatcarboxylase?

SEUBERT: *In vivo* sehen wir die Erhöhung der Pyruvatcarboxylase nach 2 Std. Allerdings haben wir kürzere Zeiten nicht gemessen, ich kann darüber also nichts sagen.

HILZ (Hamburg). EISENSTEIN und auch wir haben gezeigt, daß mit Alanin als Substrat bei Leberschnitten ein Cortisoleffekt zu beobachten ist, jedoch nicht mit Pyruvat oder Lactat als Substrat. Es sieht danach so aus, als ob in diesem System das Hormon an der Aminosäure angreift, den Stoffwechsel der anderen C_3-Körper aber nicht beeinflußt. Haben Sie dazu eine Erklärung?

SEUBERT: Ich kenne diese Arbeit von EISENSTEIN; ich habe keine Erklärung, aber einige Einwände. Man kann wohl in so einem Fall nicht nur
messen, was hereingeht und herausgeht, sondern müßte auch den Spiegel von
Enzymen oder Substraten messen, z. B. den Spiegel von Acetyl-CoA. Es
könnte ja sein, daß mit Pyruvat als Substrat die Phosphoenolpyruvat-
carboxykinase limitierend wird, und von der wissen wir, daß sie unter diesen
Bedingungen *in vitro* nicht ansteigt, daß dagegen mit Alanin als Substrat die
Pyruvatcarboxylase limitierend wird, von der wir wissen, daß sie *in vitro*
ansteigt. Versuche von KREBS sprechen weiterhin dafür, daß der Acetyl-
CoA-Spiegel eine Rolle spielen könnte. Alle diese Spiegel hat EISENSTEIN
jedoch nicht gemessen. — Mir sind jedenfalls keine experimentellen Anhaltspunkte dafür bekannt, daß zwischen Alanin und Pyruvat noch irgendwelche
kurzzeitigen Hormoneffekte existieren.

WIELAND (München): Ich möchte auch etwas zum Acetyl-CoA als mög-
lichem Regulator sagen. Sie haben ja die Beteiligung des Acetyl-CoA an den
verschiedenen Reaktionskreisen schon beschrieben, und es ist ein sehr schöner
neuer Befund, den Sie hier heute vorgestellt haben: die Hemmung der
Pyruvatkinase durch Acetyl-CoA. An sich wäre damit Acetyl-CoA ein idealer
Regulator für die drei wichtigen Schaltstellen des Stoffwechsels in Richtung
Glucosesynthese. Ich muß allerdings nach unseren eigenen letzten Untersuchungen ein wenig Mehltau streuen: Wir haben die Spiegel von Acetyl-CoA
in der perfundierten Rattenleber gemessen, und zwar unter Bedingungen,
denen wir die Bildung von Glucose aus C_3-Vorstufen stimuliert haben. Wir
haben zu unserer Überraschung gefunden, daß unter der stimulierten
Gluconeogenese die Acetyl-CoA-Spiegel keineswegs erhöht sind. Das gibt
uns etwas zu denken.

SEUBERT: Ja, Sie stehen natürlich vor dem Problem, daß Sie einen zusätzlichen Effekt bezüglich der Wirkung des Glucagons finden müssen. Es ist
wahrscheinlich, daß der Glucagoneffekt durch die Fettsäuren zustande
kommt. Das cyclische AMP scheidet meiner Ansicht nach aus, denn es ist ja
nachgewiesen, daß die Inaktivierung der Fuctose-1,6-diphosphatase durch
cyclisches AMP beschleunigt wird, es ist ferner nachgewiesen, daß die Umwandlung der Glykogensynthetase von der L- in die D-Form ebenfalls durch
cyclisches AMP beschleunigt wird; das wären alles Effekte, die der Gluconeogenese entgegenwirken. Es bleibt deshalb nur übrig, daß die Fettsäuren
für die Stimulierung der Gluconeogenese verantwortlich sind.
    Ich kenne ja Ihre Experimente, wir haben es oft diskutiert. Sie finden also
unter Glucagon einen Anstieg des Acetyl-CoA, in Gegenwart von Lactat plus
Fettsäure keinen Anstieg des Acetyl-CoA, und die Kombination von Glucagon und Lactat führt sogar zu einem Abfall. Das finde ich paradox.

WIELAND: Ganz so paradox ist es nicht. Mit Glucagon ohne Sub-
stratzusatz finden wir einen leichten Anstieg des Acetyl-CoA. Wenn wir
Lactat plus Glucagon geben, finden wir einen geringen Abfall des Acetyl-CoA;
hierbei wird aber natürlich viel mehr Glucose gebildet als im Parallelversuch.

— Wenn wir Fettsäuren plus Lactat geben, finden wir auch einen Abfall des Acetyl-CoA. Es bedarf noch weiterer Versuche, um zu sehen, ob diese Auf- bzw. Abbewegungen des Acetyl-CoA sich als signifikant erweisen werden. Eines steht jedoch schon jetzt fest: Unter den Bedingungen, unter denen maximale Glucosebildung in der perfundierten Leber stimuliert wird, steigt das Acetyl-CoA nicht an. Dabei würde man es hier gerade erwarten.

Ich möchte diese Diskussion auch zum Anlaß nehmen, um hier ein wenig zu warnen. Wenn man sich mit diesen Regulationsphänomenen befaßt und die Literatur verfolgt, dann stellt man eine enorme Häufung von Arbeiten über die Wirkung metabolischer Effektoren fest. Wenn das so weiter geht, dann werden wir in einigen Jahren so viele allosterische Effekte von Metaboliten kennen, daß man durch geschickte Kombination alles Gewünschte erklären kann, und es wird schließlich bestimmte Schulen geben, die die eine oder andere Erklärung bevorzugen.

SEUBERT: Ich kann Ihnen da nicht zustimmen. Zwar kann man auf diesem Gebiet nichts direkt beweisen, man kann nur Indizien sammeln; die bisher vorliegenden Indizien sprechen jedoch zugunsten des Acetyl-CoA als Regulator. Natürlich kann sich unser Bild noch ändern, wenn neue experimentelle Tatsachen hinzukommen.

STAIB (Düsseldorf): Herr **SEUBERT**. wenn ich Sie richtig verstanden habe, dann stützen Sie die Hypothese von **FEIGELSON** über die Freisetzung der Aminosäuren durch die Induktion der Tyrosin-Alpha-Ketoglutarattrans- aminase, Tryptophanpyrrolase und anderen. Wie stellen Sie sich zu dem frühen Effekt der Verminderung der Glucoseutilisation, der von **DRURY** u. a. beschrieben worden ist?

SEUBERT: Es war meine Aufgabe, hier einen Überblick zu geben. Da wir selbst uns nicht mit dem Aminosäurestoffwechsel und den Aminosäurespiegeln beschäftigt haben, mußte ich hierzu auf die Literatur zurückgreifen. Nach allem, was ich gelesen habe, bin ich zu der Überzeugung gekommen, daß an der Theorie von **FEIGELSON** schon etwas dran ist. Sie läuft im Prinzip darauf hinaus, daß durch das Fehlen einer essentiellen Aminosäure das Gleichgewicht in der Peripherie vom Proteinaufbau zum Proteinabbau verschoben wird; das steht durchaus in Einklang mit Fütterungsversuchen, bei denen eine essentielle Aminosäure ausgelassen wurde. Ich will aber gern einräumen, daß unsere Kenntnisse über den Mechanismus der Aminosäuremobilisation in der Peripherie noch außerordentlich lückenhaft sind.

HILZ: Ich wollte noch ein Wort zum Acetyl-CoA sagen. Es ist ja ganz sicher, daß das Acetyl-CoA bei der Oxydation der Fettsäuren zunächst in den Mitochondrien entsteht. Was Sie benötigen, ist das Acetyl-CoA im extramitochondrialen Raum. Sie müssen also noch einen stimulierten Transport hierfür annehmen. Es könnte übrigens durchaus sein, daß aus diesem Grunde auch Bilanzmessungen des Acetyl-CoA über die Konzentration im extramitochondrialen Raum nichts aussagen.

SEUBERT: Mit dieser Argumentation können Sie natürlich alle Untersuchungen über Substratspiegel in Frage stellen. Wir können experimentell einfach nicht anders vorgehen, als den Substratgehalt in der *gesamten* Leber zu bestimmen; um die *in-vivo*-Situation zu erfassen, muß man einfach einen Frierstop machen. Ich sagte schon vorhin, man kann nichts beweisen, man kann nur Indizien sammeln. — Im übrigen besteht mit Sicherheit ein Transportmechanismus für Acetyl-CoA vom mitochondrialen in den extramitochondrialen Raum.

SCHIMASSEK (Marburg): Ich wollte die von Herrn Prof. WIELAND ausgesprochene Warnung nochmals unterstreichen. Es ist einstweilen sehr schwierig, Substrate, die aus verschiedenen Metabolitketten stammen, summarisch bestimmten Schaltstellen im Sinne einer Steuerung (überhöhte oder verminderte Konzentration) zuzuordnen. Im übrigen teile ich Ihren Pessimismus bezüglich der Gehaltsbestimmung nicht. Wenn man die Werte mit der nötigen Kritik betrachtet, wird man auch aus den Bewegungen der Spiegel im steady state etwas entnehmen können.

An isoliert durchströmten Lebern haben Nebennierenrindenhormone (wir haben Prednisolon verwendet) einen doppelten Effekt: Eine Verminderung der Atmung sowie eine Verminderung der Aufnahme an Lactat und Pyruvat. Das bedeutet, daß die Leber für die Neubildung der Glucose von der Zufuhr dieser Substrate weitgehend abgeschnitten ist und daß wir deshalb in der Leber selbst nach entsprechenden Quellen suchen müssen.

Noch eine Frage zur Wirkungsdauer. Ich glaube, daß die Wirkung injizierter Nebennierenrindenhormone am Ganztier etwa 4 Std anhält. Es ist also etwa nach 4 Std mit einem Überwiegen der Gegenregulationen zu rechnen. Inwieweit kann das Ihren 6-Std-Wert von Glucose-6-Phosphat und anderen Zwischenprodukten beeinflussen?

SEUBERT: Zur Atmungshemmung kann ich nur sagen, daß wir an Nieren*schnitten* keine Atmungshemmung durch Cortisol gefunden haben. — Was den Einfluß auf den Transport betrifft, der durch Cortisol gedrosselt werden soll, so müssen wir danach in unseren Versuchen an Schnitten eine Drosselung der Gluconeogenese feststellen und keine Steigerung. — Uns kam es bei dem *in-vitro*-System vor allem darauf an, den Mechanismus der Induktion zu studieren und die Rolle des Cortisols hierbei zu ergründen. In dieser Beziehung bewerten wir es als wichtigen Befund, daß für die Enzyminduktion außer Cortisol noch Substrate, vor allem Aminosäuren, vorhanden sein müssen.

HILZ: Ich wollte nochmals auf die Frage des Substrateinstroms zurückkommen, die Herr SCHIMASSEK angeschnitten hat. An sich bedeuten ja die von ihm erwähnten Substrate Pyruvat und Lactat nichts weiter als Glucose. Sofern im Organismus ein Bedürfnis nach echter Neubildung von Glucose aus Aminosäuren besteht, und dieser Weg durch Cortisol beeinflußt wird, dann müßte man eigentlich den Einstrom von Aminosäuren in die Leber untersuchen. Sind Ihnen, Herr SCHIMASSEK, dazu Zahlen bekannt?

Schimassek: Nein, ich kann Ihnen im Augenblick keine Zahlen hierzu nennen. Nach den Versuchen von Christensen und seiner Schule in Ann Arbor müßte man allerdings eher mit einem verstärkten Einstrom von Aminosäuren unter diesen Bedingungen rechnen.

Staudinger: Da wir nachher noch die Mitgliederversammlung haben, darf ich vorschlagen, daß wir jetzt 15 min Pause machen. Ich danke Herrn Seubert sowie den Diskussionsrednern und schließe die Sitzung.

Erythropoietin. Its Nature and Mechanism of Action

By P. P. Dukes

Argonne Cancer Research Hospital and Department of Biochemistry
University of Chicago. Chicago, Illinois, USA*

With 4 Figures

Introduction

The hormone erythropoietin is a substance that regulates the formation of red cells in the hemopoietic tissues of the body. It has been shown that under conditions of anoxic stress due to anemia, low atmospheric O_2 tension or any other cause an erythropoietic factor is demonstrable in the plasma of animals. Carnot and Deflandre[1] made the first claim to this effect as long ago as 1906. It was only relatively recently however that several groups of investigators presented evidence supporting the hypothesis that erythropoietin is part of the regulatory system for red cell formation in the normal organism. For instance the presence of small amounts of erythropoietin can now be demonstrated in both the plasma[2,3] and urine[4,5,6] of normal unstressed animals and human beings. Moreover, antibodies to erythropoietin injected into normal animals cause them to become anemic[7].

The following regulatory circuit seems to exist[8]. When the oxygen tension in the body falls below a critical level a sensor signals to the sites of production of the hormone in the kidney and perhaps elsewhere. Erythropoietin appears in the circulation and interacts with susceptible multipotent stem cells in the bone marrow and other sites of erythropoiesis causing them to start development in the red cell line. After several divisions have occured the progeny of these cells appears in the blood stream. The additional oxygen carried by the new cells leads to a higher oxygen tension in the tissues and the output of erythropoietin is reduced accordingly.

* Operated by the University of Chicago for the United States Atomic Energy Commission.

Chemical nature, purification

The chemical nature of the hormone is still largely unknown. This is mainly because only extremely small amounts of it are available in a highly purified form. From an anemic plasma with about 0.007 units per mg protein to the most highly purified fraction with 6500 units per mg protein there is approximately a million-fold increase in potency but unfortunately only a 2 percent yield. This purification was achieved by Goldwasser and Kung[9] by a procedure involving adsorption of dialyzed plasma on DEAE, chromatography of the active eluate on amberlite XE-97, ammonium sulfate precipitation, chromatography of the active frac-

Table 1. *Summary of Purification of Erythropoietin[9]*

Fraction	Units/mg Protein	% Recovery	Purification Factor
Anemic Plasma	0.007	100	—
DEAE cell., XE-97 fraction (step III)	2.7	35	388
Ammonium sulfate fraction	19	20	2,700
Sulfoethyl Sephadex fraction	133	10.4	19,000
Calcium phosphate gel eluate	6500	2.2	930,000

tion on sulfoethyl sephadex gel followed by calcium phosphate gel adsorption and fractional elution (Tab. 1). Lowy and Keighley[10] have reported a considerable purification of erythropoietin from human urine, and Contrera, Gordon and Weintraub[11] have described purification of a renal erythropoietic factor.

The hormone is generally assumed to be a glycoprotein. The information on the chemistry of erythropoietin collected by various groups is based mainly on inactivation studies done on very impure preparations. These studies suggest that the erythropoietin molecule contains a polypeptide. It loses its biological activity on treatment with trypsin implying the presence of lysine and/or arginine in the peptide[12]. Fluorescein isothiocyanate, which substitutes on phenolic hydroxyl groups and on free amino groups, inactivates erythropoietin. This study by Lowy and collab.[13] suggests a possible involvement of tyrosine, lysine, or an N-terminal amino acid in a site required for activity. These workers have also shown, using anhydrous formic acid which can react with serine or threonine hydroxyl groups, that these are not involved in the hormone's activity. Sialic acid may also be a part of the erythropoietin molecule. Its removal from preparations, either with dilute acid[14] or sialidase[15], led to a total loss of activity in the *in vivo*-animal assay, but not in the *in vitro* assays[3].

Assays

The *in vivo* assays are based on the measurement of the increased incorporation of radioactive iron into red cells of rats or mice treated with the hormone. Usually the endogenous erythropoietin production in these animals is reduced by starvation[8], polycythemia[2], or some other means, prior to the assay.

The *in vitro* systems are based on the incorporation of radioactive precursors into cells in primary explant culture usually from rat bone marrow. One such system involves the increased incorporation of ^{59}Fe into heme of hemoglobin of cells when erythropoietin is added *in vitro*[16]. I shall refer to this assay system several times later on. The other system involves the stimulated incorporation of ^{14}C-glucosamine into bone marrow cells[17].

The unit of hormone is presently based on an international standard preparation. This in turn was originally standardized in the *in vivo*-^{59}Fe rat system which equated one unit to the response obtained when 5 μ Moles of $CoCl_2$ were injected into the assay rat[18].

The glucosamine incorporation system

In the following I shall mainly deal with the glucosamine incorporation system since my own work is principally concerned with it. The studies to be presented are based on collaborative efforts with Dr. GOLDWASSER, and at various times with Drs. TAKAKU and GALLIEN-LARTIGUE. The basic observation was that the addition of erythropoietin caused increased incorporation of ^{14}C-glucosamine into the acid insoluble material of bone marrow cells[19]. Most of the earlier data are based on the use of 50% medium 199 to 50% calf serum and incubation in tubes gassed with 95% air/5% Co_2. More recently standard conditions have been modified: cells are incubated in plastic culture dishes in 50% medium NCTC 109 to 50% calf serum. We also use Millipore filters to collect, wash, and when needed acid precipitate cells on them[20]. It is interesting that the system works best in the presence of calf serum and not with the homologous rat plasma. Fetal calf serum, which is optimal for ^{59}Fe incorporation into heme, seems to be inhibitory for glucosamine incorporation.

1. Specificity

The specificity of the erythropoietin effect *in vitro* was tested with respect to cells affected.

Rat bone marrow cells showed an increased incorporation of glucosamine after the second hour of culture in the presence of erythropoietin. The hormone caused equal increments of glucosamine incorporation over the control incorporation per unit time[17, 19]. I shall return to the time course of the hormone effect later.

Mouse spleen cells responded to erythropoietin in essentially the same way as rat marrow cells. This was to be expected since in the mouse the spleen is a hemopoietic organ. Both normal and polycythemic mouse spleen cells, the latter not containing any recognizable red cell precursors, were tested[21].

Rat reticulocytes incubated *in vitro* did not respond to erythropoietin. This was tested both with the incorporation of glycine into heme and of glucosamine into various unidentified cellular constituents[3]. Peripheral red cells from rats incorporated glucosamine at a level at most 1 to 2% that of marrow cells, and did not respond to erythropoietin[21].

Granulocytes from a peritoneal exudate of rats were obtained by the method of Gordon[22]. They showed very active incorporation of glucosamine when they were incubated the same way as bone marrow cells. The addition of erythropoietin, however, did not cause any increase of this incorporation over a 24 h period[21].

These data on cell specificity are all compatible with the hypothesis that erythropoietin acts on hemopoietic stem cells.

The specificity of the effector, that is, erythropoietin itself, in bringing about the increased glucosamine incorporation was tested. This was done by adding erythropoietin at various stages of purification to the marrow cell incubation system. Anemic plasma (the starting material), and various purified fractions were all active in stimulating glucosamine incorporation *in vitro* with a fair correlation to their potency as established by the *in vivo* erythropoietin assay[17]. Step III erythropoietin was the hormone standard preparation used in the experiments reported here. The activity in the glucosamine incorporation system of the highest purified fractions was about 30 to 40% of the activity predicted for them on the basis of the *in vivo* assay. This allows one to state that it is very likely that erythropoietin and the factor responsible for increased glucosamine incorporation are one and the same. If this were not so they would have had to be present in a relatively constant ratio in fractions representing a 10,000 fold range of potency.

2. Glucosamine labeling

I would like to turn now to what little is known about what the ^{14}C-glucosamine is incorporated into the bone marrow cells. On the average about 55% (range 35 to 75%) of the total radioactivity in the cell is found in the acid insoluble fraction. Water lysis of the cells followed by centrifugation yields a similar fraction with about 60 to 70% insoluble radioactivity. For simplicity we have usually used the acid precipitation procedure but we have referred to both fractions as stroma since the acid precipitable fraction certainly includes the water insoluble cell constituents. During the first 24 h of incubation the action of the hormone is similarly reflected in the radioactivity of both the soluble and stroma fractions.

The stroma fraction was investigated further as follows[17]: Stroma from cells cultured for 24 h with and without erythropoietin was obtained by the acid procedure. Samples were hydrolyzed for 2 h at 80° with 0.05 N H_2SO_4 according to the procedure of GOTTSCHALK[23] which releases neuraminic acids. Both the control and the stimulated sample had approximately 56% of their radioactivity solubilized by this procedure. After the removal of H_2SO_4 with the liquid anion exchanger LA-2 the soluble fractions were chromatographed on paper. The distribution of labeled material was similar in the chromatograms of both control and stimulated sample. Neuraminic acids accounted for about 50% of the total radioactivity applied to the paper, 40% remained at the origin. This allows one to postulate that at least 30% of the total stroma radioactivity resides in neuraminic acids. More work is in progress at this time on the fate of the glucosamine label in the cells.

3. Dose response

When the relationship between amount of erythropoietin and response in terms of increased incorporation of glucosamine is plotted on arithmetic coordinates it yields an S-shaped curve typical for hormone effects. The same kind of curve is obtained also when the data from the *in vivo* assay are plotted similarly[24]. More often however the response is expressed on a linear scale as percent increase of specific activity of the acid insoluble residue over that of the control while the dose of hormone is plotted on a logarithmic scale. This method yields a straight line relationship over part of the response range useful for assaying purposes. Different culture

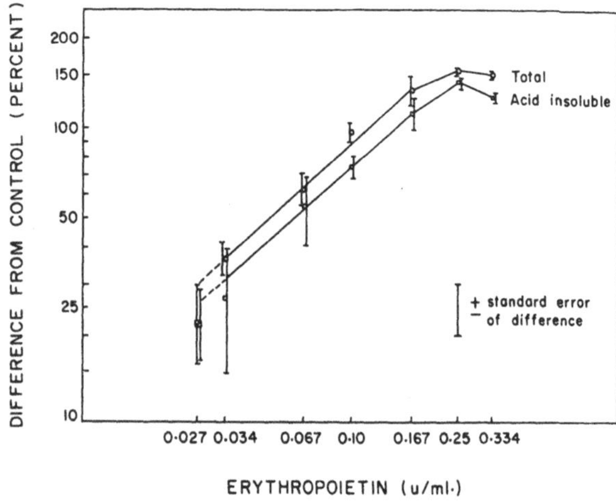

Fig. 1. Relationship of amount of erythropoietin (step III) and stimulated ^{14}C-glucosamine uptake into rat bone marrow cells in culture (18.10^6 nucleated cells/ml)

conditions gave different dose ranges within which this "quasi linear" response was obtained[17] (see also Fig. 2). It should be pointed out that this assay must be used very cautiously because of the presence of an inhibitor in the erythropoietin preparations. This causes a depression of glucosamine uptake and affects the range of response. The inhibitor seems to be largely removed from the 0.67 saturated $(NH_4)_2SO_4$ precipitate (Step IV), it is therefore concentrated in the corresponding supernatant fraction[17].

When in the ^{59}Fe *in vitro* system[18] the ^{59}Fe uptake into heme is plotted against the log of the dose, the resulting dose response curve is very similar to the one just described for glucosamine incorporation.

For the experiment described in Fig. 1 the modification of the glucosamine-system utilizing incubation in plastic culture dishes, which has already been mentioned, was used. The presentation chosen here and generally in our more recent work is a plot of % increase in incorporation on a log-scale against dose on a log-scale. This log-log presentation was adopted because it yields a more useful quasi linear range at low doses than the semilogarithmic plot. For the experiments* used in these curves, cells received erythropoietin at the start and label at an interval of from 18 to 21½ h of incubation. Both total cells washed and collected on Millipore filters, and their acid insoluble residue prepared on the filter, gave essentially paralell dose response curves, reflecting a similar erythropoietin effect on several classes of labeled compounds in the cell. In the control cells the relative proportion of acid insoluble radioactivity was slightly higher than in erythropoietin treated cells. This causes the percent increase curve for the acid insoluble fraction to be displaced below that of the total sample.

The glucosamine incorporation assay compares favorably with respect to sensitivity with the *in vivo* assay and the [59]Fe *in vitro* system[16]. They can all detect reliably about 0.05 units of hormone. The upper limit of usefulness of the glucosamine incorporation assay is set by the number of responding cells in the cultures, the culture conditions in general, and the amount, if any, of inhibitory material present in the sample to be assayed.

From the minimal detectable dose *in vitro* one can, by making some rather uncertain assumptions, obtain the number of hormone molecules per cell needed for action. Assuming a molecular weight of 50,000 for erythropoietin, a potency of the pure hormone of 10,000 u/mg and a maximum of 1% of the cells in the culture actually responding, one obtains as an order of magnitude estimate about 10^5 to 10^6 molecules required per cell. That is a large number of molecules, suggesting that perhaps only a few of them actually penetrate into the interior of a cell. This is at any rate compatible with the observation that no detectable amount of erythropoietin disappears from the medium of a 24 h culture[26].

4. Time course

The most important feature of the time course is that after a lag of about 2 h, labeled product stimulated by erythropoietin in-

---

* P. P. DUKES unpublished results.

creases linearly with time[17, 19]. There are 2 alternative explanations for this.

1. All sensitive cells start to respond to the hormone at the same time, and they incorporate glucosamine at a constant rate as long as erythropoietin is present. That the presence of the erythropoietin is required for the continued response was demonstrated in an experiment in which cells were removed from a medium containing erythropoietin after 1 h's incubation and transferred to medium without it, which had been suitably preincubated with other cells. The stimulated incorporation lasted from about $3\frac{1}{2}$ to 7 h of incubation. After that, these cells did not incorporate more glucosamine than controls not given erythropoietin from the start[20].

2. A second explanation might be that there is a constant rate of formation of sensitive cells from a pool of marrow cells that are potentially sensitive to erythropoietin. The response during the entire incubation period may be thought of as the sum of a number of short bursts of incorporation by the induced fraction of the marrow cell population. I am inclined to favor this explanation because it seems very unlikely that the same group of cells would continue to respond for 20 h with a fixed rate of glucosamine incorporation. The results of the experiment* described in Fig. 2 support the hypothesis that a constant number of cells is responding to the hormone per unit time. The accumulated response to 3 different doses of hormone was measured at an early and a late time point in the incubation. Granting that the hormone level remains essentially constant in the culture medium, and observing that the quasi linear response range remains unchanged with time, within the limits checked, it is possible to conclude that the ratio between responding cells and hormone molecules remains the same during the whole incubation period. A constant stimulated rate was established for each hormone dose. The plots of accumulated stimulated product extrapolate for all doses to the same time point for 0 effect. If the number of responding cells had varied significantly during the incubation this relationship would not have held. For instance, had the number of responding cells decreased with time then the response to the highest dose (0.5 u) would have fallen below its present value at 23 h on both plots. This would have caused a

---

* P. P. Dukes and E. Goldwasser unpublished results.

break in the semilogarithmic plot and the time point for extra-polated 0 effect would have deviated from 4 h.

If one follows the effect of erythropoietin added at 0 time to a culture past 24 h of incubation one notices that the incorporation of label from ^{14}C-glucosamine into the whole cell and its acid in-soluble fraction no longer parallel each other[27]. The stimulated in-corporation into the whole cells continues, although no longer at a constant rate, to the end of the second day and ceases then. The

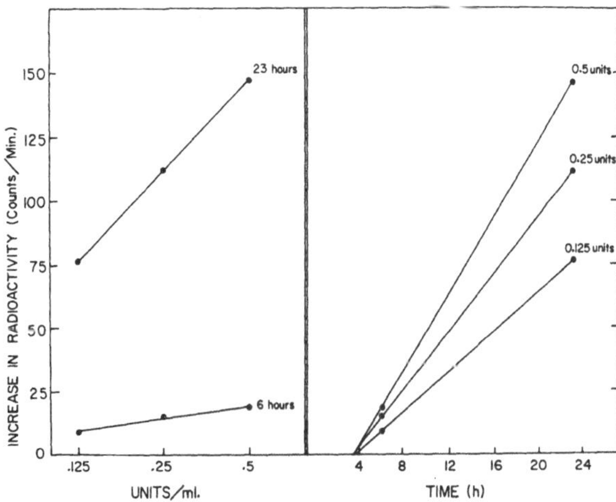

Fig. 2. Relationship of amount of erythropoietin (step III) and stimulated ^{14}C-glucosamine uptake into rat bone marrow cells at 6 and 23 h of culture (dose response on the left, time response on the right)

rate of incorporation of ^{14}C-glucosamine into the acid insoluble fraction of the cells is actually depressed in the erythropoietin samples from the second to the fourth day (Tab. 2). Perhaps a response to erythropoietin consisting of an increased influx of hexosamine and accumulation of its low molecular weight deriva-tives continues to be possible as long as there are sensitive cells in the stem cell pool. On the other hand another part of the response could depend on the availability of macromolecular acceptors for the labeled carbohydrates. These may cease to be made by the cell after 1 day's incubation perhaps due to exhaustion of the medium. It should be pointed out here that there is already a precedent for an erythropoietin mediated accumulation of a precursor in cells.

Table 2. *Response to Erythropoietin by Cells in Culture*

Incorporation of	Incubation Time				
^{14}C-glucosamine into	0—8	16—24	44—52	70—71	94—95
			hours		
1. Whole cells	+	+	+	0	0
2. Acid insoluble	+	+	—	—	—

In every instance erythropoietin was added at the start of the incubation.
+ increase over control incorporation rate.
— decrease from control incorporation rate.
0 no change due to erythropoietin.

Hrinda and Goldwasser[28] have reported recently on the increased Fe uptake of bone marrow cells caused by erythropoietin at a time before increased heme synthesis becomes measurable. This process is dependent on new protein synthesis.

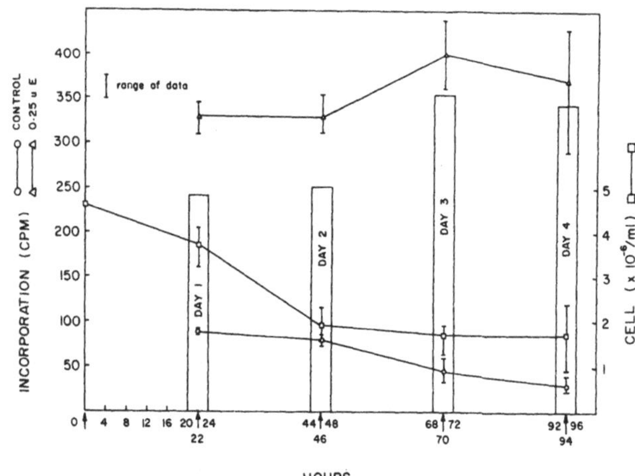

Fig. 3. Increased incorporation of ^{14}C-glucosamine into rat bone marrow cells due to 0.25 u/ml erythropoietin added after various intervals of preincubation

As has just been stated, bone marrow cells cease to respond to erythropoietin added at 0 time at the end of the second day of culture. This is not because potentially sensitive cells cannot survive past this time. Even after 4 days of incubation cells do still respond to the addition of erythropoietin by incorporating additional glucosamine*. This response is if anything greater on the fourth day than on the first, although total cell number and control incorporation decrease drastically with time (Fig. 3).

My interpretation of these findings is that the erythropoietin sensitive cells are a small fraction of the total marrow cells, which are very resistant to the unfavourable conditions that they experience during 5 days of culture. They may be in a cell cycle of about 48 h duration, and once triggered off into erythroid precursors cannot be replaced. After that length of time in the presence of erythropoietin they would have all been used up. Alternatively, they could have a shorter cell cycle, but after 48 h incubation no more replenishment of the cell pool is possible.

I must emphasize here that the responses seen after 48 h are once again only in terms of total cell uptake of label. Acid insoluble compounds could not be formed in increased amounts.

5. Inhibitor studies

I would like to turn now to the effect of inhibitors on the system. The effect of puromycin on 24 h glucosamine incorporation into whole cells was to eliminate 95% of the incorporation despite the presence of the hormone[20]. This suggests that protein synthesis is necessary for the expression of erythropoietin action in terms of glucosamine incorporation. This is also true for stimulated ^{59}Fe incorporation into hemoglobin[29] and ^{59}Fe uptake per se[28] in marrow cells.

KRANTZ, GALLIEN-LARTIGUE and GOLDWASSER[30, 31] have shown that actinomycin D inhibits the erythropoietin effect on RNA and hemoglobin synthesis. This inhibitor also prevents the stimulated uptake of glucosamine by the cells[20]. The response to actinomycin added at the same time as erythropoietin was dependent on the ratio of amount of inhibitor to the number of cells. 0.25 μg per million cells completely abolished the effect of erythropoietin. However, 0.156 μg actinomycin per million cells, (1.25 μg actinomycin/8 $\times$ 10^6 cells/ml) caused no change at all in incorporation in the first 4 h of incubation. In the next 3 h, the stimulated incorporation was reduced by about 40% in the presence of actinomycin and soon thereafter ceased altogether. I should add here that it is

---

* P. P. DUKES and E. GOLDWASSER unpublished results.

known from Krantz's[30] work that 0.16 μg actinomycin per 10^6 cells added at the beginning of the culture did not abolish all RNA synthesis immediately, but by 6 h it was down to 10% of its original value.

A study of the effect of various actinomycin concentrations on the 24 h glucosamine incorporation due to erythropoietin revealed dose response curves exhibiting an initial region of high sensitivity to low drug concentrations followed by a region of reduced further response to increasing drug concentrations[3].

It is possible to ascribe the 2 slopes observed in these curves to 2 or more processes based on DNA dependent RNA synthesis ultimately affecting glucosamine incorporation. Reich[32] has observed a similar 2 slope curve in a study of the actinomycin inhibition of purified RNA polymerase in the presence of DNA primer. One can formulate a hypothesis that the smaller amount of inhibitor stops production of inducible cells by preventing the sensitization of stem cells to erythropoietin in a process requiring RNA synthesis. The stimulation of already existing sensitized cells would continue until they were exhausted. This would lead to the intermediate degrees of inhibition already described. A large amount of actinomycin would prevent not only sensitization but also the processes involved in the response to the hormone.

In experiments in which the actinomycin was added at the 15. h of incubation, that is, at a time when the erythropoietin effect was fully established, the erythropoietin stimulated glucosamine incorporation continued unaffected for 3.5 h and then stopped abruptly[20]. If one assumes that in this case the inhibitor had an immediate effect on m-RNA synthesis, then the 3 to 4 h persistance of stimulated glucosamine incorporation suggests a similar *maximum* life time of m-RNAs related to that. The proteins coded for are probably enzymes transferring activated carbohydrate moieties, or protein acceptors for the carbohydrates, or both.

The effect of the mitotic inhibitor colchicine was tested in order to see whether cell division plays any part in erythropoietin induced glucosamine uptake[20]. Colchicine at the 10^{-7} M level had no effect whatsoever on the 23 h uptake of glucosamine. This must be contrasted with findings by Gallien-Lartigue[31] that the same amount of colchicine under the same conditions greatly reduced erythropoietin stimulated hemoglobin synthesis. These data can be reconciled by referring to the hypothesis that stimulated glucosamine incorporation involves only a relatively short burst during a part of the cell cycle[20]. It would not be affected by the arrest of mitoses as long as the generation of cells in the proper phase of the cycle continued. Hemoglobin synthesis, on the other hand, goes on more less continuously and is normally amplified by cell division. Mitotic arrest there prevents much of the stimulated hemoglobin synthesis.

Experiments were performed utilizing the incorporation of thymidine into the acid insoluble constituents of cells[27]. It was found that addition of erythropoietin to cultures caused an increased rate of DNA synthesis to appear within 3 to 4 h. The overall DNA synthesis rate fell off rapidly in control cultures; the presence of erythropoietin retarded this decrease in synthesis rates. The erythropoietin stimulated increment disappeared after about 20 h of incubation. It reappeared later in long term cultures, seemingly during periods when the control DNA synthesis was at a minimum. The increased rate of DNA synthesis due to erythropoietin could be the consequence of preparation for division by those cells that had been triggered off for differentiation by the hormone. This would be so if (as is likely) their cycle was shorter than that of undifferentiated stem cells. Alternatively, or in addition, the replenishment of the stem cells themselves could lead to increased DNA synthesis in the culture.

It may be questioned whether there is a need for active DNA synthesis in a culture to permit the initiation of erythropoietin action. Therefore the effect of erythropoietin was followed by measuring both glucosamine and thymidine incorporation.

Hydroxyurea, an inhibitor preventing the conversion of ribonucleotides to deoxyribonucleotides, was employed to reduce DNA synthesis[27].

It was established that 10^{-3} M hydroxyurea reduced DNA synthesis to 2% of its normal value in a 2 h period at the beginning of the incubation. At the same time, glucosamine incorporation was essentially unaffected.

Another sample, 24 h later, from the same group of cultures showed DNA synthesis in hydroxyurea containing samples to be 12% that of controls. The shift from 2 to 12% was due to the pronounced fall off of DNA synthesis in the controls. The erythropoietin had been added at 0 time. Whereas the rate of DNA synthesis with and without hydroxyurea was very different, the stimulated increase of glucosamine incorporation into the acid insoluble fraction was similar in terms of counts incorporated, in both sets. From this and other experiments it must also be noted that hydroxyurea did not prevent the erythropoietin stimulated increase of DNA synthesis either.

It seems then that a very low rate of DNA synthesis, such as was caused by hydroxyurea, does not preclude erythropoietin action or even seriously diminish it. This observation unfortunately does not decide whether or not DNA synthesis is needed at all. Obviously some cells are continuing to synthesize DNA, even when unable to make deoxyribonucleotides, probably by scavenging pathways utilizing preformed deoxyribonucleosides. It is impossible at this time to say whether or not these are the cells that respond to erythropoietin by increased glucosamine incorporation, but it is perhaps significant that at least some of these cells do respond with increased DNA synthesis.

Other hormones

In addition to erythropoietin, the sex hormones seem to be involved in the direct regulation of erythropoiesis. Near physiological amounts of estradiol administered to male rats caused a reduction of red cell formation suggesting a direct effect of the steroid on the hemopoietic tissues[33]. More recently Jepson and Lowenstein[34], from studies of ^{59}Fe incorporation into erythrocytes of polycythemic mice, suggested that the utilization of erythropoietin by the marrow cells was blocked by estrogen. Much higher doses of estrogen, employed by Mirand and Gordon[35], interfered also with the production of erythropoietin in the test animals.

Androgens cause an increase in erythropoiesis, but the mechanism involved is still unclear. Naets and Wittek[36], in studies of the ^{59}Fe incorporation of polycythemic mice, found that androgens either potentiate the action of erythropoietin by increasing the number of stem cells, or their sensitivity to erythropoietin, or else they influence a step in erythropoiesis after stem cell differentiation. Fried and Gurney[37] and Mirand and Gordon[35] suggest additionally that erythropoietin production itself is also increased by androgens.

Summary

Certain features of erythropoietin action[3] are recapitulated in Fig. 4. The first part of the figure shows the stem cell cycle. Cells may, after completing a full turn of the cycle, divide into 2 new stem cells, or they may, under the influence of erythropoietin, leave the cycle. It has been shown[20] that the response of an unsynchronized population seems to be made up of incremental responses distributed over a long time. This suggests that it is only during a certain phase of the cycle that stem cells can interact with erythropoietin. From experiments with actinomycin inhibition, one

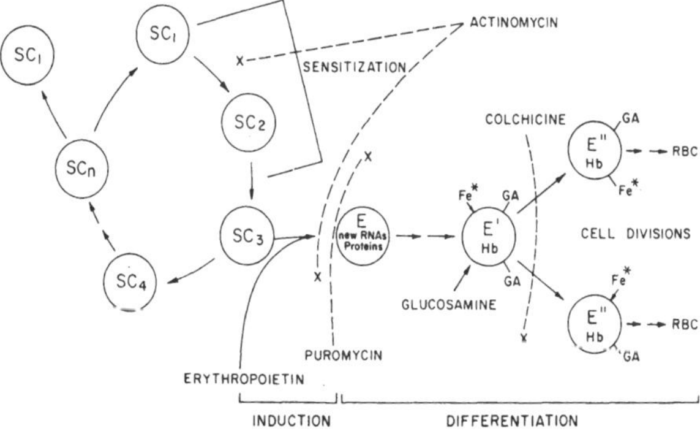

Fig. 4. Scheme of erythropoietin action. *SC* stem cells. Subscripts refer to stages in the cell cycle. *E* cells in the erythroid line. The broken lines terminating in X show steps inhibited by the compounds indicated. *GA* cellular material labeled by glucosamine. *Hb* hemoglobin

may postulate, although this is far from certain, a process of *sensitization* to erythropoietin involving DNA dependent RNA synthesis. This would be characterized by a greater actinomycin sensitivity than the rest of the cellular RNA synthesis.

The actual *induction* process follows the entry of the hormone into the cell. Erythropoietin seems to act as a derepressor in the sense of JACOB and MONOD[38]. It seems to cause the transcription of genetic information, hitherto unexpressed, from nuclear DNA into m-RNA[39] (KRANTZ[30] observed the synthesis of new m-RNA like material 15 min after adding erythropoietin to a culture). The newly initiated process of *cellular differentiation* or *specialization* includes the synthesis of RNAs related to hemoglobin, iron transport, glucosamine incorporation into various cellular materials, and

14*

probably many others. It cannot take place in the presence of actinomycin. Obviously new protein synthesis is also essential for this specialization since when it is inhibited by puromycin or actidion no increased synthesis of hemoglobin or cell constituents labeled by ^{14}C-glucosamine can take place.

Cell divisions are naturally involved before the final mature red cell can appear. But experiments with colchicine showed that the synthesis of additional stromal material is completed, and hemoglobin synthesis well on its way before the necessity for cell division arises.

References

[1] CARNOT, P., et C. DEFLANDRE: Comptes Rend. **143**, 432 (1906).

[2] JACOBSON, L. O., E. GOLDWASSER, L. F. PLZAK, and W. FRIED: Proc. Soc. exp. Biol. (N.Y.) **94**, 243 (1957).

[3] GOLDWASSER, E.: In Current Topics in Developmental Biology, p. 173. (MONROY, A., and A. A. MOSCONA eds.). New York, London: Academic Press 1966.

[4] FINNE, P. H.: Brit. med. J. **1**, 697 (1965).

[5] ALEXANIAN, R.: Blood **28**, 344 (1966).

[6] VAN DYKE, D., M. L. NOHR, and J. H. LAWRENCE: Blood **28**, 535 (1966).

[7] SCHOOLEY, J. C., and J. F. GARCIA: Proc. Soc. exp. Biol. (N.Y.) **110**, 636 (1962).

[8] FRIED, W., L. F. PLZAK, L. O. JACOBSON, and E. GOLDWASSER: Proc. Soc. exp. Biol. (N.Y.) **94**, 237 (1957).

[9] GOLDWASSER, E., and C. K. H. KUNG: Ann. N.Y. Acad. Sci. In press 1967.

[10] LOWY, P. H., and G. L. KEIGHLEY: Clin. chim. Acta **13**, 491 (1966).

[11] CONTRERA, J. F., A. S. GORDON, and A. H. WEINTRAUB: Blood **28**, 330 (1966).

[12] SLAUNWHITE, W. R., E. A. MIRAND, and T. C. PRENTICE: Proc. Soc. exp. Biol. (N.Y.) **96**, 616 (1957).

[13] LOWY, P. H., and H. BORSOOK: In Erythropoiesis, p. 33. (JACOBSON, L. O., and M. DOYLE, eds.). New York: Grune and Stratton 1962.

[14] RAMBACH, W. A., R. A. SHAW, J. A. D. COOPER, and H. A. ALT: Proc. Soc. exp. Biol. (N.Y.) **99**, 482 (1958).

[15] LOWY, P. H., G. KEIGHLEY, and H. BORSOOK: Nature (Lond.) **185**, 102 (1960).

[16] KRANTZ, S. B., O. GALLIEN-LARTIGUE, and E. GOLDWASSER: J. biol. Chem. **238**, 4085 (1963).

[17] DUKES, P. P., F. TAKAKU, and E. GOLDWASSER: Endocrinology **74**, 960 (1964).

[18] GOLDWASSER, E., and W. F. WHITE: Fed. Proc. 18, 236 (1959).

[19] DUKES, P. P., F. TAKAKU, and E. GOLDWASSER: Biochem. biophys. Res. Commun. **13**, 223 (1963).

[20] —, and E. GOLDWASSER: Biochim. biophys. Acta (Amst.) **108**, 447 (1965).

[21] TAKAKU, F., P. P. DUKES, and E. GOLDWASSER: Endocrinology **74**, 968 (1964).

[22] GORDON, A. S., R. O. NERI, C. D. SIEGEL und B. S. DORNFEST: Acta haemat. (Basel) **23**, 323 (1960).

[23] GOTTSCHALK, A., and E. R. B. GRAHAM: Biochim. biophys. Acta (Amst.) **34**, 380 (1959).

[24] WHITE, W. F., C. W. GURNEY, E. GOLDWASSER, and L. O. JACOBSON: In Recent Progress in Hormone Research, Vol. 16, p. 219 (PINCUS, G., ed.). New York: Academic Press 1960.

[25] DeGOWIN, R. L., D. HOFSTRA, and C. W. GURNEY: Proc. Soc. exp. Biol. (N.Y.) **110**, 48 (1962).

[26] KRANTZ, S. B., and E. GOLDWASSER: Biochim. biophys. Acta (Amst.) **108**, 455 (1965).

[27] DUKES, P. P.: Ann. N.Y. Acad. Sci. In press 1967.

[28] HRINDA, M. E., and E. GOLDWASSER: Fed. Proc. **25**, 284 (1966).

[29] — Ph. D. Dissertation, University of Chicago 1967.

[30] KRANTZ, S. B., and E. GOLDWASSER: Biochim. biophys. Acta (Amst.) **103**, 325 (1965).

[31] GALLIEN-LARTIGUE, O., and E. GOLDWASSER: Biochim. biophys. Acta (Amst.) **103**, 319 (1965).

[32] REICH, E.: Science **143**, 684 (1964).

[33] DUKES, P. P., and E. GOLDWASSER: Endocrinology **69**, 21 (1961).

[34] JEPSON, J. H., and L. LOWENSTEIN: Proc. Soc. exp. Biol. (N.Y.) **123**, 457 (1966).

[35] MIRAND, E. A., and A. S. GORDON: Endocrinology **78**, 325 (1966).

[36] NAETS, J. P., and M. WITTEK: Amer. J. Physiol. **210**, 315 (1966).

[37] FRIED, W., and C. W. GURNEY: Proc. Soc. exp. Biol. (N.Y.) **120**, 519 (1965).

[38] JACOB, F., and J. MONOD: Cold Spr. Harb. Symp. quant. Biol. **26**, 193 (1961).

[39] KARLSON, P.: Perspect. Biol. Med. **6**, 203 (1963).

Diskussion

ZILLIKEN (Marburg): Herr DUKES, wir danken Ihnen sehr für die interessanten Ausführungen. Ich möchte die Diskussion eröffnen.

BÜCHER (München): Brauchen Sie Transferrin oder Siderophilin, um den Eiseneinbau zu bekommen?

DUKES (Chicago): Ja. Wir benutzen Präparate, die Transferrin enthalten, um das Eisen in einem Zustand zu haben, der zum Einbau führt.

KARLSON (Marburg): Herr DUKES, das Erythropoietin ist ja ein sehr interessantes Beispiel für ein morphogenetisches Hormon; es bewirkt die Differenzierung der Reticulocyten. Ihr System scheint mir ausgezeichnet geeignet, die biochemischen Vorgänge bei einer solchen Zelldifferenzierung genauer zu analysieren. Hierzu habe ich einige Fragen: Wie weit ist die Determination

zum Erythrocyten reversibel? Ist nach Einwirkung von Erythropoietin für
z. B. einige Stunden die Zelle bereits endgültig determiniert oder nicht?

DUKES: Es scheint bei den Zellen, die wir verwenden, sehr schnell eine
endgültige Determinierung stattzufinden. Es kann da noch zu einer Abortion
kommen, aber nicht zu einer Rückentwicklung.

KARLSON: Und für die weiteren Prozesse, die sich daran anschließen
— Hämoglobinsynthese usw. — ist das Erythropoietin nicht mehr laufend
nötig? Der einmalige Anstoß genügt, um alle diese Prozesse in Gang zu
setzen?

DUKES: Ja. Es ist durchaus so, daß man nach einer mehrstündigen Einwirkung von Erythropoietin für eine lange Zeit eine gesteigerte Hämoglobinsynthese bekommt; anscheinend bekommen wir auch eine Verstärkung des
Einbaus nach der ersten Zellteilung (auch in Abwesenheit vom Erythropoietin).

KARLSON: Man muß dann doch wohl mit sehr langlebigen Messenger-
Ribonucleinsäuren rechnen, gerade für die Hämoglobinsynthese.

DUKES: Ja.

ZILLIKEN: Darf ich selbst etwas fragen, Herr DUKES. Haben Sie untersucht, ob das Erythropoietin diese schwere Messenger-RNS induzieren kann,
die von KLAUS SCHERRER in Paris nachgewiesen wurde?

DUKES: In unserem Labor hat Herr Dr. KRANTZ Versuche in dieser Richtung gemacht und folgendes gefunden: Wenn man mit den üblichen Methoden die RNS isoliert, dann findet man die schnelle Markierung nach Erythro-
poietineinwirkung bei vergleichsweise leichten Messenger-RNS-Fraktionen.
Wenn man sehr schonend isoliert, kann man sie auch in der schwereren Fraktion bekommen. Im übrigen ist es so, daß das Erythropoietin nicht spezifisch
eine bestimmte RNS-Fraktion stimuliert, sondern eigentlich alle Größen-
klassen von Ribonucleinsäuren.

REISSERT (Göttingen): Sie haben gezeigt, daß bei hohen Dosen Erythropoietin im Medium der Einbau von Glucosamin wieder abnimmt, und Sie
führen das auf einen Inhibitor zurück, der offenbar während der Inkubation
entstehen muß. *In vivo* haben wir diesen Abfall nicht gesehen. Wenn wir die
Dosis steigern, kommt es einfach zu einem Plateau. Wie würden Sie diesen
Befund erklären?

DUKES: Daß man diesen Inhibitoreffekt *in vivo* nicht sieht, dürfte darauf
zurückgehen, daß das Erythropoietin im Ganztier doch vergleichsweise
schnell inaktiviert wird. Man erreicht niemals so hohe Konzentrationen wie
in der Gewebekultur. Im übrigen scheint mir ein Mißverständnis vorzuliegen.
Der Inhibitor wird nicht während der Inkubation gebildet; er scheint ein

Begleitprotein des Erythropoietins zu sein, und wenn wir größere Mengen unseres Präparats zusetzen, dann wird die Inhibitorwirkung deutlich. Wenn man durch sorgfältige Ammonsulfatfraktionierung den Inhibitor weitgehend entfernt, dann kann man den geraden Teil der Kurve erheblich verlängern.

STAUDINGER (Gießen): Ich habe zwei Fragen, die den Anfang Ihres Vortrags betreffen. 1. Ist die Niere der einzige Bildungsort für Erythropoietin oder gibt es noch weitere? Wie ist das bewiesen worden? 2. Sauerstoffmangel führt zur vermehrten Erythropoietinsynthese. Wie groß muß der Abfall des Sauerstoffdrucks im Gewebe sein, um diesen Effekt auszulösen?

DUKES: Die ursprüngliche Ansicht, daß das Erythropoietin ausschließlich in der Niere entstehen würde, wie sie von JACOBSON vertreten wurde, beruhte auf Extirpationsexperimenten an Ratten. Wird die Niere entfernt, so kommt es nicht mehr zur Erythropoietinbildung. Das ist aber in dieser Ausschließlichkeit nicht mehr haltbar. Man hat sowohl in Versuchen mit Hunden als auch durch Beobachtungen am Menschen, denen die Nieren entfernt werden mußten, zeigen können, daß die Erythropoiese weitergeht, wenn auch auf niedrigerem Niveau. Es muß also neben der Niere, die wohl als Hauptbildungsort angesprochen werden muß, noch andere Bildungsstätten geben.

Was den Sauerstoffgrunddruck angeht, so kann ich im Augenblick den Schwellenwert für die Druckerniedrigung nicht angeben. Es ist aber sicher, daß Menschen, die längere Zeit im Hochgebirge leben, z. B. bei Expeditionen, schon bei Höhen von 3000 m einen merklich erhöhten Erythropoietinspiegel haben. Es sind auch Tierversuche durchgeführt worden, aber die genauen Zahlen kann ich Ihnen im Augenblick nicht nennen.

SENFT (Berlin): Weiß man etwas über die Halbwertzeit des Erythropoietin *in vivo*? Wann tritt z. B. nach Entfernung der Nieren eine Anämie auf?

DUKES: Die biologische Halbwertszeit beträgt etwa 2 bis 3 Std; das Hormon wird also relativ schnell aus der Blutbahn entfernt.

ZILLIKEN: Der Versuch mit Neuraminidase hat sehr schön zeigen können, daß durch Abspaltung der terminalen N-Acetylneuraminsäure (NANA) die Wirkung des Erythropoietin *in vivo* verschwindet. Für den Effekt auf den ^{59}Fe-Einbau in der Gewebskultur ist jedoch die NANA nicht erforderlich. Trifft das auch für den Glucosamineinbau zu?

DUKES: Ja, es geht völlig parallel. Während *in vivo* die N-Acetylneuraminsäure essentiell ist, sehen wir in der Gewebekultur überhaupt keinen Unterschied zwischen intaktem Hormon und Neuraminidase behandeltem Hormon, weder im Eiseneinbau noch im Glucosamineinbau.

RÓKA (Frankfurt): Ich habe eine Frage zur biologischen Rolle des Erythropoietins. Sie hatten gesagt, daß auch im Blut nichtanämischer Tiere Erythropoietin gefunden wird, wodurch sich die normale Erythropoese erklärt. Muß man nicht trotzdem noch eine Basiserythropoese annehmen, die auch ohne Hormon vor sich geht? Oder kann man z. B. mit Antikörpern gegen Erythropoietin die Erythropoese völlig unterdrücken?

DUKES: Man kann mit Antikörpern tatsächlich eine sehr weitgehende Anämie hervorrufen. Ob man die Erythropoese völlig unterdrücken kann, ist eine offene Frage. Auch die Nierenexstirpationsversuche deuten darauf hin, daß es eine Resterythropoese auch ohne Hormon geben könnte.

DUMONT (Würzburg): Bei der Leukämie wurde neben einer starken Vermehrung der weißen Blutkörperchen eine Verminderung der Erythrocyten beobachtet. Ist bekannt, ob dabei der Erythropoietinspiegel vermindert ist?

DUKES: Ich glaube, daß es sich hierbei um eine Konkurrenz handelt. Bei dieser so stark vermehrten Leukopoese werden die Stammzellen so stark beansprucht, daß für die Erythropoese gewissermaßen keine Stammzellen mehr übrig bleiben.

SEIF (Tübingen): Hat man schon versucht, durch markierte Antikörper die Bildungsstätte des Erythropoietins in der Niere zu lokalisieren?

DUKES: Soviel ich weiß, hat die Gruppe um GORDON in New York solche Versuche durchgeführt. Die Ergebnisse sind noch nicht publiziert, weil diese Versuche außerordentlich schwer auswertbar sind. Es sind noch keine entscheidenden Ergebnisse erhalten worden.

KLINGMÜLLER (Mannheim): Spielt das Erythropoietin auch eine Rolle bei der Bleivergiftung? Es sieht ja so aus, als ob das Porphyrinsystem fertig vorgebildet sein muß und daß Erythropoietin an der Stelle angreift, wo das Eisen eingefügt wird, vermutlich wohl durch die „Ferrochelatase".

DUKES: Ja, es ist durchaus möglich, daß die Ferrochelatase eines der Enzyme ist, das durch Erythropoietin in besonderem Maße induziert wird. Das ist aber noch nicht exakt bewiesen.

DUMONT: Erythropoietin wurde sowohl im Plasma wie im Urin festgestellt. Ist etwas über die quantitativen Verhältnisse bekannt, z. B. über die Clearance und Rückresorption? Wie hoch ist das Molekulargewicht des Erythropoietins, daß es in der Niere ausgeschieden wird?

DUKES: Ich habe die Angabe eines Molekulargewichts für Erythropoietin vermieden, wie Sie vielleicht gemerkt haben. Es gibt Hinweise dafür, daß es um 60000 liegt, es ist aber keineswegs gesichert. Nun, ein Molekül von dieser Größe würde vielleicht gerade noch ausgeschieden. Jedenfalls kann man sagen, daß es im Urin meistens in einem bestimmten Verhältnis zum Plasmaspiegel vorhanden ist.

OELKERS (Berlin): Wo vermutet man den Receptor für die Erythropoietinausschüttung, z. B. bei der Höhenkrankheit? Die Niere dürfte hierfür doch wohl ungeeignet sein, weil die Niere eine hohe Durchblutung hat und die arterio-venöse Sauerstoffdifferenz sehr gering ist.

Und 2.: Sie haben gezeigt, daß nach Auswaschen des Erythropoietins in Gewebekulturen der Glucosamineinbau wieder abfällt. Ist derselbe Versuch mit radioaktivem Eisen gemacht worden? Weiß man, ob der Eiseneinbau vielleicht über längere Zeit weitergeht?

DUKES: Zur Frage der Niere: Es gibt Perfusionsstudien, wonach bei Durchströmung der Niere mit sauerstoffarmem Blut oder mit anderen Perfusionsflüssigkeiten vermehrt Erythropoietin abgegeben wird. Danach dürfte die Niere doch Receptoren für eine verminderte Sauerstoffspannung besitzen. Andererseits erscheint es auch mir wahrscheinlich, daß es darüber hinaus Receptoren in anderen Organen geben müßte.

Nach Auswaschung des Erythropoietins geht der gesteigerte Eiseneinbau sogar bedeutend länger als der des Glucosamins.

KARLSON: Noch eine Frage an Herrn Kollegen OELKERS: Ist es wirklich nötig, daß im Receptororgan eine erhebliche arterio-venöse Sauerstoffdruckdifferenz vorliegt, oder reicht es nicht vielmehr aus, wenn das Receptororgan sehr empfindlich auf kleine Schwankungen in der Sauerstoffspannung anspricht?

OELKERS: Das ist durchaus möglich. Ich dachte allerdings, daß sich in der Niere auf Grund der besonderen Verhältnisse eine Anoxie am spätesten bemerkbar machen müßte. Im Fall der Panmyelopathie mit erhöhtem Erythropoietinspiegel im Plasma ist die Sauerstoffspannung im Blut auch nicht vermindert.

DEUL (Veest, Holland): Sollte man nicht annehmen, daß das Gehirn, das ja auf Sauerstoffmangel am empfindlichsten reagiert, das Organ sein sollte, das die Erythropoietinausschüttung reguliert?

DUKES: Ja, das Gehirn hat auch etwas damit zu tun. Man hat Teile des Gehirns exstirpiert und gefunden, daß bestimmte Regionen in die Regulation der Erythropoietinsynthese eingeschaltet sind.

ZILLIKEN: Wenn keine weiteren Fragen mehr sind, dann danke ich nochmals Herrn DUKES und schließe damit die Sitzung.

Hormonhemmer — Untersuchungen mit Testosteron-Antagonisten

Von F. Neumann, R. von Berswordt-Wallrabe,
W. Elger und H. Steinbeck

Hauptlaboratorium der Schering AG, Berlin

Mit 13 Abbildungen

Das Gebiet der Hormonantagonisten ist relativ jung. In den letzten 10 bis 15 Jahren wurden aber bereits große Fortschritte erzielt. Ich möchte nur an das Aldacton als Aldosteron-Antagonist erinnern. Bei den Sexualhormonen wurde vor allem nach Antioestrogenen und Antiandrogenen gesucht[22]. In meinem Vortrag beschränke ich mich auf Untersuchungen mit Antiandrogenen.

Bei den bisher in der Literatur beschriebenen Antiandrogenen handelt es sich überwiegend um Steroide, aber auch einige Nichtsteroide wurden beschrieben. Obwohl etwa 50 verschiedene Substanzen als Antiandrogene klassifiziert wurden, haben diese, vielleicht mit einer Ausnahme, bisher weder in der Klinik noch experimentell Bedeutung erlangt. Dies hat verschiedene Gründe: Die meisten Substanzen waren viel zu schwach oder nur in bestimmten Dosisbereichen antiandrogen, in höheren Dosen dagegen selbst androgen. Andere Substanzen zeigten eine antiandrogene Wirkung nur in bestimmten Testanordnungen.

Wir befassen uns seit 1963 mit Antiandrogenen. Unsere überwiegend biologisch-experimentell ausgerichteten Untersuchungen gingen dabei in zwei Richtungen: 1. dienten sie der Erforschung eventueller klinischer Anwendungsmöglichkeiten und 2. haben wir Antiandrogene als Mittel benutzt, um testosteronabhängige Vorgänge zu untersuchen.

In Abb. 1 sind die Strukturformeln einiger relativ stark wirksamer Antiandrogene dargestellt, die in unserem Hause synthetisiert wurden*. Es sind Hydroxyprogesteronderivate und einige

---

* Die meisten dieser Verbindungen wurden von Dr. Wiechert (Schering AG) hergestellt.

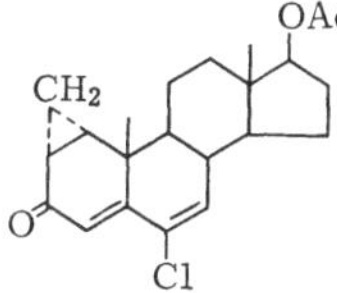
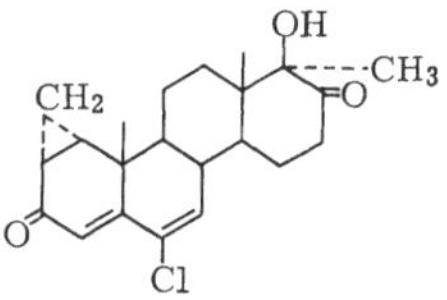
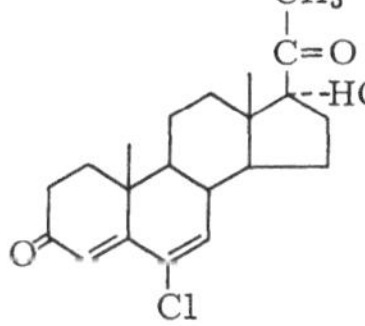
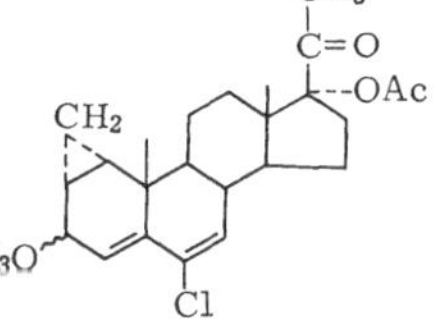
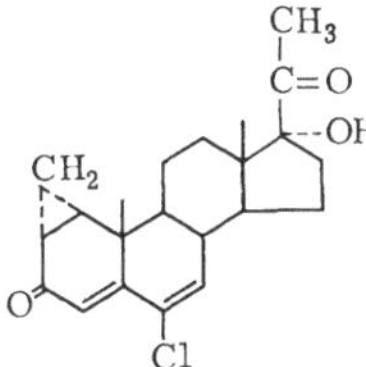

1,2α-Methylen-6-chlor-Δ^{4,6}-pregnadien-17α-ol-3,20-dion-17-acetat

1α-Methyl-6-chlor-Δ^{4,6}-pregnadien-17α-ol-3,20-dion-17-acetat

6-Chlor-Δ^{1,4,6}-pregnatrien-17α-ol-3,20-dion-17-acetat

1,2α-Methylen-6-chlor-Δ^{4,6}-pregnadien-3, 17α-diol-20-on-17-acetat

1,2α-Methylen-Δ^4-pregnen-17α-ol-3,20-dion-17-acetat

17β-Methyl-1,2α-methylen-D-homo-Δ^{4,6}-androstadien-17α-ol-3,17a-dion-17-acetat

17aα-Methyl-1,2α-methylen-D-homo-Δ^{4,6}-androstadien-17aβ-ol-3,17-dion

1,2α-Methylen-6-chlor-Δ^{4,6}-androstadien 17β-ol-3-on-17-acetat

17aα-Methyl-1.2α-methylen-6-chlor-D-homo-Δ^{4,6}-androstadien-17 aβ-ol-3,17-dion

17aα-Methyl-1,2α-methylen-6-chlor-D-homo-Δ^{4,6}-androstadien-17aβ-ol-3,17-dion-, 17a-acetat

6-Chlor-Δ^{4,6}-pregnadien-17α-ol-3,20-dion

1,2α-Methylen-6-chlor-Δ^{4,6}-pregnadien-3, 17α-diol-20-on-3-methoxy-17-acetat

1,2α-Methylen-Δ^{4,6}-pregnadien-17α-ol-3,20-dion-17-acetat

1,2α-Methylen-6-chlor-Δ^{4,6}-pregnadien-17α-ol-3,20-dion

1,2α-Methylen-6-chlor-24-nor-Δ^{4,6}-choladien-17α, 20-diol-3-on-23-säure-23→17-lacton

Abb. 1. Strukturformeln von Antiandrogenen (die meisten dieser Verbindungen wurden von Dr. WIECHERT, Schering AG, synthetisiert)

Androstenderivate. Bei letzteren ist hervorzuheben, daß der D-Ring ein Sechserring ist. Es handelt sich also um D-homo-Verbindungen. Alle am $C^{17}OH$ veresterten Verbindungen besitzen neben ihren antiandrogenen auch noch starke gestagene Eigenschaften, die freien Alkohole sind dagegen ausschließlich antiandrogen wirksam.

Nach unseren Erfahrungen kann man prinzipiell mit Antiandrogenen alle in ihrer Funktion oder Morphologie androgenabhängigen Organsysteme beeinflussen. In Tab. 1 sind einige der Funktionen des Testosterons aufgeführt.

Tabelle 1. *Funktionen des Testosterons*

Im erwachsenen Organismus	Im fetalen Organismus
1. Hoden: trophische Wirkung auf das Keimepithel	1. Differenzierung männlicher ableitender Geschlechtswege (Wolffscher Gang)
2. Accessorische Geschlechtsdrüsen: Entfaltung, Funktion	2. Entwicklung männlicher accessorischer Geschlechtsdrüsen
3. Talgdrüsen: Funktion	3. Differenzierung männlicher äußerer Geschlechtsorgane
4. Stoffwechsel: anabole Wirkung	
5. Knochen:	4. Differenzierung des männlichen Sexualzentrums (Regler der späteren Gonadotropinsekretion)
a) Schluß der Epiphysenfuge (Reifung)	
b) Verhinderung einer Osteoporose	5. Differenzierung des männlichen „Mating-Center" (Kopulationsverhalten, Triebrichtung)
6. Libido und Potenz: fördernd	
7. Rückkopplungsmechanismus: Hemmung der Gonadotropinsekretion	6. Männliche Differenzierung der Milchdrüsen

Beim erwachsenen männlichen Individuum ist Testosteron notwendig für die Aufrechterhaltung der Hodenfunktion[8], der Funktion der accessorischen Geschlechtsdrüsen und Talgdrüsen[4, 39, 40] und der Libido[45]. Ferner ist es der Mittler des Rückkopplungsmechanismus zwischen den Gonaden und dem Sexualzentrum[41]. Es hat bestimmte Stoffwechselwirkungen, z. B. eiweißanabole Eigenschaften, auch spielt es eine Rolle bei der Knochenreifung. Bei männlichen Embryonen ist Testosteron notwendig zur Differenzierung der Mehrzahl der Sexualorgane[5, 16], der Milchdrüsen[6, 17] und bestimmter nervöser Zentren[11, 12, 20, 28, 36, 37].

In meinem Vortrag werde ich vor allem über jene Versuche berichten, in denen wir Antiandrogene eingesetzt haben, um jene

Prozesse zu untersuchen, bei denen die Bedeutung des Testosterons nicht völlig geklärt ist. Vielleicht sollte eingangs noch erläutert werden, wie sich Antiandrogene von Oestrogenen in ihrem Wirkungsmechanismus grundsätzlich unterscheiden, denn vielfach herrscht auch heute noch die Meinung vor, daß sich Androgene und Oestrogene antagonistisch verhalten. Abb. 2 soll den unterschiedlichen Wirkungsmechanismus veranschaulichen.

Der Syntheseort des Testosterons ist der Hoden. Das Testosteron unterhält z. B. die Funktion der accessorischen Geschlechtsdrüsen, hier ist die Prostata dargestellt. Nun führt sowohl die Gabe von Oestrogenen als auch Antiandrogenen an intakte Tiere zu einer Atrophie der Prostata (a und c). Wie kommt nun der Effekt von Oestrogenen zustande? Oestrogene wirken hemmend auf die Produktion von ICSH[10, 21] und führen somit zu einer herabgesetzten Testosteronsynthese im Hoden, die accessorischen Geschlechtsdrüsen atrophieren (c). Antiandrogene wirken direkt am Erfolgsorgan, wie Abb. 2b veranschaulicht. Man kann das beweisen, indem man mit hypophysektomierten Tieren arbeitet, bei denen keine Testosteronsynthese mehr stattfindet. Es kommt zu einer Atrophie der accessorischen Geschlechtsdrüsen (Abb. 2f), die durch Gabe von Testosteron aufgehalten werden kann, auch wenn die Tiere gleichzeitig Oestradiol erhalten. Oestrogene können also exogenes Testosteron nicht hemmen (Abb. 2d). Dagegen ist ein Antiandrogen in der Lage, Testosteroneffekte direkt aufzuheben (Abb. 2e).

Bei kastrierten männlichen Mäusen wird das Wachstum der Samenblasen durch Gabe von Testosteronpropionat stimuliert.

Mit steigender Dosis des Antiandrogens — wir haben hauptsächlich mit Cyproteronacetat (1,2 α-Methylen-6-chlor-$\Delta^{4,6}$-pregnadien-17 α-ol-3,20-dion-17 α-acetat) und dem freien Alkohol dieser Verbindung (1,2 α-Methylen-6-chlor-$\Delta^{4,6}$-pregnadien-17 α-ol-3,20-dion = Cyproteron) gearbeitet — wird dieser Effekt progredient aufgehoben. Zur völligen Aufhebung der Testosteronpropionatwirkung ist etwa die 30fache Menge des Antiandrogens notwendig[24].

Da nach unserer Meinung die Wirkung von Cyproteronacetat *in vivo* im Sinne eines kompetitiven Antagonismus zum Testosteron verstanden werden kann, versuchten wir, ob sich die Kriterien der Enzymkinetik für den kompetitiven Antagonismus auf die *in vivo* an der Ratte erhaltenen Meßergebnisse übertragen lassen.

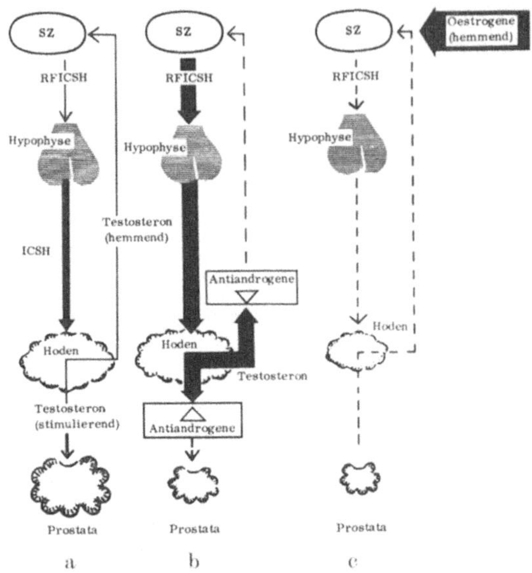

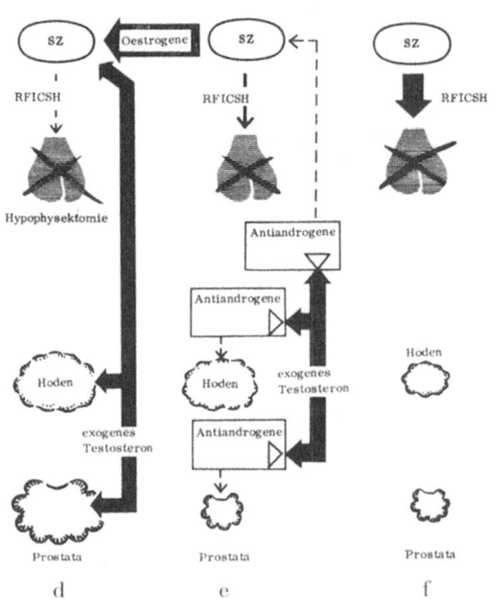

Abb. 2. Wirkungsmechanismus: Antiandrogene—Oestrogene

Als Substrat diente Testosteronpropionat, als Inhibitor das Cyproteronacetat. Unter der Annahme, daß Testosteronpropionat in der schließlich zur Gewichtszunahme der Prostata führenden Reaktionskette die langsamste und damit geschwindigkeitsbestimmende Reaktion steuert, wurde die Gewichtszunahme der Prostata in der Zeiteinheit als Meßgröße für die Reaktionsgeschwindigkeit eingesetzt.

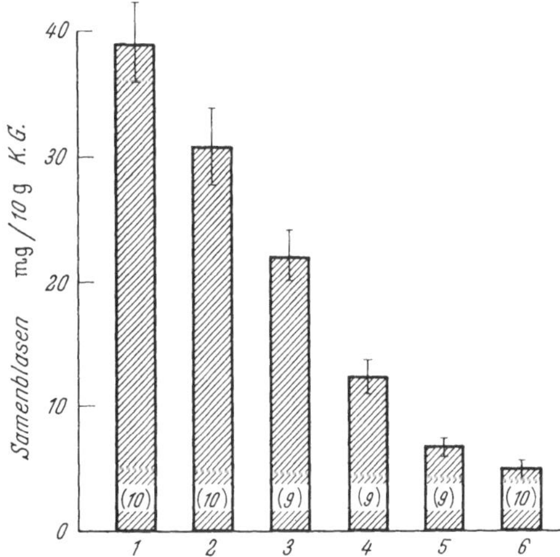

Abb. 3. Einfluß von Cyproteronacetat (1,2 α-Methylen-6-chlor-$\Delta^{4,6}$-pregnadien-17 α-ol-3,20-dion-17 α-acetat) auf das durch Testosteronpropionat stimulierte Samenblasenwachstum kastrierter männlicher Mäuse. Behandlung: täglich über 7 Tage subcutan. 1 = 003 mg Testosteronpropionat, 2 = 003 mg TP · 003 mg Cyproteronacetat, 3 = 003 mg TP · 01 mg Cyproteronacetat, 4 = 003 mg TP · 03 mg Cyproteronacetat, 5 = 003 mg TP · 1 mg Cyproteronacetat, 6 = unbehandelte kastrierte Kontrolle, I = mittlerer Fehler, ( ) = Tierzahl

Bei der graphischen Darstellung der Gewichtsveränderung in der Zeiteinheit gegen die verabreichte Testosterondosis ergab sich eine typische Substratsättigungskurve.

Bei der Darstellung nach LINEWEAVER-BURK schneiden sich die mit Testosteronpropionat (TP) bzw. mit TP + Cyproteronacetat erhaltenen Geraden auf der Ordinate.

Wir sind uns über die Fragwürdigkeit dieses Vorgehens ganz im klaren, wollen die Versuche aber fortsetzen, da es uns nicht ausgeschlossen erscheint, daß in ganz speziellen Fällen die vorstehend skizzierte Übertragung der Enzymkinetik auf die *in vivo*-Versuche erlaubt ist.

Wie bereits eingangs erwähnt, wird auch die Talgdrüsenfunktion durch Androgene beeinflußt. Abb. 4a zeigt einen histologischen Schnitt der Rückenhaut einer kastrierten Maus nach 28tägiger Be-

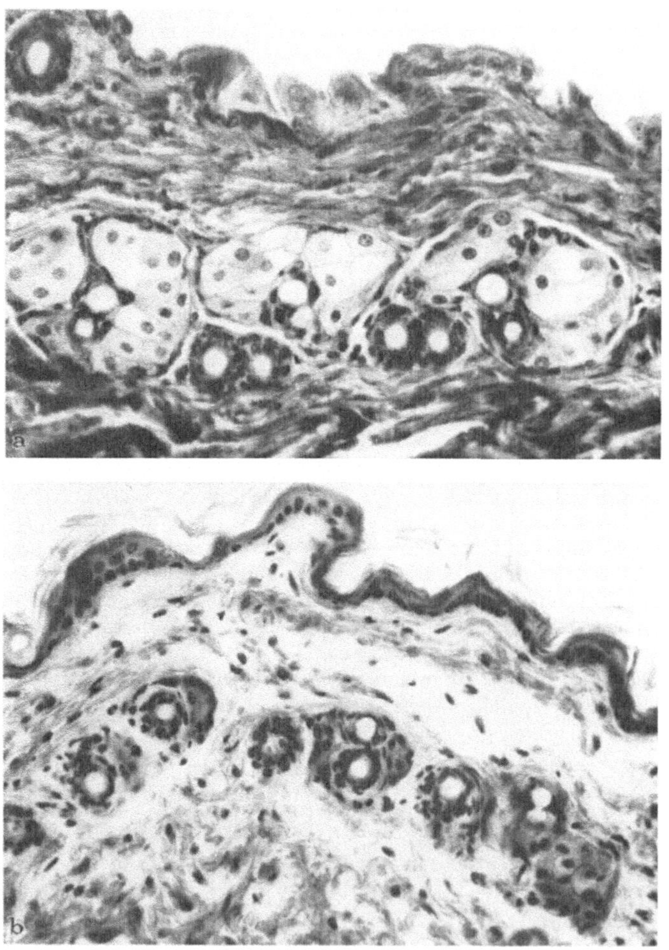

Abb. 4. Rückenhautschnitte von kastrierten männlichen Mäusen. a Nach 4wöchiger subcutaner Behandlung mit 0,5 mg/Tier Testosteronpropionat jeden 2. Tag und b nach gleichzeitiger Behandlung mit 1 mg Cyproteron-acetat. Vergr.: etwa 350mal. Färbung: Mallory

handlung mit 0,5 mg Testosteronpropionat jeden 2. Tag/Tier subcutan. Diese Behandlung führt zu einer starken Stimulierung der
Talgdrüsen, die sich zu großen Komplexen um die Haarfollikel
zusammenschließen. Bei gleichzeitiger Gabe von Cyproteronacetat
über den gleichen Zeitraum sind die Talgdrüsen fast völlig atrophisch (Abb. 4b).

Die Gabe sehr hoher Antiandrogendosen führt auch zu einer
Hemmung der Spermiogenese und damit zur Sterilität der Tiere[19].
Abb. 5a zeigt den Hoden eines Hundes nach 30tägiger Behandlung
mit täglich 10 mg/kg Cyproteronacetat. Reife Spermien sind nicht
mehr vorhanden, das Keimepithel ist atrophisch.

Ein anderes Beispiel einer Antiandrogenwirkung: Die Aktivität
der sauren Phosphatase in der Prostata einer kastrierten Maus
wird durch Behandlung mit 0,3 mg Testosteronpropionat gestei-
gert. Bei gleichzeitiger Gabe von 2,0 mg Cyproteronacetat ist die
Aktivität der sauren Phosphatase fast völlig verschwunden.

Bei kleinen Nagern, wie Ratten und Mäusen, bestehen hinsichtlich der Enzymaktivität und des Enzymmusters in den Nieren
Unterschiede bei weiblichen und männlichen Tieren. Die Gabe
eines Antiandrogens an männliche Tiere führt zu einer Änderung
der Enzymaktivität und des Verteilungsmusters in weiblichem
Sinne. Am deutlichsten wird dies bei der alkalischen Phosphatase,
der Beta-Glucuronidase und der Indoxylesteraseaktivität.

Zunächst zur Aktivität der alkalischen Phosphatase. Ohne auf
Einzelheiten des Verteilungsmusters näher eingehen zu wollen, fällt
folgendes auf: die Aktivität ist relativ gering bei männlichen Tieren.
Sie steigt nach der Kastration der Tiere an und entspricht dann
etwa derjenigen weiblicher Tiere. Bei Gabe von Cyproteronacetat
an normale männliche Mäuse steigt die Aktivität des Enzyms in
gleicher Weise an wie nach der Kastration. Bei Behandlung kastrierter männlicher Tiere mit Testosteronpropionat läßt die Aktivität
wieder nach.

Die Indoxylesterase verhält sich der alkalischen Phosphatase
genau entgegengesetzt. Die Aktivität ist bei männlichen Tieren
hoch. Sie sinkt nach der Kastration ab und entspricht jetzt der
geringen Aktivität weiblicher Tiere. Durch die Behandlung mit
einem Antiandrogen fällt die Aktivität ähnlich ab wie nach der
Kastration. Nach Behandlung kastrierter männlicher Tiere mit

Testosteronpropionat kommt es wieder zu einem Anstieg der Indoxylesteraseaktivität, bei gleichzeitiger Gabe von Cyproteronacetat bleibt dieser Effekt aus.

Auch die Beta-Glucuronidaseaktivität ist bei männlichen Mäusen oder Ratten höher als bei weiblichen Tieren. Die Aktivität dieses Emzyms und der Indoxylesterase gehen einander parallel.

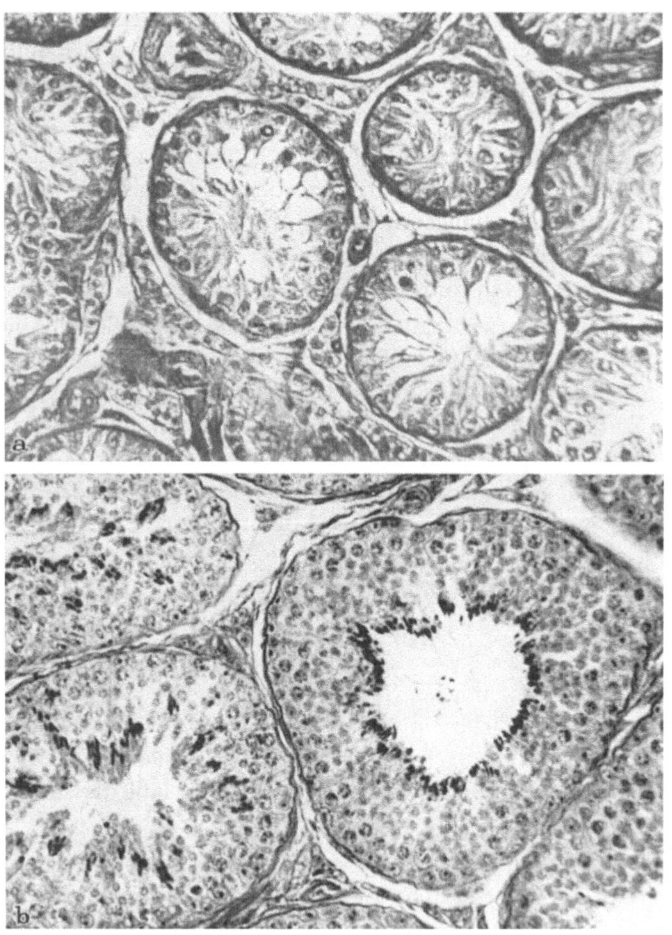

Abb. 5. Einfluß von Cyproteronacetat auf die Hodenfunktion und Spermiogenese von Hunden. a 30tägige i.m. Behandlung mit täglich 10 mg/kg.
b Kontrolle. Vergr.: 275mal. Färbung: Azan

Dementsprechend steigt die Aktivität dieses Enzyms bei einer männlichen kastrierten Maus nach Behandlung mit Testosteronpropionat (0,1 mg/d). Bei gleichzeitiger Behandlung mit Cyproteronacetat (1,0 mg/d) sinkt sie sehr stark ab.

Am Zentralnervensystem haben Antiandrogene, wie Testosteron auch, zwei Angriffspunkte: 1. hemmen sie die Libido, die bei männlichen Individuen offensichtlich durch Testosteron unterhalten wird, und 2. heben sie die Gonadotropin-hemmende Wirkung von endogenem Testosteron oder exogen zugeführten Androgenen auf[26, 32, 41]. Die Hemmung der Libido unter Antiandrogenen ist uns gleich zu Anfang unserer Experimente aufgefallen[19]. Wir haben aber diesem Gesichtspunkt nicht allzuviel Bedeutung beigemessen, obwohl Antiandrogene gerade für die Behandlung der Hypersexualität bei Männern klinisch Bedeutung erlangt und die Phantasie der Tagespresse angeregt haben.

Der zweite Gesichtspunkt interessierte uns, weil es mit Antiandrogenen möglich ist, den Rückkopplungsmechanismus zwischen Gonaden und Hypophysenzwischenhirnsystem[14] in ähnlicher Weise zu beeinflussen wie durch eine Kastration[25, 33]. Diese Beeinflussung des Rückkopplungsmechanismus ist nur mit solchen Antiandrogenen möglich, die keine gestagenen (= zentral hemmenden) Eigenschaften haben. Wir haben mit Cyproteron gearbeitet, aber auch alle anderen der in Abb. 1 dargestellten freien Alkohole (am $C^{17}OH$ unverestert) sind dazu geeignet. In Abb. 6 ist schematisch die Regulation der männlichen inkretorischen Funktionen des Hypophysenzwischenhirnsystems und der Hoden unter normalen und experimentellen Bedingungen (feed-back) dargestellt.

Unter normalen Verhältnissen besteht ein Gleichgewicht zwischen der Funktion der Gonaden und der Hypophyse (Abb. 6a). Dieser Regelkreis wird gestört durch die Kastration der Tiere (Abb. 6b). Es entfällt die Bremswirkung des Testosterons, und es resultiert daraus ein Anstieg der gonadotropen Partialfunktion der Hypophyse im Sinne einer gesteigerten Synthese und Freisetzung gonadotroper Hormone. Umgekehrt führt die Gabe von Testosteron zu einer Hemmung des Hypophysenzwischenhirnsystems (Abb .6c). Durch die Gabe von Antiandrogenen (Abb. 6d) wird die Bremswirkung des Testosterons auf das Sexualzentrum ebenfalls aufgehoben, und das Hypophysenzwischenhirnsystem reagiert mit einer

15*

Aktivitätssteigerung wie bei einem echten Mangel an Testosteron (der ja hier nicht gegeben ist).

Zunächst war das eine Hypothese, die wir aber in weiteren Versuchen dann beweisen konnten. Ein geänderter Funktionszustand der Hypophyse spiegelt sich im Zellbild dieses Organs wieder. Bei einer Steigerung der gonadotropen Funktion tritt eine Vermehrung und Hypertrophie jener Zellelemente auf, in denen Gonadotropine entstehen[7]. Abb. 7b zeigt einen histologischen Schnitt des Hypophysenvorderlappens eines männlichen Tieres, das 3 Wochen lang

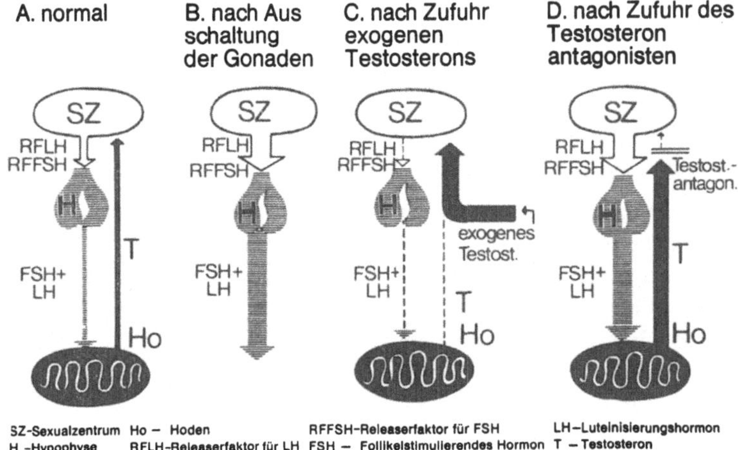

Abb. 6. „Feed-back"-Mechanismus bei männlichen Individuen

mit täglich 10 mg Cyproteron subcutan behandelt wurde, daneben (Abb. 7a) das gleiche Organ eines unbehandelten Kontrolltieres. Bei den behandelten Tieren fällt deutlich die Hypertrophie der basophilen Zellelemente auf (in der Schwarz-Weiß-Abbildung dunkel angefärbte Zellen).

Die beschriebenen morphologischen Veränderungen im Hypophysenvorderlappen sind als Ausdruck einer erhöhten Gonadotropinproduktion und -sekretion aufzufassen. Es galt nun, den direkten Nachweis der erhöhten Gonadotropinsekretion zu erbringen. Wir bedienten uns dabei zunächst der Parabiosetechnik.

Zwei infantile Tiere, ein weibliches und ein männliches, werden unter Eröffnung der Bauchhöhle operativ vereinigt. Wird das männliche Tier nicht kastriert, so bleibt das Genitale des weiblichen

Tieres in seinem infantilen Zustand bestehen. Nach der Kastration des männlichen Tieres kommt es dagegen zu einem Anstieg der Gonadotropinsekretion beim männlichen Partner und als Folge davon beim weiblichen Partner zu einer vorzeitigen Geschlechts-

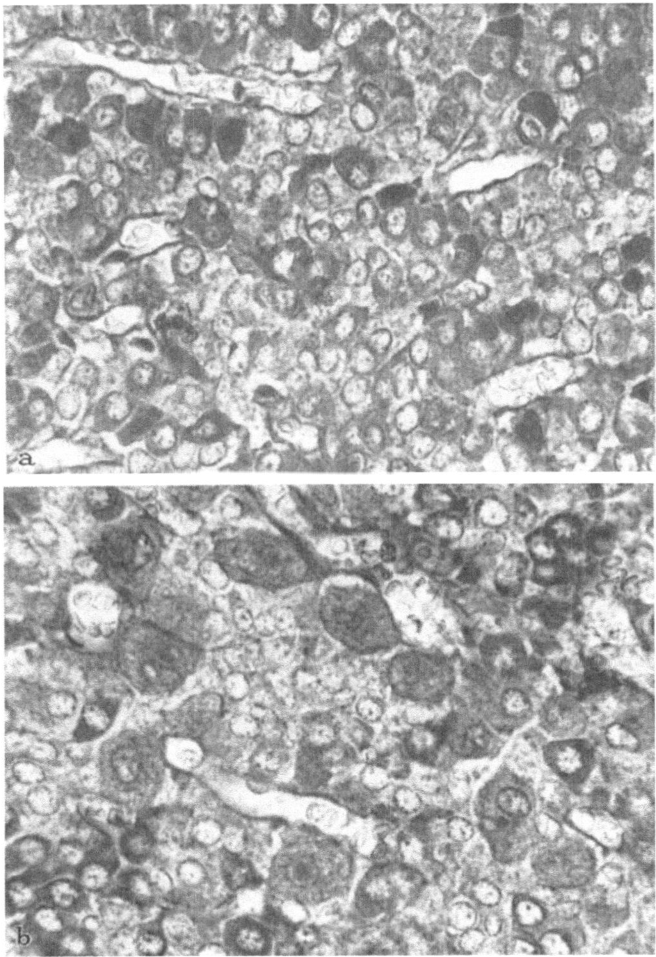

Abb. 7. Hypophysenvorderlappen erwachsener männlicher Ratten. a Kontrolle. b Nach 3wöchiger Behandlung mit täglich 10 mg Cyproteron/Tier subcutan (man beachte die starke Hypertrophie der basophilen Zellen). Vergr.: 350mal. Färbung: Trichrom

reife, erkennbar unter anderem an der Vaginalöffnung und am Auftreten eines Oestrus. — Wird ein nichtkastrierter männlicher Partner mit einem Antiandrogen behandelt, so reagiert ein hoher Prozentsatz der weiblichen Tiere mit einer vorzeitigen Geschlechtsreife, wie nach Kastration des männlichen Tieres. Dieser Versuch beweist, daß die Gonadotropinsekretion beim männlichen Partner beträchtlich erhöht worden ist.

In einem weiteren Versuch wurde dann FSH in Hypophysen und Serum von männlichen Tieren bestimmt, die 14 Tage lang mit Cyproteron behandelt worden waren. Wie erwartet, stieg der FSH-Gehalt im Serum der behandelten männlichen Tiere erheblich an (Tab. 2).

Tabelle 2. *Einfluß von Cyproteron auf den FSH-Gehalt in Hypophysen und Serum juveniler männlicher Ratten*
(m Einheiten NIH-FSH-S$_1$)

Versuchsgruppe	Hypophyse*	Serum**
Intakt	2170	keine meßbaren Mengen
Cyproteron***	875	65
Orchidektomie	78	156

* Hypophysen: 1 Äquivalent einer Hypophyse/Empfängertier
** Serum: 18 ml/Empfängertier
*** Die Spenderratten erhielten über 12 Tage täglich 10 mg Cyproteron subcutan

Vermutlich wird auch die LH-Sekretion erhöht. Wir konnten dies wegen methodischer Schwierigkeiten bisher noch nicht nachweisen, aber aus Untersuchungen von Bloch und Davidson[1], die männlichen Ratten kleinste Mengen von Cyproteron in den Hypothalamus implantierten, läßt sich dieser Schluß ziehen. Die Antiandrogenmengen reichten aus, um lokal die Bremswirkung des endogenen Testosterons auf das Sexualzentrum aufzuheben, nicht aber, um peripher an anderen testosteronabhängigen Organsystemen irgendeinen Effekt auszuüben. Unter diesen Versuchsbedingungen kam es zu einem Wachstum der accessorischen Geschlechtsdrüsen, was dafür spricht, daß Testosteron im Hoden vermehrt gebildet wurde. Schließlich fanden Voigt und Klosterhalfen[43] bei Männern unter Cyproteron eine Erhöhung der Testosteronausscheidung im Urin um den Faktor 4 und von Epitestosteron um den Faktor 3, wenn auch von diesen Untersuchern keine erhöhte Gonadotropinsekretion festgestellt wurde.

Auf Grund dieser Ergebnisse lag es nahe, zu vermuten, daß man mit Antiandrogenen auch die Hemmwirkung von exogen zugeführten Androgenen auf die Gonadotropinsekretion beeinflussen kann. Dies war dann tatsächlich der Fall, übrigens auch bei weiblichen Tieren. Bekanntlich führt die Gabe von Testosteron oder anderen Androgenen an weibliche Tiere auch zu einer Hemmung der Ovulation und des Cyclus. Durch gleichzeitige Gabe von Cyproteron, in diesem Falle 10 mg/Tier subcutan, konnte der Hemmeffekt von 0,3 mg Testosteronpropionat aufgehoben werden, die Tiere ovulierten wieder.

Mit Hilfe von Antiandrogenen war es auch möglich, zu untersuchen, welche Effekte „reines" FSH am Hoden ausübt. Die Problemstellung wird in Abb. 8 veranschaulicht. Auf die Hoden wirken das ICSH oder LH und das FSH[8]. Die Funktion des ICSH ist bekannt, es stimuliert die Testosteronsynthese in den Zwischenzellen der Hoden. Alle Effekte des ICSH sind letztlich Testosteroneffekte. Das FSH dagegen hat keinen Mittler, aber wie und wo greift es am Hoden an? Diese Frage konnte bisher deshalb nicht beantwortet werden, weil es kein FSH-Präparat gibt, das völlig frei von ICSH-Aktivitäten ist. Man hat vermutet, daß FSH die Ausbildung reifer Spermien fördert oder daß es bestimmte Strukturelemente, die nicht zum eigentlichen Keimepithel zählen, im Hoden stimuliert.

Wir haben mit hypophysektomierten Ratten und einem FSH-Präparat aus Schafshypophysen gearbeitet. Die Überlegung war folgende: Durch die gleichzeitige Gabe dieses FSH-Präparates und eines Antiandrogens ist es möglich, die Effekte des vom ICSH stimulierten Testosterons auszuschalten. Die nächsten Abbildungen zeigen die Befunde.

In Tab. 3 sind die Hodengewichte hypophysektomierter Tiere nach Behandlung mit dem FSH-Präparat allein oder in Kombination mit einem Antiandrogen dargestellt. Die Hypophysektomie führt zu einer starken Atrophie dieses Organs. Durch die Gabe des Schafshypophysen-FSH-Präparates kann diese Atrophie verhindert werden. Wenn die Tiere jedoch gleichzeitig ein Antiandrogen erhalten, so ist das FSH-Präparat nicht mehr in der Lage, die der Hypophysektomie folgende Atrophie der Hoden zu verhindern. Aus den histologischen Befunden (Abb. 9) wird folgendes ersichtlich:

Man erkennt in Abb. 9b die starke Atrophie des Hodens nach
der Hypophysektomie. Durch Behandlung mit dem FSH-(LH)-
Präparat (Abb. 9c und e) bleibt die Funktion dieses Organs mehr
oder weniger erhalten. Der Hoden in Abb. 9e zeigt sogar noch eine
Spermiogenese. Bei simultaner Gabe des Antiandrogens ähnelt das

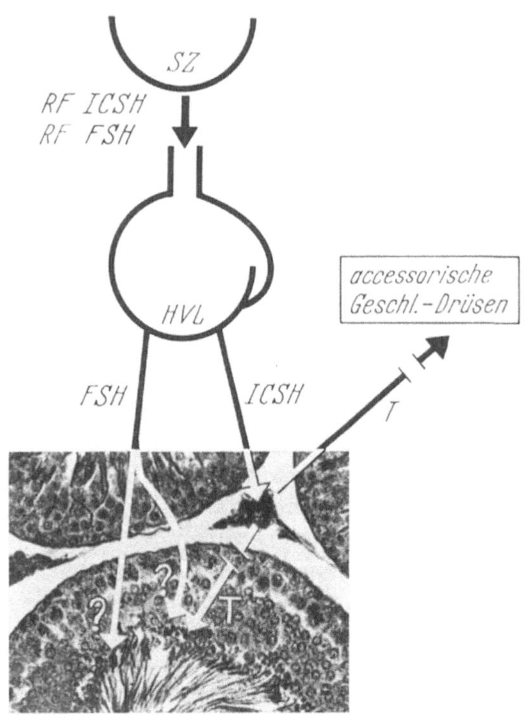

Abb. 8. Effekte des „reinen" FSH am Hoden. SZ = Sexualzentrum,
RF ICSH = Releaserfaktor für ICSH, RF FSH = Releaserfaktor für FSH,
HVL = Hypophysenvorderlappen, ICSH = Interstitial-Zell-stimulierendes
Hormon, FSH = Follikelstimulierendes Hormon, T = Testosteron

Hodenbild wieder dem unbehandelter hypophysektomierter Tiere,
die Tubuli sind vollkommen atrophisch (Abb. 9d und f).

Es läßt sich also sagen: FSH allein hat keinen Einfluß auf die
Hodenfunktion.

Ich möchte Ihnen jetzt kurz über Versuche berichten, in denen
wir Antiandrogene eingesetzt haben, um die Rolle des Testosterons

bei verschiedenen Differenzierungsvorgängen des männlichen Embryos zu studieren. Zunächst zu Sexualdifferenzierung: Die Sexualdifferenzierung findet noch in der Fetalzeit statt[15, 16, 17, 18], bei den einzelnen Spezies früher oder später. Es war dabei schon bekannt, daß Testosteron, das bereits in der Embryonalzeit gebildet wird, eine wesentliche Rolle bei der Differenzierung der männlichen Sexualorgane spielt. Mit Hilfe von Antiandrogenen konnten in den entsprechenden Differenzierungsphasen die Effekte des Testosterons ausgeschaltet werden[5, 27, 35]. Wir haben unsere Untersuchungen bei verschiedenen Spezies durchgeführt. Die Ratten wur-

Tabelle 3. *Einfluß von FSH und Cyproteronacetat auf die Hodengewichte erwachsener hypophysektomierter Ratten. Gewichtsverlust des Hodens nach 24-tägiger subcutaner Behandlung*

Substanz	Tagesdosis	Tierzahl	Gewichtsverlust der Hoden
	in mg/100 g KG		in mg/100 g KG
Kontrollen	—	10	975 ± 57
FSH	0,05	17	579 ± 37
FSH	0,50	15	134 ± 18
FSH +	0,05		
Cyproteronacetat	5,00	13	831 ± 83
FSH +	0,50		
Cyproteronacetat	5,00	17	959 ± 44

KG = Körpergewicht

den teils vom 17. bis 20. Tag, in einer anderen Serie vom 13. Tag an bis zum Ende der Schwangerschaft mit Cyproteronacetat behandelt. Kaninchen wurden vom 13. bis 24. Tag der Schwangerschaft behandelt. Die normale Tragezeit bei Ratten beträgt etwa 22 Tage, bei Kaninchen 31 bis 32 Tage. Unter der Einwirkung des Antiandrogens differenzierte sich das äußere Genitale von männlichen Feten völlig weiblich. Bei den männlichen Ratten entwickelte sich sogar eine Vagina[34]. Die Ausbildung der Vagina ist bei Ratten zum Zeitpunkt der Geburt noch nicht völlig abgeschlossen. Wenn man sowohl die Mütter in der Schwangerschaft als auch die Neugeborenen in den ersten Wochen post partum mit einem Antiandrogen behandelt, so besitzen diese Tiere später eine äußerlich in Erscheinung tretende Vagina, die auf Oestrogengaben genauso reagiert wie bei weiblichen Tieren.

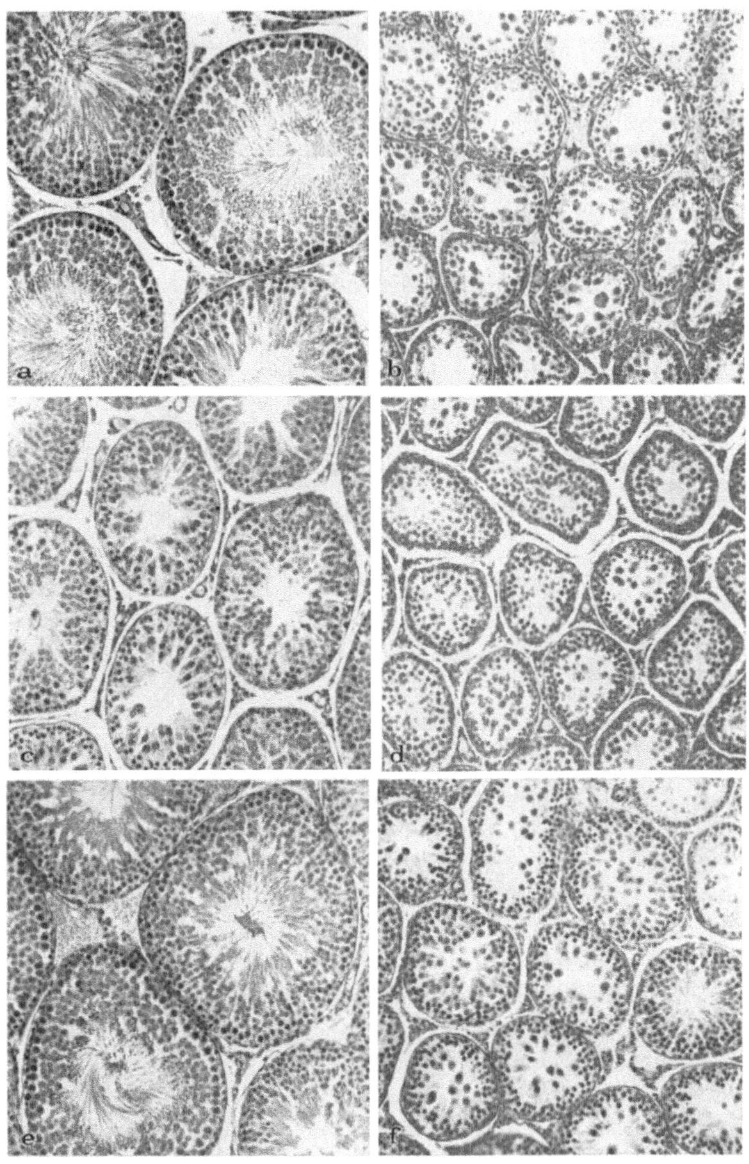

Abb. 9 (Legende siehe nächste Seite)

Antiandrogene hemmen auch die Anlagen der accessorischen Geschlechtsdrüsen. Auf die Veränderungen am inneren Genitale nach Ausschaltung des fetalen Testosterons möchte ich hier nicht näher eingehen. Vielleicht ergibt sich dazu Gelegenheit in der Diskussion. Zusammenfassend kann gesagt werden, daß die Androgene des männlichen Feten notwendig sind zur Ausbildung der männlichen ableitenden Geschlechtswege, zur Differenzierung der accessorischen Geschlechtsdrüsen sowie zur Differenzierung des männlichen äußeren Genitales. Andererseits hemmen Androgene die Ausbildung einer Vagina bei männlichen Tieren.

Wir haben Antiandrogene auch eingesetzt, um die Rolle des Testosterons für die Milchdrüsendifferenzierung zu untersuchen. Die Experimente wurden an Ratten und Mäusen durchgeführt[6, 29]. Normalerweise entwickeln sich bei männlichen Tieren dieser Spezies keine Saugwarzen, wohl aber nach Behandlung mit Antiandrogenen in der Fetalzeit. Die Saugwarzen männlicher Tiere sind dann von denen weiblicher Tiere nicht zu unterscheiden. Bei den Feten der Maus ist die Milchdrüsenanlage zunächst bei beiden Geschlechtern gleichermaßen ausgeprägt, aber am 15. Tag der Embryonalentwicklung tritt eine Zerstörung der Drüsenanlage bei männlichen Tieren auf. In Abb. 10 wird dies veranschaulicht.

Das den primären Drüsensproß umgebende Mesenchym ist bei den männlichen Tieren sehr stark entwickelt und scheint den Drüsensproß regelrecht zu strangulieren. Die Drüsenzellen werden pyknotisch. Dieser Prozeß wird offensichtlich durch Testosteron ausgelöst, denn wenn man die Mütter mit einem Antiandrogen über diese Phase der Differenzierung behandelt, so unterbleibt dieser Zerstörungsprozeß, und der Drüsensproß entwickelt sich in gleicher Weise weiter wie bei weiblichen Tieren[6]. Wir haben die Milchdrüsenentwicklung auch in der postnatalen Phase bei Ratten verfolgt. Die einmal angelegten Saugwarzen wachsen nach der Geburt

Abb. 9. Histologische Hodenbilder hypophysektomierter erwachsener Ratten 22 Tage nach der Hypophysektomie. a Nichthypophysektomierte unbehandelte Kontrolle (normale Spermiogenese und Leydig-Zellfunktion). b Hypophysektomierte Kontrolle. c Täglich 0,05 mg FSH*/100 g Körpergewicht. d Wie c, zusätzlich täglich Behandlung mit 5 mg/100 g Körpergewicht Cyproteronacetat. e Täglich 0,5 mg FSH*/100 g Körpergewicht. f Wie e, zusätzlich täglich Behandlung mit 5 mg/100 g Körpergewicht Cyproteronacetat. * 10 mg des Schafshypophysen-FSH = 1 mg NIH-FSH-S_1. Vergr.: etwa 140mal. Färbung: Hämatoxilin-Eosin

bei feminisierten männlichen Tieren in gleicher Weise weiter wie bei normalen weiblichen Tieren[29]. Das gleiche gilt auch für die Ausbildung des Drüsengewebes. Später war es sogar möglich, durch entsprechende Hormonbehandlung bei feminisierten männlichen Ratten im Erwachsenenalter eine Laktation hervorzurufen[31]. Aus

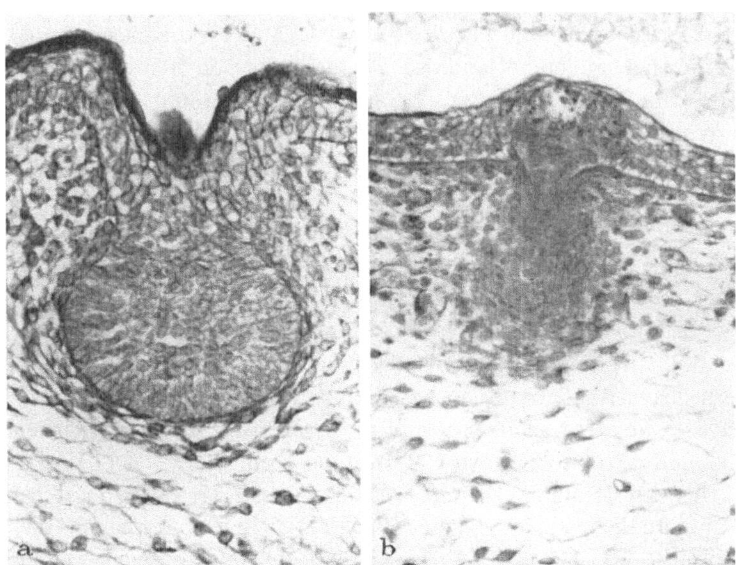

Abb. 10. Differenzierung der Milchdrüse bei Mäusen am 15. Tag der Embryonalentwicklung. a Männlicher Embryo. Die Mutter wurde vom 12. bis 14. Tag der Schwangerschaft mit täglich 3 mg/Tier Cyproteronacetat behandelt (der primäre Milchdrüsensproß ist völlig intakt und entwickelt sich wie bei weiblichen Tieren). b Männlicher Embryo, die Mutter blieb unbehandelt (der Milchdrüsensproß ist bereits weitgehend zerstört, seine Zellen zeigen verschiedene Anzeichen von Degenerationen). Vergr.: etwa 300mal. Färbung: Azan

diesen Befunden konnte nun geschlossen werden, daß für die geringe Entwicklung des Drüsengewebes und die unvollständige Entwicklung des Ausführungsgangsystems bei männlichen Ratten und Mäusen allein die Androgene des Fetus verantwortlich zu machen sind. Wird, wie in unserem Experiment, durch einen Testosteron-Antagonisten die Wirkung der fetalen Androgene gehemmt, so läuft eine der weiblichen Organogenese entsprechende Differenzierung dieses Organs ab. Man kann weiterhin schließen, daß für die Reifung der einmal angelegten Organstrukturen weibliche Sexualhormone, Oestrogene und Gestagene, bis zum Eintritt der Pubertät ohne Bedeutung sind.

Ein anderer Punkt ist die Differenzierung derjenigen nervösen Zentren, die Höhe und Modus der Gonadotropinsekretion regeln. Dazu sei nur soviel gesagt, daß die Prägung dieses Zentrums zum männlichen acyclischen Funktionstyp unter der Einwirkung von Testosteron bei Ratten in den ersten Lebenstagen erfolgt[11, 36]. Werden die Tiere in diesem Zeitraum mit einem Testosteron-Antagonisten behandelt, so arbeitet das Sexualzentrum später potentiell cyclisch. Es ist also im weiblichen Sinne differenziert worden. Man kann das nachweisen, wenn man feminisierte männliche Tiere im Erwachsenenalter kastriert und ihnen Ovarien implantiert. Dann laufen bei den Tieren ähnlich cyclische Veränderungen ab, wie sonst bei weiblichen Tieren[30]. Wir konnten das, da die Tiere ja eine Vagina hatten, schon an den Veränderungen im Vaginalabstrich erkennen. Außerdem fanden sich in den Ovarimplantaten jüngere und ältere Corpora lutea als Beweis einer vorangegangenen Ovulation. Es muß betont werden, daß unter gleichen Versuchsbedingungen normale männliche Tiere niemals ovulieren. Es entwickeln sich in Ovarimplantaten nur Follikel. Das Sexualzentrum hat offensichtlich die potentielle Eigenschaft zur cyclischen Funktionsweise, und nur unter dem Einfluß des Testosterons wird es definitiv zum männlichen acyclischen Funktionstyp geprägt. Das gleiche gilt auch für ein ebenfalls im Hypothalamus lokalisiertes, im anglo-amerikanischen Schrifttum als „Mating Center“ bezeichnetes Zentrum. Durch Gabe eines Testosteron-Antagonisten in der kritischen Phase der Differenzierung wird auch dieses Zentrum zum weiblichen Funktionstyp geprägt. Ich möchte Ihnen an einer Abbildung unsere Befunde demonstrieren.

Dargestellt sind die cyclischen Veränderungen, wie wir sie durch Abstrich am Vaginalepithel festgestellt haben, dazu oben einige Originalabstrichbilder. Ich hoffe, sie sind noch zu erkennen. Im Stadium des Oestrus ausschließlich verhornte Schollen, im Dioestrus Leukocyten, im Metoestrus ein gemischtes Bild. Das Sexualverhalten dieser Tiere gegenüber normalen männlichen Tieren war vollständig weiblich, dem Cyclusstadium entsprechend. Im Zustand des Oestrus ließen sich die feminisierten Tiere immer von normalen

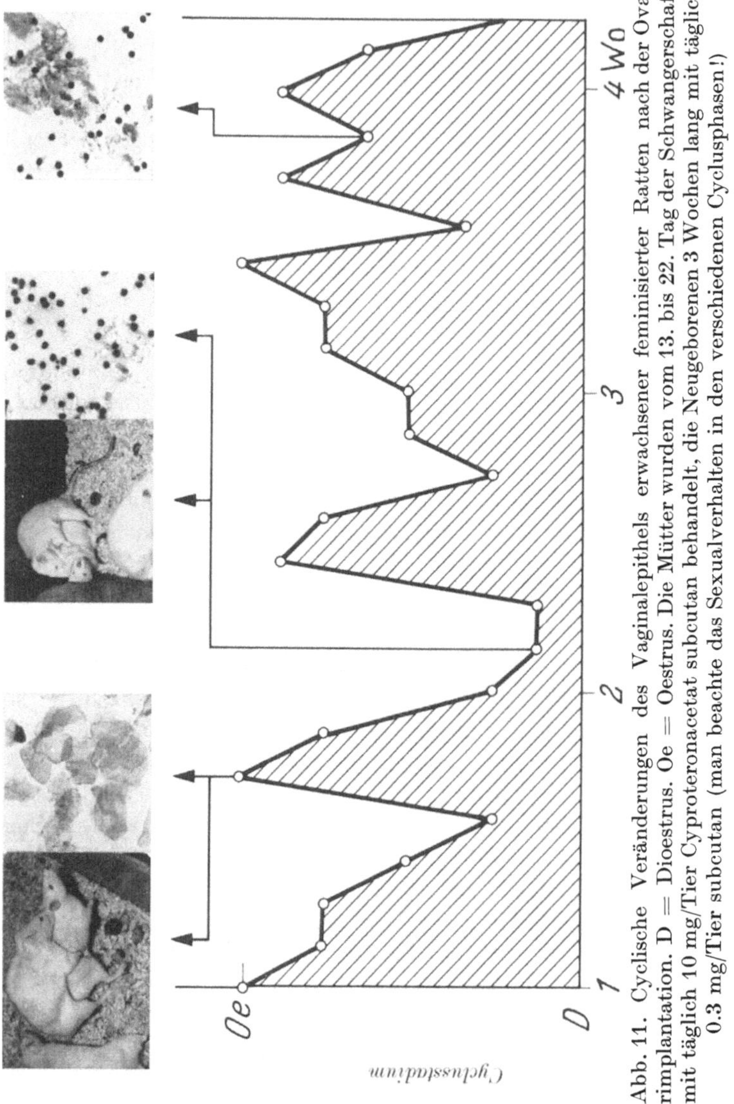

Abb. 11. Cyclische Veränderungen des Vaginalepithels erwachsener feminisierter Ratten nach der Ovarimplantation. D = Dioestrus. Oe = Oestrus. Die Mütter wurden vom 13. bis 22. Tag der Schwangerschaft mit täglich 10 mg/Tier Cyproteronacetat subcutan behandelt, die Neugeborenen 3 Wochen lang mit täglich 0.3 mg/Tier subcutan (man beachte das Sexualverhalten in den verschiedenen Cyclusphasen!)

männlichen Tieren decken, in anderen Stadien überwog dagegen die Abwehr.

Primär, unabhängig vom genetischen Geschlecht, geht die Differenzierung der Sexualorgane in weibliche Richtung. Zur männlichen Differenzierung vieler Strukturen (aber nicht aller) ist Testosteron notwendig. Abb. 12 zeigt am Beispiel der Ratte noch einmal zusammenfassend den zeitlichen Ablauf der besprochenen Sexualdifferenzierungsvorgänge.

Unter Zuhilfenahme dieser Ergebnisse und der Ergebnisse anderer Untersucher war es nun relativ leicht möglich, die kausale Genese verschiedenster Formen der Intersexualität zu klären.

Inkubationsversuche von Hoden, die in Zusammenarbeit mit SOMMERVILLE u. Mitarb.[2] vom Chelsea Hospital in London durch-

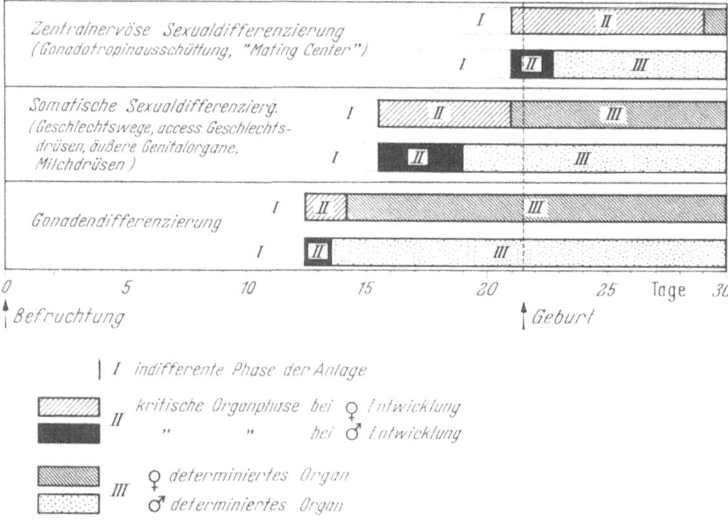

Abb. 12. Zeitplan der Sexualdifferenzierung (Ratte)

geführt wurden, haben gezeigt, daß sich nach der Antiandrogenbehandlung der graviden Ratten in den Hoden der feminisierten männlichen Nachkommen bleibende Enzymdefekte nachweisen lassen. Ich darf darauf hinweisen, daß die nun zu beschreibenden Befunde nicht ohne weiteres mit unseren bisherigen Vorstellungen über den Wirkungsmechanismus der Antiandrogene übereinstimmen.

Es gibt zwei Hauptwege der Testosteron-Biosynthese: 1. den Weg über die Δ^5-3-Hydroxy-Steroide und 2. den Weg über die Δ^4-3-Keto-Verbindungen. Die nächste Abbildung zeigt die Befunde von SOMMERVILLE u. Mitarb. (Tab. 4).

Tabelle 4. *Steroidbiosynthese in Hoden normaler und feminisierter männlicher Ratten*

Angabe in % der initial zugesetzten Radioaktivität, Tierzahl immer 5

Inkubation mit Δ^4-Androstendion	Kontrolle	Feminisiert
Δ^4-Androstendion	19,91	17,03
Testosteron	9,33	41,22
Oestron	3,58	2,90
Oestradiol	10,37	4,53
wäßrige Fraktion	4,32	3,92

Inkubation mit Dehydroepiandrosteron		
Dehydroepiandrosteron	1,13	3,61
Δ^5-Androstendiol	7,66	4,11
Δ^4-Androstendiol	6,58	7,37
Testosteron	8,08	22,70
Oestron	0,16	0,087
Oestradiol	0,13	0,112
wäßrige Fraktion	4,14	3,09

Inkubation mit Progesteron		
Progesteron	1,41	18,24
17-Alpha-Hydroxyprogesteron	5,47	10,02
Δ^4-Androstendion	2,98	3,28
Testosteron	36,05	5,76
Oestron	0,078	0,034
Oestradiol	0,111	0,019
wäßrige Fraktion	3,23	5,22

Inkubation mit Δ^5-Pregnenolon		
Δ^5-Pregnenolon	0,45	3,86
17-Alpha-Hydroxy-Δ^5-Pregnenolon	1,11	1,28
Dehydroepiandrosteron	0,61	0,83
Δ^5-Androsten-3-Beta, 17-Beta-diol	2,89	0,56
Progesteron	0,60	7,86
17-Alpha-Hydroxyprogesteron	0,91	6,27
Δ^4-Androstendion	2,24	2,92
Testosteron	11,29	5,33
Oestron	0,051	0,035
Oestradiol	0,070	0,013
wäßrige Fraktion	3,73	3,69

In SOMMERVILLES Versuchen wurden die Hodenschnitte mit vier in C_4 markierten Vorstufen des Testosterons, Δ^5-Pregnenolon, Progesteron, Dehydroepiandrosteron und Δ^4-Androstendion, inkubiert.

Die Ergebnisse werden bei den einzelnen Stoffwechselprodukten in % der initial zugesetzten Aktivität ausgedrückt. In den Hoden der feminisierten Tiere wird bei der Inkubation mit Δ^5-Pregnenolon und Progesteron erheblich weniger Substrat umgesetzt als bei den Kontrollen. Die Oestrogensynthese war in den Hoden der feminisierten Tiere gegenüber Kontrollen nicht erhöht. Bei der Inkubation mit Δ^4-Androstendion ist die Oestrogensynthese in den Hoden feminisierter Tiere sogar deutlich erniedrigt. In den Hodenschnitten feminisierter Tiere, die mit Δ^5-Pregnenolon inkubiert wurden, wurde mehr 17-Alpha-Hydroxyprogesteron und Progesteron gebildet als bei den Kontrollen und sehr viel weniger Testosteron. Auch bei der Inkubation mit Progesteron wird sehr viel mehr 17-Alpha-Hydroxyprogesteron und weniger Testosteron gebildet. Hinweise auf die bei der Testosteronbiosynthese eine Rolle spielenden Enzyme der Hoden feminisierter Tiere wurden durch Anwendung der von NEHER und KAHNT[23] entwickelten Formeln erhalten. Tab. 5 zeigt, wie die Berechnung vorgenommen wurde.

Es werden die in dieser Berechnung gewonnenen Ergebnisse veranschaulicht.

Interessant sind die Punkte 3 und 4, nämlich die Seitenkettenabspaltung und die 17-Hydroxylierung. Es wird erkennbar, daß in den Hoden feminisierter Tiere sowohl die Seitenkettenabspaltung als auch die 17-Hydroxylierung gehemmt ist. Die Antiandrogenbehandlung der Mütter muß demnach zu einem bleibenden Defekt der Desmolase- und 17-Hydroxylaseaktivität geführt haben. In die gleiche Richtung weisen neueste Untersuchungen von HOFFMANN und BREUER[13], die bei *in-vitro*-Mikrosomen-Inkubationsversuchen unter akuter Antiandrogenwirkung ebenfalls eine Hemmung der Desmolaseaktivität unter Cyproteron fanden.

Auch histochemische Befunde an Rattenhoden sprechen dafür, daß die 3-Beta-Hydroxy-Steroid-Dehydrierung nach 14tägiger Behandlung mit Cyproteron nicht beeinflußt wird. Hodenschnitte wurden mit Dehydroepiandrosteron inkubiert. Die Aktivität der 3-Beta-Hydroxy-Dehydrogenase war bei den behandelten Tieren nicht vermindert, eher etwas vermehrt.

Wacker, Chandra und Feller[44] fanden, daß der Antagonismus von Cyproteron und Testosteron auf einer Konkurrenz am Receptormolekül beruht. An isolierten Prostataribosomen und bei einem Mikroorganismus, dem Pseudomonas testosteroni, konnte durch Anwendung des Cyproterons die zellfreie durch Testosteron induzierte Proteinsynthese gehemmt werden. Auch die Testosteron-induzierte Aktivität der Ketosteroidisomerase konnte bei Pseudomonas testosteroni durch Cyproteron gehemmt werden.

Tabelle 5. *Berechnung der Enzymaktivitäten*
Nach Neher und Kahnt (1966)

	Inkubation mit							
	Dehydro-epiandro-steron		Δ^4-Andro-stendion		Δ^5-Pregne-nolon		Proge-steron	
	norm.	fem.	norm.	fem.	norm.	fem.	norm.	fem.
1. 17-Ketosteroid-reduktion	2,00	2,42	0,84	2,30	4,91	1,56	11,79	1,74
2. 3-Beta-Hydroxy-steroid-Dehydrie-rung und Δ^5-Δ^4-Isomerisation	1,67	3,90	—	—	2,97	3,43	—	—
3. Seitenketten-abspaltung	—	—	—	—	8,43	1,28	7,14	0,90
4. 17-Hydroxylierung	—	—	—	—	18,14	1,47	31,56	1,04

1. 17-Ketosteroidreduktion
$$= \frac{\%\ 17\text{-Beta-Hydroxy-}C_{19}\text{-Steroide}}{\%\ 17\text{-Ketosteroide}}$$

2. 3-Beta-Hydroxysteroid-Dehydrierung und Δ^5-Δ^4-Isomerisation
$$= \frac{\%\ \Delta^4\text{-3-Ketosteroide}}{\%\ \Delta^5\text{-3-Beta-Hydroxysteroide}}$$

3. Seitenkettenabspaltung
$$= \frac{\%\ C_{19}\text{-Steroide}}{\%\ 17\text{-Hydroxy-}C_{21}\text{-Steroide}}$$

4. 17-Hydroxylierung
$$= \frac{\%\ 17\text{-Hydroxy-}C_{21}\text{-Steroide} + C_{19}\text{-Steroide}}{\%\ 17\text{-Desoxy-}C_{21}\text{-Steroide (inklusive wiedergefundene Vorstufen)}}$$

norm. = Kontrolle
fem. = feminisiert

TRÄGER und WACKER gelangten bei der Untersuchung der testosteronbedingten Enzyminduktionen bei Streptomyces hydrogenans und Pseudomonas testosteroni ebenfalls zur Annahme eines kompetitiven Antagonismus von Testosteron und Cyproteron. Im Falle des kompetitiven Antagonismus wird Testosteron aus der Zelle verdrängt, im Gegensatz zum nichtkompetitiven Antagonismus (z. B. durch Actinomycin D).

In unserem Hause werden von GERHARDS und HART[9] Untersuchungen durchgeführt zu der Frage, ob sich bei der Ratte die Proteinsynthese in Erfolgsorganen des Testosterons durch Cyproteronacetat hemmen läßt.

Der Leucineinbau wurde *in vitro* an Gewebeschnitten der ventralen Prostata, der Samenblase, am Diaphragma sowie in einem — nach WILLIAMS-ASHMAN präparierten — zellfreien System der Prostata untersucht.

Nach den bisher vorliegenden Ergebnissen wird der Leucineinbau pro mg Protein der Prostata durch 9tägige subcutane Verabreichung von 30 mg Cyproteronacetat um 30% gesenkt, im zellfreien System um 20%.

Im Anschluß an die 9tägige Cyproteronacetatbehandlung führt die 5tägige subcutane Verabreichung von 3 mg Testosteronpropionat (TP) zu einer Normalisierung des Leucineinbaues oder zu einer leichten Steigerung über die Kontrollwerte.

Der Proteingehalt pro 100 mg Prostatagewebe wurde durch Cyproteronacetatverabreichung von 8,7 mg bei Kontrollen auf 7,9 mg gesenkt, durch TP auf 11,5 mg erhöht.

Bei Umrechnung der Impulse/mg Protein auf die Einbaurate pro 100 mg Gewebe ergibt sich, daß die alleinige Verabreichung von Cyproteronacetat den Leucineinbau gegenüber Kontrollen um 35% herabsetzt. Daran anschließend steigert TP den Leucineinbau wieder um 50% über die Kontrollen.

In der Samenblase führte die alleinige Verabreichung von Cyproteronacetat im Gegensatz zur Prostata nicht zu einer Hemmung, eher zu einer Steigerung des Leucineinbaues.

Auch am Diaphragma ließ sich nach Cyproteronacetatverabreichung keine Beeinflussung des Leucineinbaues pro mg Protein nachweisen.

Wie schon erwähnt, tritt unter anderem auch Progesteron im Biosyntheseweg des Testosterons auf. Es gelingt an hypophysekto-

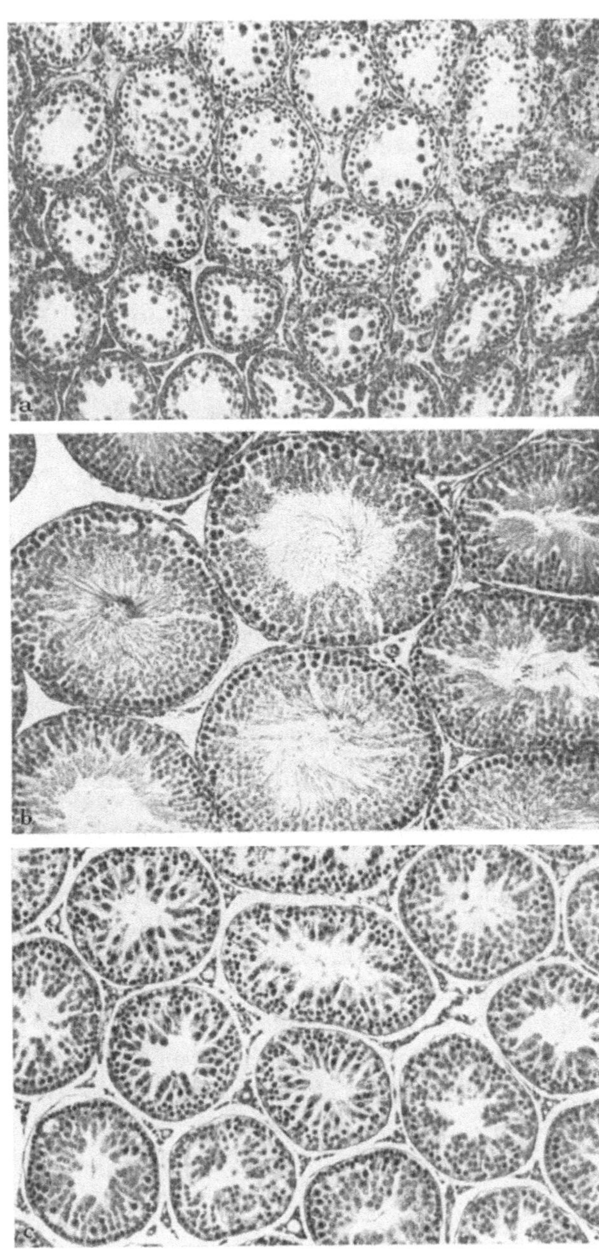

Abb. 13 (Legende siehe nächste Seite)

mierten Tieren, z. B. Ratten, mit Progesteron die Spermiogenese im Hoden aufrechtzuerhalten. Auch mit 17-Alpha-Hydroxyprogesteron ist dies natürlich möglich. Als noch nicht bekannt war, daß der Hoden in der Lage ist, aus Progesteron oder 17-Alpha-Hydroxyprogesteron Testosteron zu synthetisieren, sprach man daher von der sog. spermatogenen Wirkung dieser Gestagene. Das Ergebnis eines solchen Versuches wird in Tab. 6 veranschaulicht.

Tabelle 6. *Gewichtsverlust des Hodens bei erwachsenen hypophysektomierten Ratten*
(21tägige subcutane Behandlung)

Substanz	Tagesdosis in mg/100 g KG	Tierzahl	Gewichtsverlust der Hoden in mg/100 g KG
Testosteronpropionat	1,5	23	16 ± 27
Progesteron	5,0	17	90 ± 22
Progesteron +	5,0		
Cyproteronacetat	5,0	17	260 ± 16
Kontrollen	—	10	455 ± 28

KG = Körpergewicht

Diese Tabelle zeigt das Verhalten der Hodengewichte von unbehandelten hypophysektomierten Tieren nach Behandlung mit Testosteronpropionat, Progesteron bzw. einer Kombination von Progesteron und Cyproteronacetat.

Nach der Hypophysektomie kommt es zu einer starken Atrophie der Hoden. Die Atrophie bleibt aus, wenn die Tiere mit Testosteronpropionat oder Progesteron behandelt werden. Interessant ist nun, daß durch die gleichzeitige Gabe eines Antiandrogens diese Progesteronwirkung dosisabhängig aufgehoben werden kann. Da Antiandrogene lediglich Effekte von Androgenen aufheben können, ist damit auch biologisch der Beweis dafür erbracht, daß der Hoden auch in Abwesenheit gonadotroper Hormone in der Lage ist, aus

Abb. 13. Histologische Hodenbilder hypophysektomierter Ratten 22 Tage nach der Hypophysektomie. a Hypophysektomierte Kontrolle. b Subcutane Behandlung mit täglich 5 mg Progesteron/Tier. c Wie b, zusätzlich behandelt mit täglich 5 mg Cyproteronacetat/Tier subcutan. Vergr.: 140mal. Färbung: Hämatoxilin-Eosin

246 F. Neumann et al.:

Progesteron Androgene zu synthetisieren. Das Progesteron selbst ist natürlich nicht spermatogen, zumal andere Wirkungen des Progesterons durch die gleichzeitige Gabe eines Antiandrogens nicht beeinflußt werden[32].

Ich habe Ihnen an einigen Beispielen demonstriert, wie man Testosteron-Antagonisten sinnvoll zum Studium testosteronabhängiger Vorgänge einsetzen kann. Ich hoffe, Sie verzeihen einem Nicht-Biochemiker, daß die Biochemie in meinem Vortrag zu kurz gekommen ist.

Herrn Dr. Dr. E. Gerhards und Herrn Dr. H. Gibian danken wir für einige wertvolle Anregungen.

Literatur

[1] Bloch, J. G., and J. M. Davidson: Antiandrogen implanted in brain stimulates male reproductive system. Endocrinology. In press (1967).

[2] Bottiglioni, F., W. P. Collins, I. F. Sommerville, and F. Neumann: Steroid transformations in rat testicular tissue. III. Experimental feminisation. Acta endocr. (Kbh.). In press (1967).

[3] Callow, R. K., and R. Deanesley: Effect of androsterone and of male hormone concentrates on the accessory reproductive organs of castrated rats, mice and guinea-pigs. Biochem. J. **29**, 1424 (1935).

[4] Ebling, F. J.: Sebaceous glands; effect of sex hormones on sebaceous glands of female albino rat. J. Endocr. **5**, 297 (1948).

[5] Elger, W.: Die Rolle der fetalen Androgene in der Sexualdifferenzierung des Kaninchens und ihre Abgrenzung gegen andere hormonale und somatische Faktoren durch Anwendung eines starken Antiandrogens. Arch. Anat. micr. Morph. exp. In press (1967).

[6] —, and F. Neumann (introduced by R. A. Edgren): The role of androgens in the differentiation of the mammary gland in male mouse fetuses. Proc. Soc. exp. Biol. (N.Y.) **123**, 637 (1966).

[7] Everett, J. W.: Central neural control of reproductive functions of the adenohypophysis. Physiol. Rev. **44**, 373 (1964).

[8] Gaarenstroom, J. H., and S. E. de Jongh: Contribution to the knowledge of the influence of gonadotropins and sex hormones on the gonads of rats. New York—Amsterdam: Elsevier Publ. Company 1946.

[9] Gerhards, E., u. Ch. Hart: Unveröffentlicht.

[10] Greep, R. O., and J. Chester Jones: Steroid control of pituitary function. Recent Progr. Hormone Res. **5**, 197 (1950).

[11] Harris, G. W.: Sex hormones, brain development and brain function. Endocrinology **75**, 627 (1964).

[12] —, and S. Levine: Sexual differentiation of the brain and its experimental control. J. Physiol. (Lond.) **163**, 42 (1962).

[13] Hoffmann, W., and H. Breuer: Wirkung von 1,2-Alpha-Methylen-6-chlor-$\Delta^{4,6}$-pregnadien-17-Alpha-ol-3,20-dion (Cyproteron) auf die Biogenese von C^{19}-Steroiden in Rattentestes. Acta endocr. (Kbh.). In press (1967).

[14] HOHLWEG, W., u. K. JUNKMANN: Die hormonal-nervöse Regulierung der Funktion des Hypophysenvorderlappens. Klin. Wschr. II, 321 (1932).

[15] JOST, A.: Recherches sur la différenciation sexuelle de l'embryon de lapin. (I. Partie: Introduction et embryologie génitale normale). Arch. Anat. micr. Morph. exp. 36, 151 (1947).

[16] — Recherches sur la différenciation sexuelle de l'embryon de lapin. (II. Partie: Action des androgénes de synthèse sur l'histogenèse génitale). Arch. Anat. micr. Morph. exp. 36, 242 (1947).

[17] — Recherches sur la différenciation sexuelle de l'embryon de lapin. (III. Partie: Rôle des gonades foetales dans la différenciation sexuelle somatique). Arch. Anat. micr. Morph. exp. 36, 271 (1947).

[18] — Problems of fetal endocrinology: The gonadal and hypophyseal hormones: Recent Progr. Hormone Res. 8, 379 (1953).

[19] JUNKMANN, K., u. F. NEUMANN: Zum Wirkungsmechanismus von an Ratten antimaskulin wirksamen Gestagenen. Acta endocr. (Kbh.) Suppl. 90, 139 (1964).

[20] KINCL, F. A., A. FOLCH PI, M. MAQUEO, L. HERRERA LASSO, A. ORIOL, and R. I. DORFMAN: Inhibition of sexual development in female rats treated with various steroids at the age of fife days. Acta endocr. (Kbh.) 49, 193 (1965).

[21] LEAVITT, W. W.: Relative effectiveness of estradiol and coumestrol on the reversal of castration changes in the anterior pituitary of mice. Endocrinology 77, 247 (1965).

[22] LERNER, L. J.: Hormone antagonists: Inhibitors of specific activities of estrogen and androgen. Recent Progr. Hormone Res. 20, 435 (1964).

[23] NEHER, R., and F. W. KAHNT: In Androgens, Ed. A. VERMEULEN. Excerpta med. Found., Amsterdam, p. 130 (1966).

[24] NEUMANN, F.: Methods for evaluating anti-sexual hormones. Symposium on Methods in Drug Evaluation, Mailand 1965, North-Holland Publ. Company, Amsterdam, p. 548 (1966).

[25] — Auftreten von Kastrationszellen im Hypophysenvorderlappen männlicher Ratten nach Behandlung mit einem Antiandrogen. Acta endocr. (Kbh.) 53, 53 (1966).

[26] — Antagonismus von Testosteron und 1,2 α-Methylen-6-chlor $\Delta^{4,6}$-pregnadien-17 α-ol-3,20-dion (Cyproteron) an den die Gonadotropinsekretion regulierenden Zentren bei männlichen Ratten. Acta endocr. (Kbh.) 53, 382 (1966).

[27] —, and W. ELGER: Physiological and psychical intersexuality of male rats by early treatment with an antiandrogenic agent (1,2 α-Methylene-6-chloro-Δ^6-hydroxy-progesterone-acetate). Acta endocr. (Kbh.) Suppl. 100, 174 (1965).

[28] — — Proof of the activity of androgenic agents on the differentiation of the external genitalia, the mammary gland and the hypothalamus-pituitary-gland system in rats: IInd Symposium on Steroid Hormones, Ghent 1965. Excerpta med. Int. Congr. Series 101, 168.

[29] — — The effect of the antiandrogen 1,2 α-Methylene-6-chloro-$\Delta^{4,6}$-pregnadiene-17 α-ol-3,20-dione-17 α-acetate (cyproterone acetate) on the development of the mammary glands of male fetal rats. J. Endocr. 36, 347 (1966).

[30] — — Permanent changes in gonadal function and sexual behavior as a result of early feminization of male rats by treatment with an antiandrogenic steroid. Endokrinologie **50**, 209 (1967).

[31] — —, and R. von Berswordt-Wallrabe: Structure of mammary glands and lactogenesis in feminized male rats. J. Endocr. **36**, 353 (1966).

[32] — — — Aufhebung der Testosteronpropionat-induzierten Unterdrückung des Zyklus und der Ovulation durch ein antiandrogen wirksames Steroid an Ratten. Acta endocr. (Kbh.) **52**, 63 (1966).

[33] — — — und M. Kramer: Beeinflussung der Regelmechanismen des Hypophysenzwischenhirnsystems von Ratten durch einen Testosteron-Antagonisten, Cyproteron (1,2 α-Methylen-6-chlor-$\Delta^{4,6}$-pregnadien-17 α-ol-3,20-dion). Naunyn-Schmiedebergs Arch. Pharmak. exp. Path. **255**, 221 (1966).

[34] — —, and M. Kramer: Development of a vagina in male rats by inhibiting the androgen receptors with an antiandrogen during the critical phase of organogenesis. Endocrinology **78**, 628 (1966).

[35] —, u. H. Hamada: Intrauterine Feminisierung männlicher Rattenfeten durch das stark gestagen wirksame 6-Chlor-Δ^6-1,2-methylen-17 α-hydroxy-progesteron-acetat. 10. Symp. Dtsch. Ges. für Endokrinologie. Wien, S. 301. Berlin-Göttingen-Heidelberg: Springer 1963.

[36] Pfeiffer, C. A.: Sexual differences of the hypophysis and their determination by the gonads. Amer. J. Anat. **58**, 195 (1936).

[37] Phoenix, C. H., R. W. Goy, A. A. Gerall, and W. C. Young: Organizing action of prenatally administered testosterone propionate on the tissues mediating behaviour in the female guinea pig. Endocrinology **65**, 369 (1959).

[38] Sachs, L.: Haupttendenzen und Regulationsprinzipien der Steroidhormon-Biogenese. Arzneimittel-Forsch. **12**, 244 (1962).

[39] Strauss, J. S., A. M. Kligman, and P. E. Pochi: The effect of androgens and estrogens on human sebaceous glands. J. invest. Derm. **39**, 139 (1962).

[40] —, and P. E. Pochi: The human sebaceous gland: its regulation by steroidal hormones and its use as an end organ for assaying androgenicity in vivo. Recent Progr. Hormone Res. **19**, 385 (1963).

[41] Szentágothai, J., B. Flerko, B. Mess, and B. Halász: Hypothalamic control of the anterior pituitary. Budapest: Académiai Kiédo 1962.

[42] Träger, L., u. A. Wacker: Pers. Mittlg.

[43] Voigt, K.-D., u. H. Klosterhalfen: Einfluß des Antiandrogens Cyproteron auf die Ausscheidung von Testosteron, Epitestosteron, Oestron, Oestradiol, Oestriol und der gesamten gonadotropen Aktivität im Urin sowie auf einige Enzymaktivitäten im Serum bei normalen Männern. Ford Foundation Colloquium, Venedig 1966.

[44] Wacker, A., P. Chandra und H. Feller: Einwirkung von Antiandrogen auf die Testosteron-induzierten Prozesse bei Bakterien und Tieren. Naturwissenschaften (Im Druck).

[45] Young, W. C., R. W. Goy, and C. H. Phoenix: Hormones and sexual behaviour. Science **143**, 212 (1964).

Diskussion

CHANDRA (Frankfurt): The demonstration by NEUMANN that Cyproterone is a potent antiandrogen has contributed to the study of fundamental hormonal action-mechanisms. His results show a competitive antagonism between Testosterone and cyproterone at the biological level. We have done some experiments to elucidate the biochemical mechanism of cyproterone action.

Table 1. *Effect of cyproterone on the RNA-polymerase activity in prostate gland of the rat*

Group	System	AMP-^{3}H-Incorporation (mμmoles/g prostate)	
Non-castrated	Complete	1,19	
	Complete-(UTP, CTP, GTP)	0,19	
	Complete-Mn^{++}	0,08	
	Complete-Enzyme	0,06	
		Expt. I	Expt. II
Castrated	Complete	0,62	0,47
Castrated + Testosterone	Complete	1,09	1,15
Castrated + Testosterone + Cyproterone	Complete	0,74	0,40

Tab. 1 shows the effect of cyproterone on RNA-polymerase activity of the prostate gland from castrated rats treated with testosterone. As reported by WILLIAMS-ASHMAN and others, we too found a marked decrease of the polymerase activity in castrated rats which could be restored almost to normal by the administration of testosterone. If however, both the hormones are administered simultaneously then the polymerase activity decreases again to the control level.

In the recent years a remarkable progress in our understanding of the role of various forms of ribonucleic acids in protein synthesis has been made. We, therefore, studied the effect of cyproterone on the activity of prostatic ribosomes in a cell-free system.

Table 2. *Effect of Cyproterone on the activity of prostatic ribosomes in a cell-free system*

Rats treated with	Lysine-[1-^{14}C]-Incorporation cpm/ml	
	— Poly A	+ Poly A
Castrated	192	614
Castrated + Testosterone	518	496
Castrated + Testosterone + Cyproterone	299	409

Each incubation flask with the microsomal system contained in a final volume of 1,0 ml the following (in µmoles unless otherwise stated): sucrose 35; tris-HCl buffer (pH 7,8) 5,6; KCl 52; MgCl$_2$ 13; beta-mercaptoethanol 11; ATPl, GTP 1; PEP 10, PEP-kinase 50 µg radioactive lysine 0,05; s-RNA 1,5 mg and poly-A 200 µg. The reaction was started by adding 2,2 mg protein of the supernatant fraction and 0,7 mg protein of ribosomal fraction. All incubations were carried out at 35 C for 40 min.

Tab. 2 shows the effect of cyproterone administration on the incorporation of lysine by prostatic ribosomes from castrated and testosterone-treated rats. The incorporation of lysine is about thrice as great with ribosomal particles isolated from testosterone-treated castrates. Similiar results have been reported by LIAO and WILLIAMS-ASHMAN, and others, except that the magnitude of the stimulation caused by testosterone was about twice. This difference may be due to the fact that their experiments were performed soon after castration, while we began the treatment one week after castration. The increase in the fresh weight of prostate gland caused by testosterone propionate was more than 3 times. It is interesting to note that the prostatic ribosomes isolated from testosterone treated castrates, which received in addition cyproterone, lose their capacity to incorporate lysine. These experiments were done in the presence of naturally occuring m-RNA. The addition of polyadenylic acid showed a marked stimulation of lysine incorporation by the prostatic ribosomes from castrated animals. However, this stimulation due to Polyadenylic acid was not observed in the prostatic ribosomes of testosterone-treated castrates. This effect is most probably due to less m-RNA in the ribosomes from castrated rats so that they can bind greater amounts of the synthetic polynucleotide. The response of prostatic ribosomes isolated from testosterone and cyproterone-treated castrates to the added polynucleotide is less than observed with ribosomes from castrated rats. One should have expected a response more than exhibited by our results.

In the presence of an added polynucleotide, however, we are measuring the incorporation directed by both endogenous m-RNA and by the added m-RNA. If the endogenous m-RNA could be removed, then the relative number of m-RNA-acceptor sites could be measured by the addition of saturating amounts of m-RNA. The removal of the endogenous m-RNA was done by a preincubation of ribosomes in the presence of an energy source, GTP and MG^{++} ions.

Tab. 3 shows the effect of cyproterone on phenylalanine incorporation by the preincubated prostatic ribosomes. One notices an over all decrease in the phenylalanine incorporation by prostatic ribosomes. This loss of activity is not due to an irreversible inactivation of ribosomes but mainly due to release of endogenous m-RNA. This is proved by the fact that the microsomes so incubated, can be reactivated by the addition of energy source and m-RNA. The addition of poly-U stimulates phenylalanine incorporation to a much greater extent in the prearations from testosterone-treated castrates than in normal castrated animals. The prostatic ribosomes from cyproterone and testosterone-treated castrates respond also to the added polynucleotide but to a lesser extent than from testosterone-treated rats.

Table 3. *Effect of hormones on the amino acid-incorporating system with pre-incubated prostatic ribosomes*

Rats treated with	Phenylalanine-[1-^{14}C]-Incorporation cpm/ml	
	— Poly U	+ Poly U
Castrated	49	340
Castrated + Testosterone	83	538
Castrated + Testosterone + Cyproterone	73	449

The experimental results suggest that the antagonistic action of the anti-androgenic agent may be due to a competition for the receptor molecule, responsible for the primary action of testosterone in the target organ.

NEUMANN (Berlin): Mir waren diese Ergebnisse natürlich bekannt, und es hat mich sehr gefreut, daß auch in diesem System ein kompetitiver Antagonismus aufgezeigt werden konnte.

DIRSCHERL (Bonn): Man kann Herrn NEUMANN nur gratulieren zu seinem schönen Vortrag und zu seinen aufregenden Ergebnissen. — Was Adam und Eva angeht, so stimmt es biochemisch wohl doch mit dem Adam als ersten Menschen: das Oestron geht ja aus Testosteron hervor, und nicht umgekehrt.

In der Gruppe der Anabolika findet man doch Substanzen, die nur noch eine sehr abgeschwächte andogene Wirkung, aber eine starke proteinanabole Wirkung haben. Mit anderen Worten, die beiden Wirkungen können dissoziiert werden. Gilt das nun auch für die Antiandrogene? Kann man auch hier durch Variation des Moleküls die verschiedenen Wirkungen verschieden beeinflussen?

NEUMANN: Bezüglich der Biogenese des Oestrons haben Sie natürlich recht. Aber Oestrogene sind ja gar nicht nötig zur Differenzierung, sie spielen dabei gar keine Rolle.

Zur Frage der dissozierten Wirkung: Prinzipiell hemmen Antiandrogene auch die anabole Wirkung der Androgene und der Anabolika. Das ist gezeigt worden von GEYER in Wien in Bilanzversuchen am Menschen. Es ist auch in Tierversuchen gezeigt worden. Wie weit bei den Hemmstoffen eine Dissoziation zwischen antiandrogener und antianaboler Wirkung möglich ist, kann ich Ihnen nicht beantworten. Ich persönlich glaube nicht daran.

DIRSCHERL: VERA DANSCHAKOFF hat ja in den 30iger Jahren in Berlin viele Experimente zur Differenzierung der Geschlechtsorgane bei Vögeln, Kaltblütern und kleinen Säugetieren gemacht. Sie hat sich vor allem dafür interessiert, ob es möglich ist, das Geschlecht umzustimmen, also aus einem Hoden ein Ovar zu machen und umgekehrt. Das gelang auch in gewissem Umfange unter der Hormongabe; nach Absetzen der Androgene bzw. Oestrogene bildeten sich die Veränderungen aber langsam zurück. Danach schien es aber doch so zu sein, daß die Oestrogene auch eine gewisse Wirkung auf die Entwicklung haben.

NEUMANN: Vögel und Kaltblüter verhalten sich ganz anders als Säugetiere in dieser Beziehung. Man kann da sogar eine vollständige Umstimmung mit Sexualhormonen erreichen, z. B. bei Kaulquappen, bei denen man das Hormon dem Wasser zusetzen kann. Es genügen auch schon ganz kleine Konzentrationen in der Größenordnung von 0,5 γ/l Aquarienwasser. — Vögel verhalten sich genau umgekehrt wie Säugetiere, bei Vögeln ist das männliche Geschlecht dominant.

STAIB (Düsseldorf): Wird der Steroidstoffwechsel durch Cyproteron beeinflußt?

NEUMANN: Untersuchungen hierzu wurden von HOFFMANN und BREUER in Bonn durchgeführt. Sie haben bei Hodenschnitten die Biosynthese mit und ohne Cytroteronzusatz verfolgt und gefunden, daß *in vitro* die Aktivität der Desmolase, d. h. des Enzyms, das die Seitenkette abspaltet, gehemmt wird. Dieser Befund paßt eigentlich nicht in unsere Vorstellung hinein, und Prof. BREUER war auch sehr vorsichtig in der Interpretation seiner Ergebnisse. Es läuft jetzt noch ein anderer Versuch, bei dem die Tiere mit Antiandrogen vorbehandelt wurden; in die Inkubationslösung kommt kein Hemmstoff. Die Ergebnisse dieser Versuche liegen allerdings noch nicht vor. Über den Einfluß auf den Aufbau der Androgene ist noch nichts bekannt.

SENFT (Berlin): Wie Sie sagten, wird die 17-Hydroxylase und die Desmolase durch Cyproteron gehemmt, nicht aber die 3-Beta-ol-Hydrogenase. Waren das vergleichbare Versuchsbedingungen?

NEUMANN: Nein, das waren verschiedene Versuche. Sie sind nicht unmittelbar vergleichbar.

REISERT (Göttingen): Haben die Antiandrogene einen Einfluß auf die adrenotrope Funktion der Hypophyse?

NEUMANN: Ja, sie scheinen alle die adrenocorticotrope Funktion der Hypophyse zu beeinflussen im Sinne einer Hemmung. Wir haben keine Erklärung für diesen Effekt, denn die Verbindungen haben sonst keine Corticoideigenschaften.

WIELAND (München): Wurden diese Verbindungen bei der Behandlung des Prostatacarcinoms eingesetzt und mit welchem Erfolg?

NEUMANN: Ich habe in diesem Gremium die bisherigen klinischen Erfahrungen bewußt ausgeklammert. Kurz zusammengefaßt läßt sich folgendes sagen: Am besten anzusprechen scheint die Bremsung der Sexualität bei Männern. Über die Behandlung des Prostatacarcinoms mit Cyproteron läßt sich noch kein abschließendes Urteil geben, obwohl schon einige klinische

Berichte vorliegen — es war neulich in New York ein Symposium über dieses Thema. Die Behandlung mit Antiandrogenen scheint zumindest nicht schlechter zu sein als die bisher übliche Behandlung mit Oestrogenen. Sie bietet außerdem den Vorteil, daß die Nebenwirkungen der Oestrogenbehandlung (Gynäkomastie, Psychische Wirkungen) wegfallen.

KLINGMÜLLER (Mannheim): Läßt sich das adrenogenitale Syndrom mit Cyproteron behandeln und hemmt das Cyproteron die Bildung von Androgenen auch im Ovar? Die letztere Frage wäre für den klinischen Chemiker wichtig.

NEUMANN: Zur zweiten Frage zuerst: In der Regel kommt es nicht zu einer Bremsung der Androgenproduktion, sondern es wird nur der Effekt der Androgene am Erfolgsorgan unterdrückt. — Zur Behandlung des adrenogenitalen Syndroms: es sind einige Fälle in der klinischen Prüfung, ich kann aber noch nichts weiter dazu sagen.

SCHMIDT-THOME (Frankfurt-Höchst): Nach Ihrem Hypophysen-Schema bremst das Cyproteron auch die Rückkopplung der Androgene auf die Hypophyse. Dadurch müßte es zu einer erhöhten Gonadotropinbildung kommen, die wiederum die Testosteronproduktion so stark anregt, daß auf längere Sicht der Cyproteroneffekt wieder aufgehoben werden müßte. Ist das richtig?

NEUMANN: Das stimmt ganz genau. Es gibt auch Beobachtungen am Menschen von VOGT und KLOST, die einen Anstieg des Testosterons um den Faktor 4 gefunden haben.

SUMM (Frankfurt-Höchst): Wir haben im Verlauf der Lactation bei der Ratte die Enzymmuster in Milchdrüsengewebe und in der Leber untersucht. Man findet nach 18 Tagen ein Maximum der Induktion einer Reihe von Enzymen, z. B. Glucose-6-Phosphat-Dehydrogenase u. a. Ist zu erwarten, daß Antiandrogene auf den Verlauf dieser Induktion einen Einfluß nehmen?

NEUMANN: Beim erwachsenen Tier (wenn die Milchdrüse bereits voll ausdifferenziert ist) ist mit Antiandrogenen überhaupt kein Effekt mehr zu erwarten. Der Einfluß besteht nur in der Phase, in der die Milchdrüse angelegt wird.

STÖTTER (Augsburg): Ich habe zwei Fragen: Werden die in der Nebennierenrinde gebildeten Androgene auch unterdrückt, und zweitens, gibt es einen Unterschied zwischen Cyproteronacetat und dem freien Alkohol auf die Gonadotropinbildung?

NEUMANN: Ja, die Androgene der Nebennierenrinde werden auch gehemmt. Die Hypophyseneffekte, über die ich berichtet habe, erhält man nur mit dem freien Alkohol. Mit dem Cyproteronacetat sieht man keine Wirkung, ja man kann sogar den umgekehrten Effekt erzielen.

FRIEDRICH (Köln): Frage zu Mechanismus: 1. Wurde der Metabolismus der sog. Antihormone untersucht. Ich halte die Cyclopropyl- bzw. Cyclopropengruppe für sehr reaktiv. 2. Halten Sie „Ihre Antiandrogene" für *sehr gezielte* Methyl- bzw. Äthyldonatoren? 3. Wurden die Stoffe auf ihre Cancerogenität untersucht?

NEUMANN: Untersuchungen zum Metabolismus sind noch nicht abgeschlossen; sie werden in unserem Hause von Dr. GERHARDS u. Mitarb. durchgeführt. — Zur Frage, ob das Cyproteron ein guter Methyldonator ist, kann ich gar nichts sagen. Dazu müßten sich die Chemiker äußern. — Zu Ihrer dritten Frage: auf Cancerogenität ist Cyproteron noch nicht eingehend geprüft worden.

WIELAND: Damit möchte ich die Diskussion schließen und Herrn KARLSON um sein angekündigtes Schlußwort bitten.

Schlußwort

Von P. Karlson

Meine Damen und Herren, es ist meine Aufgabe, die Ergebnisse
dieses Colloquiums noch einmal kurz zusammenzufassen. Ich glaube,
wir alle haben viel gelernt.

Zunächst darf ich kurz auf die Methoden eingehen, die hier vor-
getragen wurden.

Neben den klassischen Methoden der Hormonforschung sind
einige ältere Methoden verbessert und erst neuerdings auf Hormon-
probleme angewendet worden. Wir haben von der Leberperfusion
gehört, die viel Interessantes erbracht hat, von Organkulturen etwa
vom Typus der Kaulquappenschwänze, über die Herr Tata berichtet hat, sowie von Zellkulturen: Die Entwicklung der Erythrocyten in der Gewebekultur, die Leberzellkulturen, die Herr Tomkins für seine Probleme verwendet. Besonders herausstellen möchte
ich das interessante System von Herrn Sekeris, die isolierten Zellkerne. Hier kann man zwar nur noch ganz begrenzte Funktionen
studieren, diese aber sehr im Detail. Zu den methodischen Hilfs-
mitteln werden ganz sicher noch die Antihormone gehören. Wir
haben von Herrn Neumann gehört, daß in vielen biologischen Bereichen durch die Antihormone wichtige Klärungen über Entwick-
lungs- und Differenzierungsprozesse möglich gewesen sind. Man
kann, glaube ich, erwarten, daß man ähnliche Phänomene auch im
molekularbiologischen Bereich beobachten wird und dort die Antihormone als Forschungsmittel wird einsetzen können. Über An-
fänge in der Richtung ist von Herrn Dr. Chandra berichtet
worden.

Unser Thema war der Wirkungsmechanismus der Hormone.
Hormone sind, nach den klassischen Definitionen von Bayliss und
Starling, chemische Sendboten. Die Frage, die wir uns hier auf
diesem Symposium vorgelegt haben, ist diese: Wer ist der Empfänger dieser Botschaft? Nun, gleich beim ersten Vortrag haben wir
gelernt, daß es nicht unbedingt nur *einen* Empfänger geben muß,

und daß der Empfänger auch gleichzeitig wieder Botschafter sein
kann. Herr Sutherland hat uns berichtet, wie aus dem ersten
„Messenger“, dem ursprünglichen Hormon, durch eine komplizierte
Reaktion ein zweiter Messenger — 3.′5′Adenosinmonophosphat —
wird, und daß dieses dann eine Reihe von Funktionen übernimmt:
das ist gewissermaßen ein biologischer Verstärkermechanismus. Es
scheint mir im Augenblick allerdings noch nicht klar, wie weit alle
Funktionen der Hormone, die nach diesem Prinzip wirken, über
diesen Mechanismus zu erklären sind oder ob wir mehrere Wirkungsmechanismen nebeneinander anzunehmen haben.

Als Empfänger der Botschaft kann man wohl in erster Linie die
Hormonreceptoren ansehen, über die Herr Jungblut berichtet hat.
Es ist meines Erachtens ein ganz wesentlicher Fortschritt, daß
diese Receptoren, die bisher nur in der Theorie gefordert wurden,
jetzt allmählich Gestalt gewinnen und greifbare chemische Substanzen werden: Substanzen von Proteincharakter, wie man ja
wohl immer angenommen hat. Allerdings, so wichtig die Kenntnis
dieser Receptoren ist, sie sagt uns noch nichts darüber aus, wie
diese Receptoren wirken. Hierzu sind noch weitere Untersuchungen
nötig; sie werden mit isolierten Receptoren möglich sein.

Als wichtiger Bindungs- und Wirkungsort der Hormone ist in
den Vorträgen immer wieder der Zellkern herausgestellt worden.
Das gilt für die Hormonreceptoren, die wahrscheinlich dort lokalisiert sind, es gilt für die Untersuchungen, über die Herr Sekeris
und Herr Tata berichtet haben; auch die Untersuchungen von
Dukes und Goldwasser scheinen in die Richtung zu gehen, obwohl dort die Experimente wohl noch weitergeführt werden müssen.
Schließlich geht die Kontrolle auf dem „translational level“, also
auf der Ebene der Übersetzung der Nucleinsäure-Sprache in die
Protein-Sprache, im Grunde ja auch über den Zellkern: Der Repressor, der für die Hemmung dieser Übersetzung verantwortlich
ist, wird nach Herrn Tomkins wohl im Zellkern gebildet. Man sieht
hier auch wieder, wie wichtig das bedeutende Denkschema von
Jacob und Monod für diese Untersuchungen geworden ist.

Sekundär zu diesen Primärprozessen, die im Zellkern stattfinden,
sehen wir dann die Enzyminduktion und damit die Veränderungen
im Stoffwechsel, über die die Herren Schimassek und Seubert so
ausführlich berichtet haben. Hier kommen wir allerdings in das
Gebiet komplizierter Zusammenhänge von Regulation und Gegen-

regulation. Es ist in der Diskussion schon angeklungen, daß die metabolischen Effektoren von Enzymsystemen gerade in letzter Zeit sehr viel Interesse gefunden haben, daß aber dort sehr komplizierte Verhältnisse vorliegen. Bezüglich der Regulation im Stoffwechsel durch Metabolite und durch Hormone bleibt noch ein sehr weites Feld zu untersuchen.

Es bleibt mir nur noch übrig, den Rednern zu danken, die die Aufgabe übernommen haben, hier die Referate zu halten, sowie den Diskussionsrednern, die diese Veranstaltung erst zu einem Colloquium gemacht haben. Nicht minder möchte ich aber denjenigen danken, die die Organisation hier am Orte gemacht haben und uns hier so schön geholfen haben, vor allem Herrn Oberarzt Dr. WINTER, der die Übertragungs- und Tontechnik betreut hat, Herrn AUHAGEN und seinem Stab wie auch dem Herrn Bürgermeister von Mosbach und allen Stellen, die hier in Mosbach für unser Unterkommen und unser Wohlbefinden gesorgt haben.